T0419253

PHYSICS RESEARCH AND TECHNOLOGY

LASER-INDUCED PLASMAS: THEORY AND APPLICATIONS

PHYSICS RESEARCH AND TECHNOLOGY

Additional books in this series can be found on Nova's website
under the Series tab.

Additional E-books in this series can be found on Nova's website
under the E-books tab.

PHYSICS RESEARCH AND TECHNOLOGY

LASER-INDUCED PLASMAS: THEORY AND APPLICATIONS

ETHAN J. HEMSWORTH
EDITOR

Nova Science Publishers, Inc.
New York

For permission to use material from this book please contact us:
Telephone 631-231-7269; Fax 631-231-8175
Web Site: http://www.novapublishers.com

Library of Congress Cataloging-in-Publication Data

Laser-induced plasmas : theory and applications / editor, Ethan J. Hemsworth.
p. cm.
Includes bibliographical references and index.
ISBN 978-1-61324-851-5 (hardcover : alk. paper) 1. Laser plasmas. I. Hemsworth, Ethan J.
QC718.5.L3L365 2011
530.4'4--dc23

2011017309

Published by Nova Science Publishers, Inc. † New York

CONTENTS

PREFACE

This book examines the theory and applications of laser-induced plasmas. Topics discussed include the application of laser-induced plasma expansion models for thin film deposition; cluster-containing plasma fumes for high-order harmonic generation laser radiation; pulsed laser deposition of nanocrystalline V2O5 thin films; nanosecond and femtosecond laser ablation of TeO2 crystals; resonant harmonic generation of short pulse laser in plasma and the influence of the heterogeneous nature of laser ablation on near-surface plasma formation and propagation.

Chapter 1 - This chapter presents a study that combines several plasma expansion models and plasma characterization techniques to answer the following question, can the optimum conditions for the growth of good quality ZnO thin films be predicted from models or at least bounded between some limits?

Using ion signals, acquired with a Langmuir probe, it is corroborated that the Anisimov model describes successfully the expansion of laser produced plasma in vacuum. Furthermore, plasma properties such as initial energy and specific heats ratio obtained from it can be used as input for the initial conditions of the plasma expansion in a background gas.

For plasma expansion in a background gas, firstly, a full characterization of the expansion is performed using intensified charged coupled device (ICCD) photography and ion signals. The commonly used shock wave model (Sedov-Tylor model) is applied and it is shown that it describes the general features of the expansion, i.e. front expansion $\sim t^{2/5}$; yet it failed in a full description of the plasma properties. However a different approach of another shock wave model, the Freiwald-Axford model, shows to be more appropriate. The proposed approach describes the plasma contact front instead of the commonly described shock front. This allowed using the plasma specific heats ratio as obtained from Anisimov model instead of approximated values of the same physical magnitude but for the gas background. A detailed comparison between the two models and the correctness of the plasma expansion velocities, positions and pressures corroborated that such an approach is correct. It is shown that the use of ion signals as extracted using the Langmuir probe is not correct for describing the plasma front. On the other hand the use of ICCD photography is correct. It is corroborated that in a background gas the plasma reaches a maximum in its expansion which is well predicted by the Predtechensky model. Beyond this distance the plasma enters in a diffusive regime which is well described by a 3-D diffusion model.

The results obtained from the thin film depositions are presented. Structure, surface smoothness and optical properties of the ZnO thin films are studied and shown to present good quality. It was shown that oxygen incorporation into the growing thin film occurs, preferentially, during the material condensation on the substrate.

Finally, it is presented a discussion that answers the question posed above.

Chapter 2 - A review of recent studies and applications of cluster-containing plasmas as the nonlinear optical media for frequency conversion of ultrashort laser pulses towards the extreme ultraviolet range is presented. The studies of conditions when the plasma producing on the surface of laser-ablated targets contains nanoparticles, clusters, and nanotubes are presented. The results are presented which show that nanoparticle formation in the plasma plumes can be accomplished using relatively long laser pulses under tight focusing condition. The nanoparticles were produced by the interaction of 300 ps, 20 mJ laser pulses with bulk Ag target at an intensity of $\sim 1\times 10^{13}$ W cm^{-2}. The laser ablation of chromium in vacuum at the tight and weak focusing conditions of a Ti:sapphire laser radiation allowed to synthesis the 60 nm particles. The studies show that nanoparticle formation during ablation of metals by laser pulses strongly depends on the concentration of surrounding gas. At moderate laser intensity of the 300 ps pulses on the surface of nanoparticle-containing materials ($<5\times 10^{9}$ W cm^{-2}), the deposited material remains approximately the same as the initial nanoparticles. The ablation of carbon nanotubes and C_{60} powders has been proved the effective technique to create the plasma plumes containing these species. The integrity of carbon nanotubes and C_{60} within the plasma plume was confirmed by structural studies of plasma debris. These cluster-containing plasma plumes proved to be the effective media for high-order harmonic generation of femtosecond laser pulses. The advantages of clustered plasmas are shown allowing the enhancement of low-order harmonic yield compared with atomic/ionic plasmas.

Chapter 3 - This chapter presents results of pulsed laser deposition of nanocrystalline V_2O_5 thin films along with a time-of-flight (TOF) analysis of the laser-induced plasma. Additionally, a brief historical overview of this deposition method is provided. In this study, the deposited films were characterized by X-ray diffraction (XRD), X-ray photoelectron spectroscopy (XPS), high-resolution transmission electron microscopy (HRTEM), and by optical transmission spectroscopy. The films were deposited on amorphous glass substrates while keeping the O_2 partial pressure at 13.33 Pa and the substrate temperature at 220 °C. The characteristics of the films were changed by varying the laser fluence and repetition rate. XRD revealed that films are nanocrystalline with an orthorhombic structure. XPS shows the substoichiometry of the films that generally relies on the fact that during the formation process of V_2O_5 films lower valence oxides are also created. From the HRTEM images, the size evolution and distribution characteristics of the clusters as a function of the laser fluence were observed. From the spectral transmittance, the absorption edge using the Tauc plot was determined. Time-of-flight (TOF) optical spectroscopy was used to analyze the elemental composition, and temporal- and spatial distributions of vanadium atoms and ions in the laser-induced V_2O_5 plasma. Neutral atoms and singly charged V ions were detected in the TOF spectra. The time evolution of emission lines is discussed along with the velocity of the ablated species. The data are discussed and conclusions are drawn for the elemental analysis by laser ablation of solid vanadia samples.

Chapter 4 - Near-IR femtosecond (fs) (pulse duration = 150 fs, wavelength = 775 nm, repetition rate 1 kHz) and VUV nanosecond (ns) (pulse duration = 20 ns, wavelength = 157 nm, repetition rate 1 to 5 Hz) laser pulse ablation of single-crystalline TeO_2 (c-TeO_2) surfaces was performed in air using the direct focusing technique. A multi-method characterization using optical microscopy, atomic force microscopy and scanning electron microscopy revealed the surface morphology of the ablated craters. This allowed us at each irradiation site to characterize precisely the lateral and vertical dimensions of the laser-ablated craters for different laser pulse energies and number of laser pulses per spot. Based on the obtained information, we quantitatively determined the ablation threshold fluence for the fs laser irradiation when different pulse numbers were applied to the same spot using two independent extrapolation techniques. We found that in the case of NIR fs laser pulse irradiation, the ablation threshold significantly depends on the number of laser pulses applied to the same spot indicating that incubation effects play an important role in this material. In the case of VUV ns laser pulses, the ablation rate is significantly higher due to the high photon energy and the predominantly linear absorption in the material. These results are discussed on the basis of recent models of the interaction of laser pulses with dielectrics. In the second part of this chapter, we use time-of-flight mass spectrometry (TOFMS) to analyze the elemental composition of the ablation products generated upon laser irradiation of c-TeO_2 with single fs- (pulse duration ~ 200 fs, wavelength 398 nm) and ns-pulses (pulse duration 4 ns, wavelength 355 nm). Due to the three order of magnitude different peak intensities of the ns- and fs laser pulses, significant differences were observed regarding the laser-induced species in the plasma plume. Positive singly, doubly and triply charged Te ions (Te^+, Te_2^+, Te_3^+) in the form of many different isotopes were observed in case of both irradiations. In the case of the ns-laser ablation, the TeO^+ formation was negligible compared to the fs case and there was no Te trimer (Te_3^+) formation observed. It was found that the amplitude of Te ion signals strongly depends on the applied laser pulse energy. Singly charged oxygen ions (O^+) are always present as a byproduct in both kinds of laser ablation.

Chapter 5 - The process of second harmonic generation of an intense short pulse laser in a plasma is resonantly enhanced by the application of a magnetic wiggler. The wiggler of suitable wave number k_0 provides necessary momentum to second harmonic photons to make harmonic generation a resonant process. Harmonic generation can also be made resonant in the presence of density ripple $(0, \vec{k}_0)$. Density ripple also provides necessary momentum to second harmonic photons to make harmonic generation a resonant process. In both the processes resonant second harmonic is produced. However, the group velocity of the second harmonic wave is greater than that of the fundamental wave, hence, the generated pulse slips out of the fundamental laser pulse and its amplitude saturates.

The process of third harmonic generation of an intense short pulse laser in plasma is resonantly enhanced by the application of a magnetic wiggler. The laser exerts a ponderomotive force at second harmonic driving density oscillations. The second harmonic oscillations coupled with electron velocity at the laser frequency, produces a non-linear current, driving the third harmonic. Third harmonic pulse generates in the fundamental pulse domain. However, the group velocity of the third harmonic wave is greater than the fundamental wave. Hence, the third harmonic pulse saturates strongly and moves forward from the fundamental pulse at shorter distance than the second harmonic pulse.

Chapter 6 - Experiments investigations and numerical simulation of thermophysical and hydrogasdynamic processes in near surface layer of metalls, irradiated by pulsed laser, allowed to expose low-threshold character of initial destruction of materials, to establish mechanisms responsible for lowering of plasma ignition thresholds.

Chapter 7 - Laser-induced breakdown spectrometry (LIBS) is one of the most promising and powerful methods for direct spectral analysis of different materials. The method is based on the use of the emission spectrum of laser-induced plasma on the surface or in the bulk of the analyzed sample. Processing of spectral data serves to quickly and effectively obtain information on the elemental composition of the target. The main advantages of this technique are the relatively simple instrumentation, the fast qualitative and quantitative determination of the light and heavy elements, the possibility of local and remote analysis etc. This chapter presents an overview of the actual state of art, main achievements and problems of LIBS over the past years. Progress of the laser systems, spectral devices and detectors for performance of LIBS are shortly discussed with respect to the future LIBS applications. Up to date the weak point of LIBS is insufficient sensitivity of the LIBS determination of microcomponents, particularly, in environmental objects. Methods for LIBS sensitivity enhancement, such as double-pulse ablation, a combination of LIBS with laser-induced fluorescence, the use of additional sources of excitation (spark) and confinement of plasma by magnetic field or shock wave and other specific approaches (microchip lasers, microwave excitation etc), are compared with respect to figures-of-merit. Minimal achievable detection limits of the most elements obtained until now are critically considered. The Achilles' heel of all analytical application of laser plasma is bad reproducibility due to laser energy instability and absence of local thermodynamic equilibrium. Modern approaches for improvement of LIBS reproducibility are illustrated and thoroughly discussed.

Chapter 8 - Laser light propagation in plasma is known to be subject to parametric instabilities, where the laser energy decays into scattered light and plasma modes giving rise to stimulated Raman (Brillouin) scattering and related phenomena. The presence of instabilities such as SRS, SBS and self-focusing in laser plasma interaction can result in significant losses of the incident laser energy leading to poor laser plasma coupling. For the ultra-short laser pulses, the most important instabilities are electron plasma instabilities. In nonlinear regime the relativistic self-focusing has affect on SRS and SBS, hereby becomes important to investigate and understand them. In the present work we investigate nonlinear interaction of a high power intense Gaussian electromagnetic beam with the electron and ion waves in an unmagnetized plasma. The resulting stimulated Raman and Brillouin scattering phenomena are studied analytically and numerically. Based on WKB and paraxial theory the propagation characteristics and regimes of an intense laser pulse is completely determined by the degree of diffraction, nonlinear defocusing and self-focusing suffered by the beam as it traverses through the plasma. When the laser power exceeds the critical power, the laser beam can undergo periodic self-focusing due to relativistic nonlinearity. The effect of nonlinear coupling between the pump laser and scattered laser beam has been incorporated. The effect of finite laser beam size and that of scattered beam and relativistic self-focusing of the pump laser beam on SRS and SBS back-reflectivity have been illustrated. Numerical calculations are made for typical parameters of relativistic laser-plasma interaction processes applicable for wide range of arbitrary pump strength and background plasma density.

Chapter 9 - The CO_2 laser of 10.6 μm in wavelength is an inexpensive, rapid and flexible one for the soft polymer and hard glass and ceramic related materials processing. It has been widely applied to the fabrication of microchannel ablation, cutting, microhole drilling, material annealing and modification in the categories of MEMS, bio-chip, optical/optoelectronic devices, displays and laser dentistry. The basic CO_2 laser physics is photo-thermal mechanism for material removal therefore some defects of debris, bulges, cracks and scorches around ablated microstructure are formed during laser processing in air which degrades the device yield and quality for bonding. In this article, some advanced laser processing methods have been proposed for improving the microstructure quality of fabrication including Liquid Assisted Laser Processing (LALP), cover-layer protection processing and Glass Assisted CO_2 LAser Processing (GACLAP) for eliminating the cracks and scorches defects, diminishing bulges height and reducing feature size, even making the transparent-in-nature silicon material to be etched, drilled and cut. LALP can effectively reduce the temperature, heat-affected zone, thermal gradient and stress via water for hindering the crack and scorch formation together with the laser heating induced stronger natural convection in water for carrying debris away to reduce bulge height. The feature size can be reduced from 400–500 μm via traditional processing in air to 150–200 μm even smaller via LALP. Combing LALP and low-temperature bonding techniques have been used for the fabrication of capillary-driven glass-based microfluidic chip for the application of low-to-high viscosity fluid actuating and biomedical blood coagulation testing. GACLAP can change light absorption behavior of Si and make Si be etched from the top surface toward the interface whose new mechanism is discussed in viewpoint of the variation of electronic band structure, surface oxidation and light absorption of Si at high temperature. A simple thermal model and ANSYS software are adopted for the analysis of thermal and stress distribution on specimen during the laser irradiation in air and water ambient. Also, a simple CO_2 laser annealing process for titanium dioxide treatment has also been developed instead of conventional expensive short wavelength laser annealing and non-selective high-temperature furnace annealing. Both crystalline rutile and anatase titanium dioxide transformation from amorphous titanium oxide can be controlled by the sol-gel composition and laser annealing parameters for material property adjustment which is potentially used for the photocatalyst and optoelectronic application. The relationship between process, microstructure and phase transformation of titanium oxide is discussed and established.

In: Laser-Induced Plasmas
Editor: Ethan J. Hemsworth, pp. 1-33
ISBN 978-1-61324-851-5

Chapter 1

PRACTICAL APPLICATION OF LASER-INDUCED PLASMA EXPANSION MODELS FOR THIN FILM DEPOSITION

E. de Posada[*,1]***, J. G. Lunney***[2]***, T. Flores***[1]***, L. Ponce***[1]***, M. Arronte***[1] ***and E. Rodríguez***[1]

[1]CICATA-Instituto Politécnico Nacional, Altamira 89600, Tamaulipas, México
[2]School of Physics, Trinity College, Dublin 2, Ireland

ABSTRACT

This chapter presents a study that combines several plasma expansion models and plasma characterization techniques to answer the following question, can the optimum conditions for the growth of good quality ZnO thin films be predicted from models or at least bounded between some limits?

Using ion signals, acquired with a Langmuir probe, it is corroborated that the Anisimov model describes successfully the expansion of laser produced plasma in vacuum. Furthermore, plasma properties such as initial energy and specific heats ratio obtained from it can be used as input for the initial conditions of the plasma expansion in a background gas.

For plasma expansion in a background gas, firstly, a full characterization of the expansion is performed using intensified charged coupled device (ICCD) photography and ion signals. The commonly used shock wave model (Sedov-Tylor model) is applied and it is shown that it describes the general features of the expansion, i.e. front expansion $\sim t^{2/5}$; yet it failed in a full description of the plasma properties. However a different approach of another shock wave model, the Freiwald-Axford model, shows to be more appropriate. The proposed approach describes the plasma contact front instead of the commonly described shock front. This allowed using the plasma specific heats ratio as obtained from Anisimov model instead of approximated values of the same physical magnitude but for the gas background. A detailed comparison between the two models and the correctness of the plasma expansion velocities, positions and pressures corroborated that such an approach is correct. It is shown that the use of ion signals as

[*] E-mail: edeposada@ipn.mx

extracted using the Langmuir probe is not correct for describing the plasma front. On the other hand the use of ICCD photography is correct. It is corroborated that in a background gas the plasma reaches a maximum in its expansion which is well predicted by the Predtechensky model. Beyond this distance the plasma enters in a diffusive regime which is well described by a 3-D diffusion model.

The results obtained from the thin film depositions are presented. Structure, surface smoothness and optical properties of the ZnO thin films are studied and shown to present good quality. It was shown that oxygen incorporation into the growing thin film occurs, preferentially, during the material condensation on the substrate.

Finally, it is presented a discussion that answers the question posed above.

INTRODUCTION

The development of the electronic industry and the requirement to produce thin films of new materials has opened a wider field for the laser application known as Pulsed Laser Deposition (PLD). The relative simplicity of the technique, coupled with comparatively inexpensive equipment requirements, as well as its success in growing high quality high temperature superconducting materials, are the main reasons for interest in the study of this technique. PLD involves a rather complex group of physical and chemical processes which in general can be divided into three main regimes. Firstly, photons are absorbed by the material and energy released in a small volume ($\sim 3\times10^{-7}$ cm^3 in the experiments described in this thesis). This allows temperatures of the order of the melting or vaporization temperatures for the material to be reached. The next regime is plasma formation. A vapour layer on top of the material surface increases its degree of ionization due to both interactions with the laser pulse and particle collisions. Finally, the plasma expands away from the material surface. In general the main goal in the process is to obtain a thin film with good stoichiometry. Although PLD has proved adequate in achieving good stoichiometry, problems arise when growing oxide or nitride compounds. For these materials a lack of oxygen or nitrogen in the as-grown film is detected. Scattering during expansion, re-evaporation and sputtering from the substrate, due to particles with high kinetic energies, are the main reasons for such behaviour. The introduction of a background gas during the deposition process has been used to overcome these problems. In addition to the experimental achievements of the technique, great attention has being given to the development of the theoretical basis.

Plasma expansion is of particular importance to the PLD technique for the following reasons. The initial conditions or properties of the plasma are related, to some extent, to the way the vaporization processes takes place and on the other hand, the behaviour of the plasma expansion plays an important role in the final properties of the thin film (film thickness, homogeneity, stoichiometry, etc.). Studying the plasma properties and its evolution in both time and space gives a better insight into the overall process. Great attention has been given to this topic. As a result several models have been developed and a wide range of characterization techniques have been applied. Despite the available theory, it is no secret for those who work in this field that models stand in a second place during the thin film growth. A trial and error process takes place until a film with the desired characteristics is obtained. In addition, particularly for oxides, post-deposition annealing is common in order to obtain the final stoichiometry. As a result, a simple review of the literature results in different reports in which good quality thin films of the same material have been obtained while grown under considerably different deposition conditions [Dinescu, et al., 1996; Ohtomo, et al., 1998; Ryu, et al., 2000; Sankur, et al., 1983; Shim, et al., 2002; Tang, et al., 1998; Wu, et al., 2001].

The question that arises is whether the optimum conditions for the growth of good quality thin films can be predicted from models or at least bounded between some limits.

EXPERIMENTAL SET UP

Figure 1 shows a schematic of the experimental apparatus used for the work presented in this chapter.

Laser ablation was performed using a KrF (248 nm, 26 ns) excimer laser. The work was carried out in a stainless steel chamber with a base pressure of $8x10^{-6}$ mbar using a Pfeiffer-Balzers TSH 050 turbo molecular system (50 l/s). When working in a background gas, 5.0 grade oxygen (Air Products) was fed through a Whitey needle-valve into the chamber. A Balzers Penning was used as pressure gauge for the 10^{-5} to10^{-3} mbar range. For the region of 0.2-1 mbar an Edwards Pirani gauge was used. The latter was calibrated using a Hastings HPM-2002-OBE wide range gauge, which consists of a dual sensor unit containing a piezoresistive direct force sensor and a thin film Pirani type sensor. Substrate heating was accomplished by using a resistive heater able to reach up to 1000 ^{0}C. The temperature was controlled using a Eurotherm 94 PID controller and a k-type thermocouple as a sensor. An Ultimax UX-20 optical pyrometer was used for temperature corroboration. The laser pulses were focused on the rotating target using a 50 cm focal length lens obtaining a rectangular spot of 0.6 x 0.1 cm dimensions. The laser pulse was incident at an angle of 45^{0} to the normal of the target.

The spot dimensions were kept constant and the fluence was varied by varying the energy of the pulse. The average energy per pulse was measured using a Scientech thermopile joulemeter.

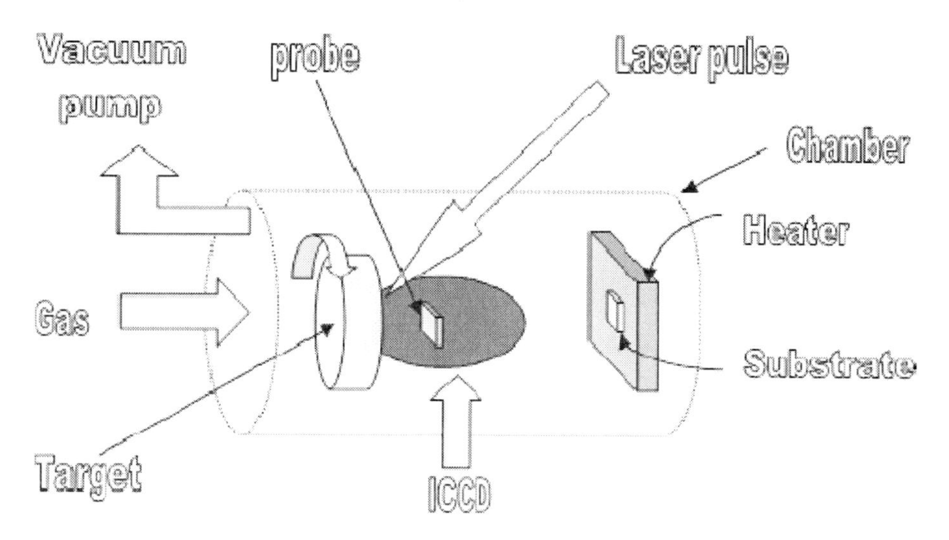

Figure 1. General scheme of the used pulsed laser deposition system.

The ZnO thin films were deposited on (001) sapphire substrates at 400 ^{0}C, using two different oxygen pressures and two different target-substrate distances. A 99.99% purity ZnO ceramic target was used, while the deposition time was 10 min running the laser at 10 Hz. Some of the films were post-annealed in the deposition chamber at the same substrate temperature and oxygen pressure as during the growth to see if there was any improvement of the crystal quality or the state of oxidation.

The surface morphology of the films was examined scanning electron microscopy (SEM) technique. Profilometer measurements of the film thickness were in the range ~ (150-200) nm, implying a growth rate of 0.03 nm/pulse. X-ray diffraction (XRD), with a Cu K_α (1.54056 A) source and θ-2θ configuration, was used to study the crystalline properties of the films.

The photoluminescence of the films was measured with a HeCd (λ=325 nm) laser, a monochromator and a photomultiplier tube cooled to -23^{0}C. All the spectra were acquired for samples at room temperature using step size of 1 nm and an integration time of 1 s.

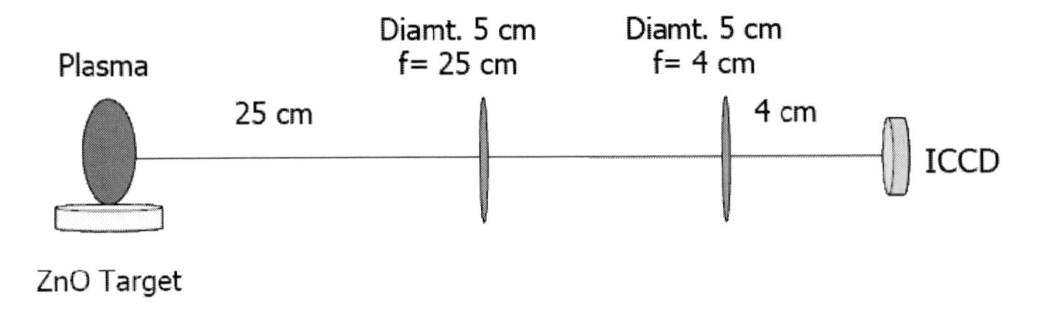

Figure 2. Delivery system used for the imaging of the plasma expansion using fast ICCD photography.

Images of the plasma were acquired by placing the ICCD camera (ANDOR Technology) parallel to the plasma expansion axis and using a combination of two lenses to image the plasma onto the detector of the camera. Figure 2 shows a schematic of the system used. A 25 cm focal length and 50 mm diameter lens was used to collect the total emission of the plasma. A lens of 4 cm focal length and 50 mm diameter was the used to produce the image. The axis of the delivery system cut the plasma expansion axis at a distance of 2 cm from the target surface in order to cover a total distance of 4 cm which was the maximum substrate position used for film growth.

The demagnification was 1/6 making a pixel in the ICCD equivalent to 0.165 mm along the plasma expansion axis. All the measurements were performed using an acquisition time of 3.8 ns and at different delay times after the arrival of the laser pulse to the target. The acquisition time used guaranteed that even for the in vacuum expansions velocity of 2.3×10^6 cm/s obtained in the experiments the displacement was about 0.09 mm. This corresponds to less than a pixel error (0.165 mm). Both delay and acquisition times were established using a Stanford delay generator.

The ionic currents were acquired using a planar Langmuir probe [Langmuir, et al., 1926] of 0.053 cm^2 of area and a Koopman circuit [Koopman, 1971], whose schematic is presented in the figure 3. The bias voltage was – 20 V.

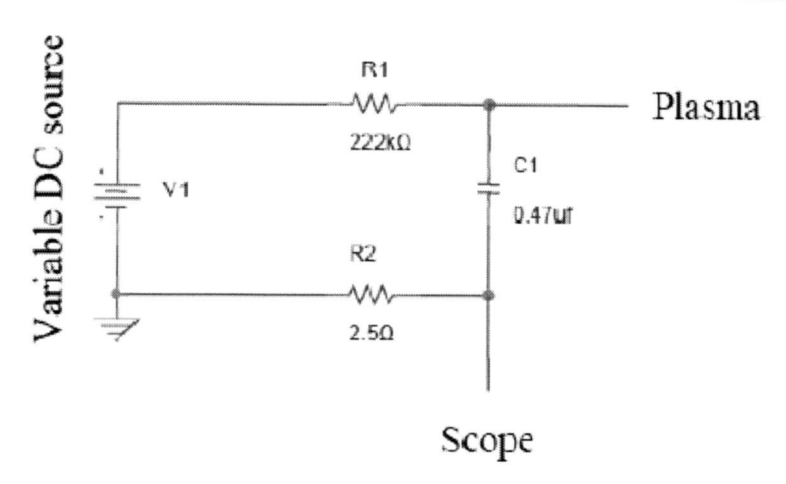

Figure 3. Koopman circuit used for acquiring ionic currents.

Table 1 summarises the maximum errors for several of the parameters values quoted throughout the chapter.

Table 1. Estimated maximum errors

Parameter	Maximum error
Averaged laser energy per pulse	± 0.005 J
Laser spot area	± 0.007 cm^2
Fluence	± 0.4 J/cm^2
Distance to target	± 0.2 cm
Background gas pressure	± 0.004 mbar
Target mass loss per pulse	± 0.1 μg
Substrate temperature	± 15 ^{0}C
Radiance and Irradiance	± 5 %

Plasma Expansion

Studies on the expansion of laser produced plasma can be traced back to as early as the 60s, where both the in vacuum [Linlor, 1963] and in gas background [Bobin, et al., 1968] expansion was analysed. From these early studies, it was realised that the expansion was strongly directed along the target normal. Cowin et al. [Cowin, et al., 1978] used TOF measurements to correlate the speed of the plasma components to the surface temperature of the target. By performing angular scanning, it was found that the extracted temperatures were too high for perpendicular emitted particles, while for the obliquely emitted particles the temperatures were too low. Similar behaviour was reported by [Namiki, et al., 1985]. These reports suggested that part or all of this behaviour was due to near-surface gas phase collisions. The region of the vapour where such collisions take place was termed the "Knudsen layer" and it was shown that evaporation of even 0.8 monolayer/pulse was enough for this layer to be developed [Kelly, et al., 1988]. Within the Knudsen layer particles tend to reach equilibrium and develop a common center of mass with a flow velocity directed normally to the target. This speed equals the sound speed in the layer [Kelly, et al., 1988], which is dependent of the temperature in the Knudsen layer. Two possibilities exist for the flow of particles beyond the boundaries of the Knudsen layer. If there are no collisions, a free expansion will develop, whereas if collisions persist, an adiabatic expansion takes place [Kelly, 1989]. For most of the conditions under which the PLD process is carried out an adiabatic expansion is present.

Of great importance for PLD is knowledge of the plasma distribution during its expansion as both film homogeneity and stoichiometry are related to this issue. Again, different models have been proposed to describe these expansions [Bulgakov, et al., 1995; Jeong, et al., 1998a; Leboeuf, et al., 1996], most of them using a gas-dynamic approach and describing both temporal and spatial behaviour of macroscopic parameters, such as density, temperature and expansion velocity. The usual high particle concentration in laser produced plasmas allows the use of gas-dynamic formulation. Although most thin films are grown in the presence of a gas background, PLD of some materials is still carried out in vacuum. In any case knowledge of the characteristics of the in vacuum expansion can be used as a reference or as input data for analysis of the expansion in a gas background. The following two sections present an analysis of the plasma expansion in the experiments performed for this chapter. The vacuum expansion will be treated separately from the expansion in background gas. For the in vacuum case the Anisimov model will be presented [Anisimov, et al., 1996]. For gas background different models will be assessed and compared.

Expansion in Vacuum

Anisimov Model

Anisimov et al. treated the adiabatic expansion of a one component vapour cloud into vacuum using a particular solution of the gas-dynamic equations, which applies when describing flows with self similar expansion. It is assumed that the formation time of the vapour cloud is much less than its expansion time and that the focal spot of the laser has an elliptical shape with semi-axes X_0 and Y_0. The expansion is modelled as a triaxial gaseous semi-ellipsoid (Figure 4) whose semi-axes are initially equal to X_0, Y_0 and $Z_0 \approx v_{sound}t_p$, where t_p is the duration of the laser pulse and v_{sound} is the sound speed in the vaporised material.

The gasdynamic equations are:

$$\begin{aligned} &\frac{\partial \rho}{\partial t} + div(\rho \mathrm{v}) = 0 \\ &\frac{\partial \mathrm{v}}{\partial t} + (\mathrm{v}\nabla)\mathrm{v} + \frac{1}{\rho}\nabla p = 0 \\ &\frac{\partial S}{\partial t} + \mathrm{v}\nabla S = 0 \end{aligned} \qquad (1)$$

where ρ, p, v and S are the density, pressure, velocity and entropy of the gas respectively. It is assumed that the flow parameters are constant on ellipsoidal surfaces and the density and pressure profiles can be written as:

$$\begin{aligned}\rho(x,y,z,t) &= \frac{M}{I_1(\gamma)XYZ}\left[1-\frac{x^2}{X^2}-\frac{y^2}{Y^2}-\frac{z^2}{Z^2}\right]^{\alpha} \\ p(x,y,z,t) &= \frac{E}{I_2(\gamma)XYZ}\left[\frac{X_0Y_0Z_0}{XYZ}\right]^{\gamma-1}\left[1-\frac{x^2}{X^2}-\frac{y^2}{Y^2}-\frac{z^2}{Z^2}\right]^{\alpha+1}\end{aligned} \tag{2}$$

where $M = \int \rho(r,t)dV$ is the mass and $E = (\gamma-1)^{-1}\int p(r,0)dV$ is the initial energy of the vapour cloud and γ is the ratio of the plasma specific heats.

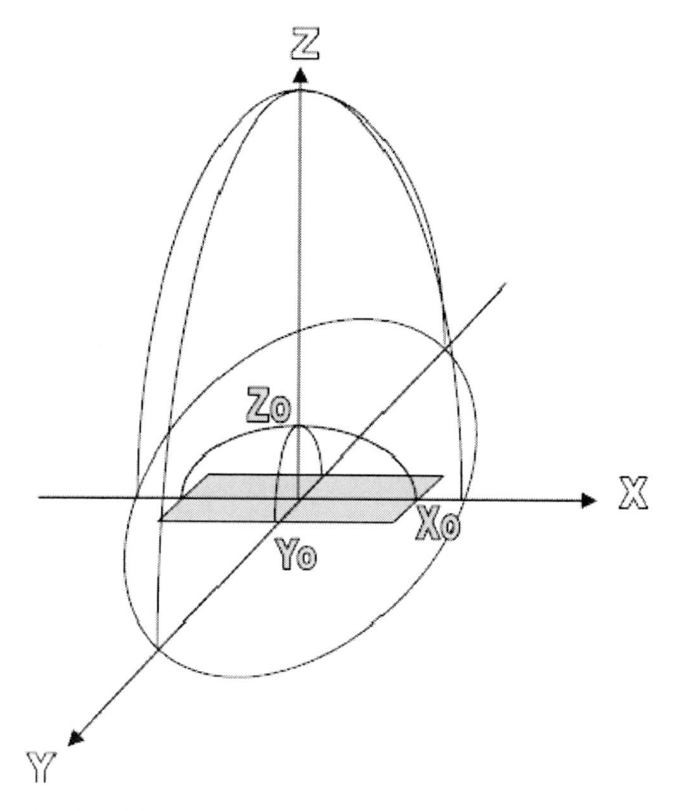

Figure 4. Scheme of the ellipsoid assumed for the in vacuum plasma expansion. Grey rectangle represents the laser spot.

The constants I_1 and I_2 are equal to:

$$\begin{aligned}I_1(\gamma) &= \frac{\pi^{\frac{3}{2}}\Gamma(\alpha+1)}{2\Gamma\left(\alpha+\frac{5}{2}\right)} \\ I_2(\gamma) &= \frac{\pi^{\frac{3}{2}}\Gamma(\alpha+2)}{2(\gamma-1)\Gamma\left(\alpha+\frac{7}{2}\right)}\end{aligned} \tag{3}$$

where Γ(z) is the Gamma-function. For the pressure and density profiles defined by equations (2) and when the initial vapour cloud is isentropic (α=1/γ-1), the entropy depends on the coordinates and time as:

$$S = \frac{1}{\gamma - 1}\ln\left\{\frac{E}{I_2(\gamma)X_0Y_0Z_0}\left(\frac{I_1(\gamma)X_0Y_0Z_0}{M}\right)^{\gamma}\left[1 - \frac{x^2}{X^2} - \frac{y^2}{Y^2} - \frac{z^2}{Z^2}\right]^{\frac{\gamma-2}{\gamma-1}}(\gamma - 1)\right\} + const \quad (4)$$

Using equations (2) and equations (3), the gas-dynamic equations can be reduced to a set of ordinary differential equations:

$$\begin{aligned} \ddot{X} &= -\frac{\partial U}{\partial X} \\ \ddot{Y} &= -\frac{\partial U}{\partial Y} \\ \ddot{Z} &= -\frac{\partial U}{\partial Z} \\ U &= \frac{5\gamma - 3}{\gamma - 1}\frac{E}{M}\left[\frac{X_0Y_0Z_0}{XYZ}\right]^{\gamma-1} \end{aligned} \quad (5)$$

The initial conditions for equations (5) are set as:

$$\begin{aligned} & X(0) = X_0, Y(0) = Y_0, Z(0) = Z_0 \\ & \dot{X}(0) = \dot{Y}(0) = \dot{Z}(0) = 0 \end{aligned} \quad (6)$$

In order to integrate equations (5) numerically, the following dimensionless quantities are introduced, using X_0 as a spatial scale length:

$$\xi\ddot{\xi} = \eta\ddot{\eta} = \zeta\ddot{\zeta} = \left[\frac{\eta_0\zeta_0}{\xi\eta\zeta}\right]^{\gamma-1} \quad (7)$$

with initial conditions:

$$\begin{aligned} & \xi(0) = 1, \eta(0) = \eta_0, \zeta(0) = \zeta_0 \\ & \dot{\xi}(0) = \dot{\eta}(0) = \dot{\zeta}(0) \end{aligned} \quad (8)$$

where

$$\xi = \frac{X}{X_0}, \eta = \frac{Y}{X_0}, \zeta = \frac{Z}{X_0}$$
$$\tau = \frac{t\beta^{1/2}}{X_0}, \eta_0 = \frac{Y_0}{X_0} \qquad (9)$$
$$\zeta_0 = \frac{Z_0}{X_0}, \beta = (5\gamma - 3)\frac{E}{M}$$

Equation (7) is solved using a fourth order Runge-Kutta-Nyström method. Details of the program and the mathematical method used can be found in Hansen [Hansen, et al., 1999]. From equations (2) we can obtain the mass flux in the Z direction:

$$j(z,t) = \rho(x,y,z,t)\mathrm{v}_z(x,y,z,t)$$
$$\mathrm{v}_z(x,y,z,t) = \frac{z}{Z}\dot{Z} \qquad (10)$$

The angular dependence of the number of particles arriving per unit area normal to the flow, $J(\theta)$, can be found by integrating equations (10) from the arrival time t_s to infinity [Hansen, et al., 1999]. The arrival time t_s is the time at which the expansion front reaches the position of the probe. If $t_s >> X_0/\beta^{1/2}$ the analytical expression for $J(\theta)/J(0)$ in the plane y-z will be:

$$\frac{J(\theta)}{J(0)} = \left(1 + \tan^2\theta\right)^{\frac{3}{2}} \left[1 + \left(\frac{Z_{\inf}}{Y_{\inf}}\right)^2 \tan^2\theta\right]^{-\frac{3}{2}} \qquad (11)$$

From equations (9) it can be seen that the model depends on two pure geometric parameters (X_0 and Y_0), which are related to the shape of the laser spot, and a third parameter related to the dimension of the plasma perpendicular to the target surface after the laser pulse has elapsed (Z_0). This parameter depends on the physical properties of the plasma through γ. For a sudden expansion of a gas into vacuum the velocity of the expansion front (v_f) is related to the sound speed through the following expression [Kelly, 1989] $v_f = 2(\gamma-1)^{-1}v_s$. The expansion front velocity was calculated from the measured arrival time of the plasma front to different positions using the ion signals from Langmuir probe measurements. The expansion velocity was calculated to be ($v_f = 2.3\times10^6$ cm/s). In all cases the time at which 10% of the maximum signal is reached was used in the calculations. Numerical calculations were performed for different γ values. Each γ value was used to calculate different v_s and Z_0. Then using equations (10) the current flow was calculated at different positions along the plasma expansion axis. The calculated current flows were compared in each case to the experimentally obtained ion currents in order to obtain the best γ value. Figure 5 shows examples of the results. The best fit for both time of arrival and signal shape was obtained for γ=1.1. In all cases both the calculated and the ion signals were normalised to their respective maximum. Figure 6 shows the time (dimensionless) dependence of the Z/Y, Z/X and Y/X values for γ=1.1.

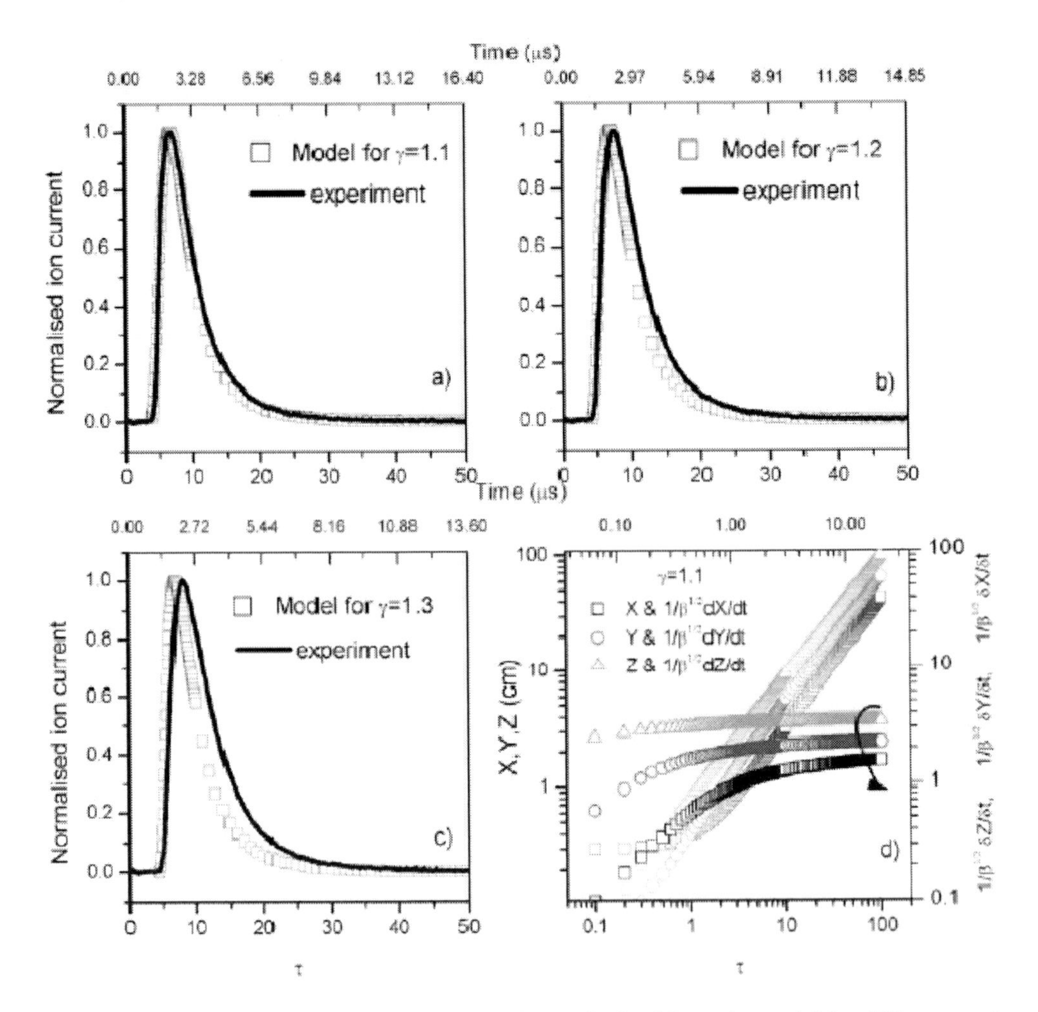

Figure 5. In a),b) and c) comparison of the particle flows obtained from the model for different γ values and the ion current measured experimentally is shown. In all cases the ion current was measured at 2.6 cm from the target surface in vacuum. Top axes are the real time and bottom dimensionless time. In d) the results of the numeric solution of equation (7) are presented.

The plasma angular distributions in both the X and Y directions were measured by rotating the probe in the x-z and y-z planes respectively. Figure 7 shows the normalised ion angular distributions. For the y-z plane the radius of the described circumference was 4 cm from the target, while for the x-z plane it was 7.8 cm. Both data were fitted, using equation (11), giving the ratios Z/X=2.4 and Z/Y=1.8. Comparison of these values with those predicted by Anisimov model shows good agreement (see Figure 6). This result is another way of corroborating the used γ value.

Using this γ value, the total initial energy of the plasma was estimated from equation (9), obtaining E=0.053 J, which is a reasonable result if we take into account the energy losses during the evaporation of the material.

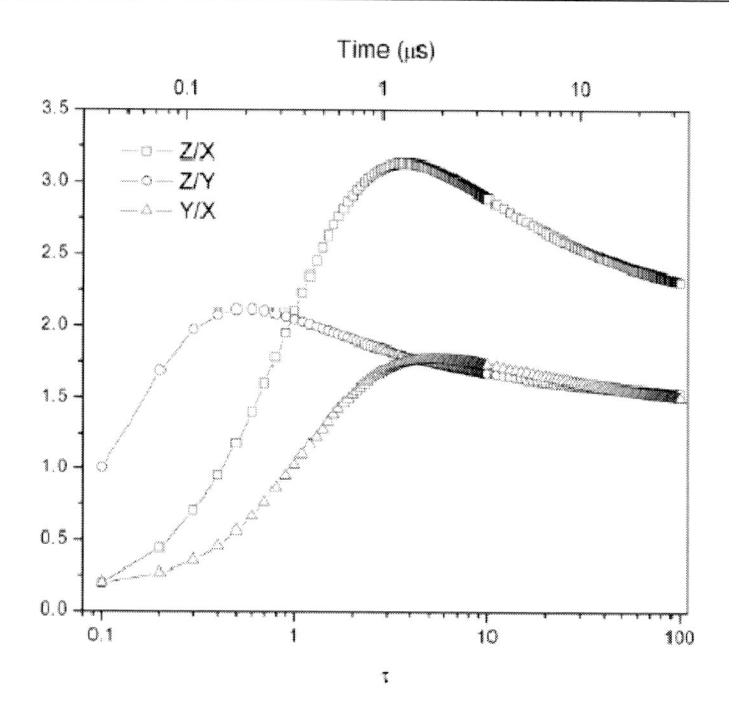

Figure 6. The time dependence of the Z/X, Z/Y and Y/X values. Note that at about 0.3 μs, the dimension of the plasma in the Y direction is greater than the dimension in the X direction i.e. there is a rotation of the semi-axes of the ellipsoid.

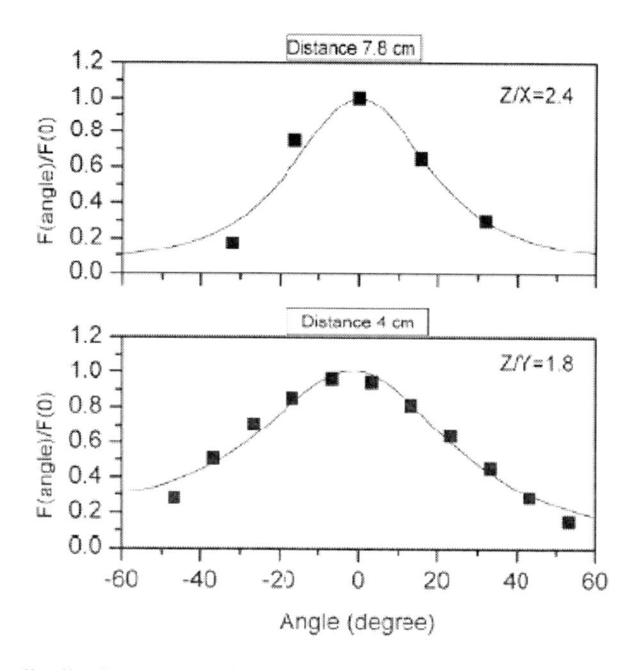

Figure 7. Ion angular distribution, measured by a Langmuir probe at different angles with respect to the plasma expansion axis. Top: at a distance from target of 7.8 cm in the x-z plane. Bottom: at a distance of 4 cm and in the y-z plane. Squares are experimental values and solid lines are the fits obtained using equation (11).

Here, it was assumed that the energy required for evaporation of the target material is the sum of the energy expended in heating up the material E_{heat} and the vaporization energy E_{vap}. Thousands of degrees Kelvin are temperatures typically reached on the target material during PLD process. For ZnO at the pressures used in this thesis, vaporization temperatures are in the range of 1000 to 1500 K [Searcy, et al., 1970].

The energy needed to bring a certain mass of material at room temperature to the above mentioned temperatures can be estimated as $E_{heat} = mc\Delta T$; with $c = 0.5\ JK^{-1}g^{-1}$ and a mass of 1.6 μg, which was the average evaporated mass per pulse, a value of $E_{heat} = 0.001$ J is obtained. Using 464 kJ mol^{-1} as the vaporization energy per mol [Searcy, et al., 1970] and knowing the number of moles in the evaporated mass (1.23×10^{-8} mol of ZnO), the energy expended for the vaporization will be $E_{vap} \sim 0.006$ J. The total energy required for the evaporation is then ~ 0.007 J, adding this value to the energy obtained from the model (0.053 J) gives a total energy of 0.06 J which was the average energy used in the experiments. In both Figures 6 and 7 the so called "flip over" can be observed. After a period of time, hundreds of nanoseconds for this case, following the initiation of the plasma expansion, the dimension of the plasma in the Y-Z plane is greater than in the X-Z plane. The angular distribution is broader in the Y-Z plane, which is the plane of the semi-minor axis of the laser spot (see Figure 4).

Expansion in a Gas Environment

Under this condition the analysis of the plasma expansion is definitively more complex than the vacuum case. Now the expanding plasma particles will interact with the molecules of the gas, new processes of energy exchange appear speeding up the reduction of the total energy of the expanding plasma. Both elastic and inelastic interaction probabilities are related to the relative kinetic energy of the interacting particles; for steady plasmas in equilibrium and when the most important interactions are those in which electrons are involved, it will suffice to know the electron temperature. None of the latter conditions are completely fulfilled in a PLD process. Not only have the transient properties of the laser-produced plasma compromised the possibility of reaching equilibrium over the entire expansion but now other important interactions are present, i.e. chemical reactions involving ions and the background gas species (molecules, atoms, ions, etc.). However interactions of electrons with the background gas play an important role in producing ionised or exited species which increases the probability of the desired reaction with ions. The characterization of the electron temperature will be dealt with in the next chapter.

In supersonic expanding plasmas, as developed during PLD, translational energies are greater than the internal energies of the ions; from this the importance of being able to characterise the plasma expansion is evident. Depending on the pressure of the gas, plume splitting and total braking of the expansion occurs. Plume splitting is characterised by the detection of two peaks in TOF studies of the plasma flow. In general terms the explanation for such a splitting has been attributed to the existence of an energetic component of the plasma particles which travel through the gas background suffering little or no collisions while a slow component is formed after it couples with the gas background losing momentum and decelerating its flow. After a period of time the two components will be spatially separated. However, in terms of a theoretical explanation things are not so clear. Wood and Geohegan [Wood, et al., 1998] explained the process using a simple theory in which the particles were distributed within the plasma plume depending on their energy after scattering by elastic collisions with the gas background. Similar results using the collisional approach in Monte-Carlo simulation were reported by Sushmita et al. [Sushmita, et al., 2001]. In the development of a model for the dynamics of laser produced plasmas into ambient gases, Bulgakov et al. [Bulgakov, et al., 1995] related the splitting to oscillations that occur in the plasma as different secondary shock waves travel within it. However, while splitting due to such a process was detected at distances of 2-4 cm from the target for a pressure of 40 Pa, the time scale was in the order of a hundreds of microseconds, which is much greater than the results presented here.

In general both features, plume splitting and plasma deceleration, are related with the appearance of both a shock wave and a contact front. The shock wave is the propagation of a perturbation created by the sudden impact of the expanding plasma with the gas background. It defines a boundary layer separating the unperturbed gas in front of the shock wave from the gas that went through this layer and is perturbed. The contact front, then, establishes a boundary between this perturbed gas and the plasma.

The formation and propagation of shock waves has been widely studied [Sedov, 1959; Zel'dovich, et al., 1966]. It is well known that the formation of such a feature is related to the ratio of the pressure (density) of the perturber and the gas and that the speed at which it propagates through the gas is defined by the thermodynamic properties of the gas. On the other hand, a full formulation has been given in which parameters such as pressure, density and temperature of the perturbed gas behind the shock wave are expressed as a function of the speed at which this wave propagates through the gas. A more specific work on shock waves developed for the modelling of strong explosions [Sedov, 1959] has been widely used for the description of the laser produced plasmas [Geohegan, 1992a; Gupta, et al., 1991; Harilal, et al., 2003; Jeong, et al., 1998b; Misra, et al., 1999; Gonzalo, et al., 1997]. The justification for such an approach can be understood by analysing the basic assumptions in the strong explosion theory [Sedov, 1959]: (i) a large amount of energy E_s is released instantaneously from a small (negligible) volume, (ii) the mass of the energy source m_s is negligible compared with the mass of the gas background swept by the shock wave and (iii) the pressure exerted by the explosion over the gas should be greater than the pressure in the unperturbed background gas.

The first assumption is generally fulfilled for the energies and laser spot size used in laser ablation processes. However, the second, strictly speaking, is not fulfilled unless a very small amount of the ablated material is removed or the plasma has expanded over a volume for which the mass of the swept gas is greater than the ablated mass. A more fundamental problem arises from the fact that this model describes the propagation of a shock wave, which experimentally is difficult to measure.

The remainder of this section will be devoted to the analysis of the use of shock wave theory for laser produced plasmas, identifying both regions of applicability and plasma measurable parameters which could be used not only for a post experiment plasma characterization but also in predicting its characteristics.

Sedov-Taylor Model

The Sedov-Taylor (S-T) theory defines the shock position R by:

$$R = \varepsilon_0 \left(\frac{E_s}{\rho_0} \right)^{\frac{1}{(2+\upsilon)}} t^{\frac{2}{(2+\upsilon)}}$$

$$\varepsilon_0 = 1.08 \left(\frac{\gamma+1}{2} \right)^{\frac{2}{(2+\upsilon)}} \tag{12}$$

where ρ_0 is the density of the gas background and γ its specific heats ratio. The parameter υ will take values of 1, 2 or 3 for expansions with planar, cylindrical or spherical symmetry. The flow speed U_{bs}, density ρ_{bs}, peak pressure P_{bs} and temperature T_{bs} behind the shock (bs) are given by:

$$U_{bs} = \frac{2}{\gamma+1} \frac{dR}{dt}$$

$$\rho_{bs} = \frac{\gamma+1}{\gamma-1} \rho_0$$

$$P_{bs} = \frac{2}{\gamma+1} \rho_0 \left(\frac{dR}{dt} \right)^2 \tag{13}$$

$$T_{bs} = \frac{2\gamma}{(\gamma+1)} \left[\frac{(\gamma-1)}{(\gamma+1)} M^2 + 1 \right] T_0$$

where $M = \frac{dR/dt}{v_{sound}}$ is the Mach number with $v_{sound} = \left(\gamma \frac{P_0}{\rho_0} \right)^{\frac{1}{2}}$.

Until this point the existence of the shock wave has been assumed. A shock wave is formed only if the shock wave thickness (Δ) is greater than the diffusion length (l_{Diff}) of the gas molecules through the perturbed region. These two parameters are defined as follows:

$$\Delta = \frac{\gamma-1}{\gamma+1} \frac{R}{3}$$

$$l_{Diff} = 2\sqrt{Dt} \tag{14}$$

where D is the diffusion coefficient within the shock region and t is the diffusion time. The diffusion coefficient is defined as follows [Dyer, et al., 1990]:

$$D = D_0 \left(\frac{T_{bs}}{T_0} \right)^{0.75} \left(\frac{\rho_{atm}}{\rho_{bs}} \right) \quad (15)$$

In practice this model is commonly used to extract the plasma expansion velocity by fitting curves of shock position versus time measured during the experiment. Langmuir probe and ICCD images are the main characterization techniques used to measure the experimental values.

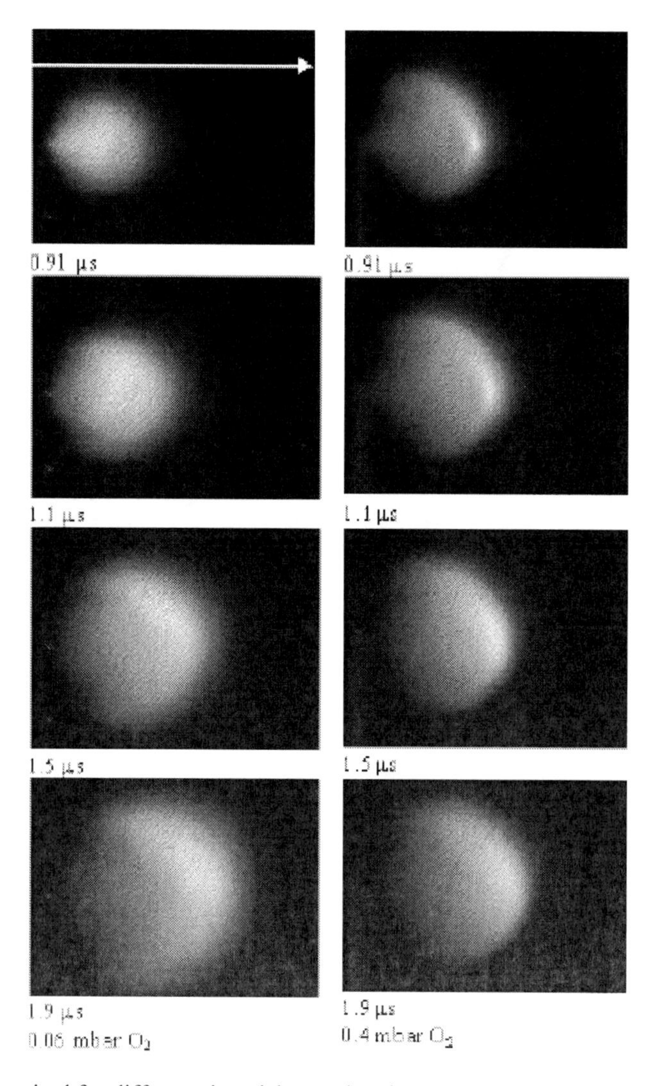

Figure 8. Images acquired for different time delay as the plasma expands in gas background at 0.06 mbar and 0.4 mbar. The direction of the expansion is denoted on the top-left image with length scale of 4 cm. The target position is the left border in all cases. Images are normalised to their maximum intensity.

Figure 8 shows time resolved images. The images were recorded for two different pressures of oxygen, 0.06 mbar and 0.4 mbar. The low pressure case is characterised by a broader and homogeneous distribution of the hotter regions in the plasma plume. As the plasma expands, the hotter region starts concentrating toward the leading edge of the plume. The higher pressure case shows the presence of the confined region occurring earlier in time during the expansion. For longer periods of time this region begins to broaden. For both cases the appearance of such a region defines the formation of a contact front, which in turn is related to the formation of a shock wave.

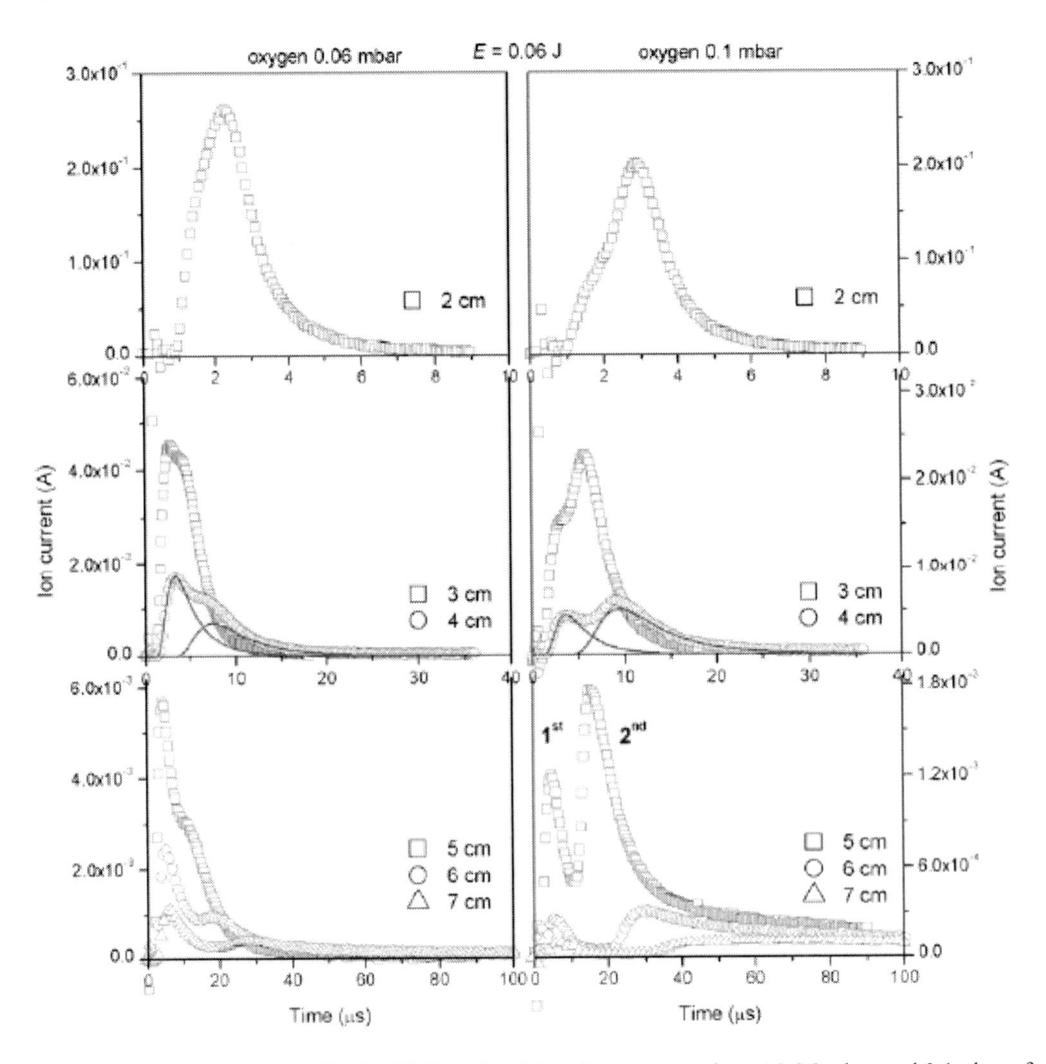

Figure 9. Ion currents measured in the TOF study of the plasma expansion at 0.06 mbar and 0.1mbar of oxygen. Two peaks are detected. The first peak expands through the gas background with a vacuum-like constant velocity, while the second peak decelerates. First and second peaks are labelled in the right-bottom graph for clarity. Examples of the fits (red lines) are presented for TOF measured at 4 cm for both pressures.

Figure 9 shows the measured TOF using the Langmuir probe for the 0.06 and 0.1 mbar oxygen pressures. The plume splitting is clearly present at all distances and at the two pressures, though its resolution increases with the increase of both distance and pressure. For clarity the peaks are labelled in the bottom-right graph. Figure 10 shows the TOF of the ion signals as measured using the Langmuir probe at a distance of 2, 3 and 4 cm from the target surface and at a pressure of 0.4 mbar of oxygen. At this pressure plume splitting is no longer detected. There is a stronger peak value reduction with increasing distance and a broadening of the distribution.

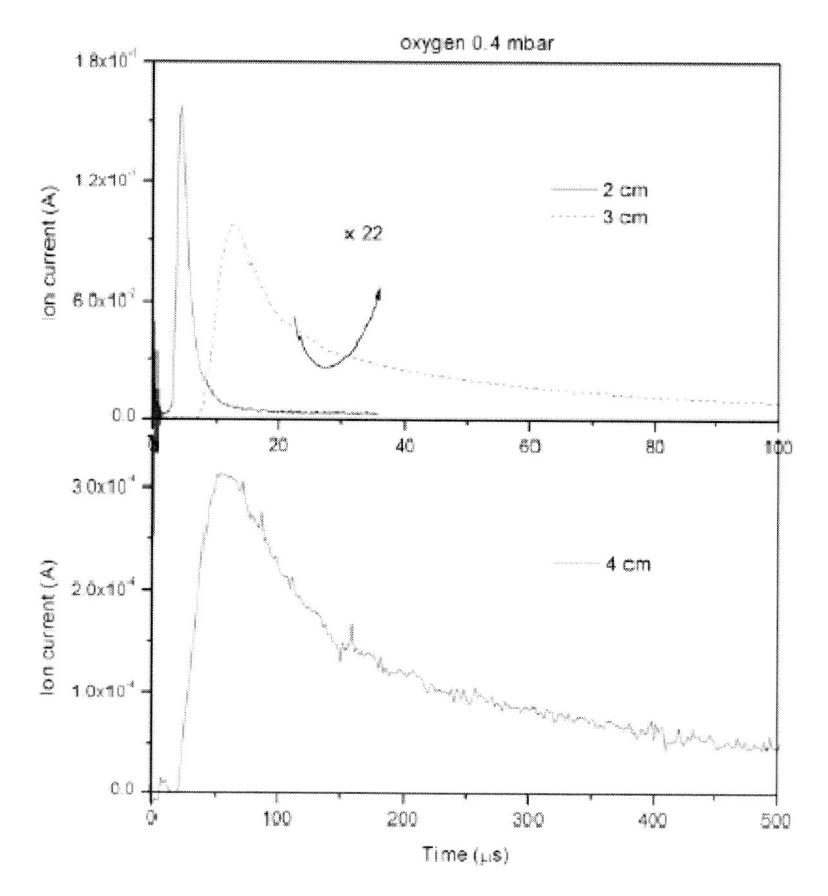

Figure 10. Ion currents measured in the TOF study of the plasma expansion at 0.4 mbar of oxygen. The expansion is characterised by a strong deceleration and a broadening of the signals. Note that the signal corresponding to 3 cm has been increased for clarity in the plots.

In order to extract the profiles of the curves which convolve to form the observed TOF, the curves were fitted using shifted half Maxwellians (Figure 9) and then both front and peak arrival time were extracted. The plots of the arrival time at each position for both peaks at the three different pressures are shown in Figure 11. As mentioned before, the first peak shows a TOF which indicates a linear vacuum-like flow. The velocities, in each case extracted from the slope of the plots, were similar for both pressures and half the value obtained for the in-vacuum expansion in the previous section. The second peak shows a decelerated expansion with a greater deceleration at the higher pressure.

The Sedov-Taylor model, equation (13), was used to fit the data extracted from both the ion signals and the ICCD photographs.

As mentioned above the Figure 11 corresponds to the data extracted using the Langmuir probe. Dashed and dotted lines were obtained by setting fixed values of both E_p and ρ_0 to 0.053 J and $9x10^{-5}$ kg/m^3 respectively allowing the fitting procedure to find the best ε_0 and υ. The values of E_p and ρ_0 correspond to the energy predicted by the Anisimov model in the previous section and the background gas density for a pressure of 0.06 mbar respectively. The solid lines were obtained by setting fixed values of parameter ε_0, choosing $\gamma = 1.28$, and $\upsilon = 3$ corresponding to a spherical symmetry as explained before.

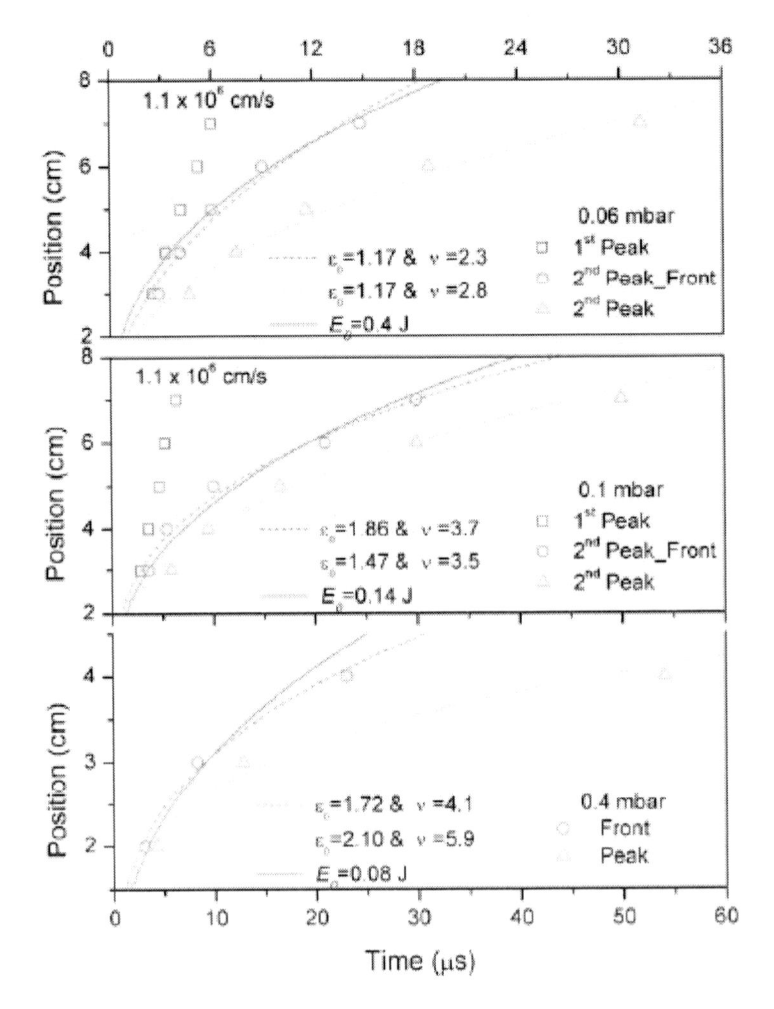

Figure 11. Plots of the arrival time of ion TOF at each position for both peaks at 0.06, 0.1 and 0.4 mbar of oxygen. For 0.06 and 0.1 mbar, where plume splitting is observed, the first peak shows a vacuum-like expansion. Both front and peak arrival times at each position show a decelerated expansion. Dashed, dotted and solid curves represent the fits performed using equation (12).

The analysis performed to the data extracted using the ICCD camera is presented in the Figure 12. The data represents the position of the front edge of the images presented in Figure 8. The fitted curves were obtained following the same idea as for the fit of the data presented in Figure 11.

In an overview of the results presented in both Figure 11 and 12 it can be corroborated that in general the expansion of laser produced plasmas is satisfactorily reproduced by a general shock wave law of the type of equation (12). However a close comparison reveals that different values of the parameters of the equation are obtained for the same experimental data. At 0.06 mbar of oxygen background, Figure 11, the fit of the front position (dashed line) suggests that the plasma expands with a cylindrical symmetry while the fit of the peak position (dotted line) is closer to a spherical one. Similar problems arise if comparing the previous mentioned results with the 0.06 mbar of oxygen case but for the data measured using ICCD images (Figure 12). These last results pose the question as to whether the reason for these inconsistencies is due to the non fulfilment of the basis of this theory, presented earlier in this section, or simply that, strictly, none of the used techniques are able to characterise the developed shock front. In trying to answer this question another model was used.

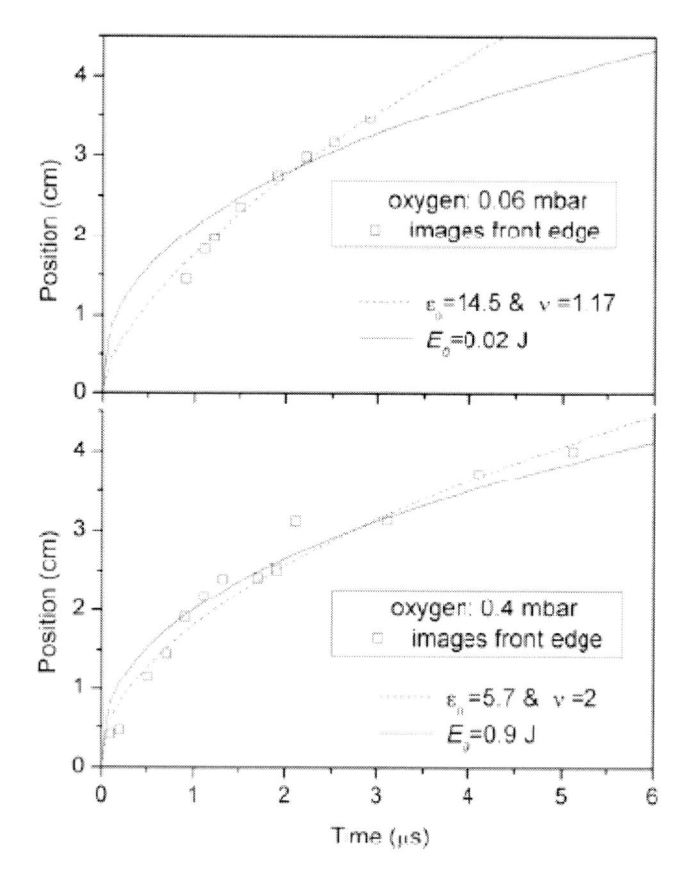

Figure 12. Plots of the arrival time of the images front edge at each position for both pressures 0.06 and 0.4 mbar of oxygen. For the two background pressures the expansion is decelerated. Both dashed and dotted fit lines were obtained following the same procedure as explained for Figure 11.

Freiwald-Axford Model

This model, a modification of the Sedov-Taylor theory, was developed by Freiwald and Axford [Freiwald, et al., 1975] with the aim of taking into account the mass of the explosion source. Their article showed that this modification is of particular importance at early stages of the plasma expansion. In what is, at least from the literature review performed for this thesis, the only work referring to this model, Kapitan and Coutts [Kapitan, et al., 2002] used it to explain the deviations of the S-T theory in describing shock waves produced in laser ablation processes. One of their main conclusions was that the model could be applied to describe processes carried out in background pressures >1 mbar. However, while their work followed the traditional approach of relating the plasma expansion to the properties of the gas background, in this chapter another approach is proposed.

Freiwald et al. took into account that a portion of the total energy E_p, released in the explosion is converted into both kinetic and internal energy of the source mass and not only in perturbing the gas through a shock wave. During the formulation an average source mass density of the following form was defined, $\bar{\rho}_p \equiv \rho_p \left(\frac{R_p^3}{R^3} \right)$, which implies the assumption that the radius of the plasma $R_p \approx R$, the shock wave radius. This assumption supports the idea of firstly, using the model to describe the propagation of the plasma itself and secondly to use the plasma gamma value instead of that of the gas background. As a result, an equation relating the energy of the source (initial plasma energy E_p), its density ρ_{p0}, plasma front position R_p and time was obtained:

$$E_p^{1/2} t = C_1^{1/2} \left(\frac{C_1}{C_2} \right)^{1/2} \left(0.455 F(\phi, 75^o) + \frac{2}{5} R_p \left(\frac{C_2}{C_1} \right)^{1/3} \left(1 + R_p^3 \frac{C_2}{C_1} \right)^{1/2} - 0.842 \right) \quad (16\text{ a})$$

$$C_1 = \frac{8.37}{(\gamma+1)^2} \rho_{p0} R_{p0}^3 \qquad C_2 = 8.37 \rho_0 \left(\frac{2}{(\gamma+1)^2} + \frac{1}{\gamma^2 - 1} \right)$$

$$\phi = \cos^{-1} \left(\frac{0.73 - R_p \left(\frac{C_2}{C_1} \right)^{1/3}}{2.73 + R_p \left(\frac{C_2}{C_1} \right)^{1/3}} \right) \quad (16\text{ b})$$

where R_{p0} is the radius of the volume of the plasma before the explosion and the function $F(\Phi, 75^o)$ is the elliptic integral of first kind. Knowing the values of the energy, density and radius of the plasma before starting the expansion the position of the plasma front as a function of time can be obtained by numerical evaluation. The plasma front velocity is obtained as follows:

$$\left(\frac{dR}{dt} \right)^2 = \frac{E_p}{\frac{16.75}{(\gamma+1)^2} \rho_0 R^3 + \frac{8.37}{(\gamma+1)^2} \rho_p R_p^3 + \frac{8.37}{\gamma^2 - 1} \rho_0 R^3} \quad (17)$$

Once the plasma front velocity is known the rest of the parameters can be calculated using equation (13).

Figure 13 shows the results (solid lines) of the numerical calculation performed using equation (16) for both 0.06 and 0.4 mbar pressures. In the calculation the values of the energy and γ obtained from the Anisimov model (see previous section) were used as input data. Both results show good agreement in describing the plasma front position as extracted from the ICCD images.

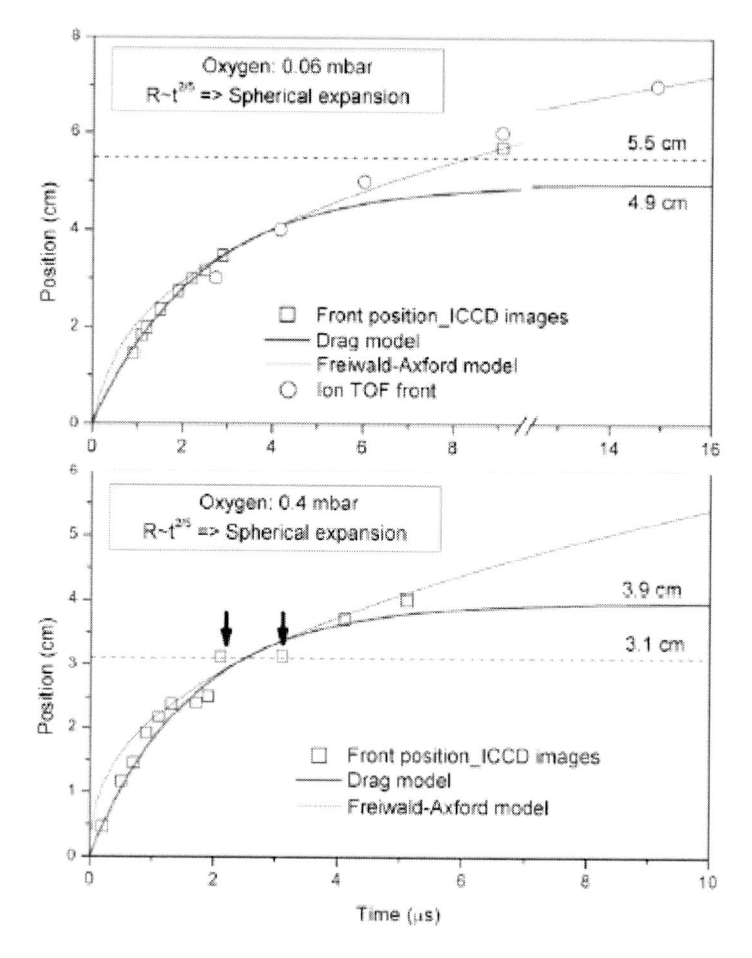

Figure 13. Solid lines represent the results of the numerical calculation preformed using equation (16). Red lines represent the Drag model and dashed lines maximum position as predicted by equation (19). Squares represent the front edge in the ICCD images presented in Figure 8. Circles (top figure) represent the arrival times of the 2nd peak fronts from the ion TOF at 0.06 mbar as in Figure 11.

In corroborating the correctness of the previous results the following phenomenological analysis was used. As has been reported [Predtechensky, et al., 1993] when the plasma expands in the presence of a gas background there exists a maximum distance which it will be able to reach. This is due to the reduction of its internal energy as a consequence of the work developed in displacing the surrounding gas. At this distance the pressure exerted by the plasma equals that of the gas background.

Two different models were used to estimate the maximum distance for the experimental conditions presented in this work, the so called drag model developed by Goehegan et al [Geohegan, 1992b] and a model developed by Predtechensky et al. [Predtechensky, et al., 1993].

The drag model is defined by:

$$R = R_{max}\left(1 - e^{-\beta t}\right)$$
$$R_{max} = \frac{v_0}{\beta} \qquad (18)$$

where v_0 is the initial velocity of the plasma expansion and β is a slowing coefficient. Predtechensky model is defined by:

$$R_{max} = \left(\frac{2ME_p}{\pi^2 \rho_g P_g}\right)^{1/6} \qquad (19)$$

The values for the maximum distance as predicted from each model are indicated in Figure 13. It can be observed that the drag model allows not only the prediction of the maximum travel distance; it also describes the plasma expansion. From the Figure 13 it seems that the Predtechensky model gives more accurate results in the calculation of the maximum position. In this figure the two arrows mark two data points which for the same distance were measured at different times with an interval of 1 μs. The data points correspond to a distance of ~ 3 cm which coincides with the value predicted by this model.

Figure 14 shows the values of the front kinetic energy and pressure exerted over the gas background as predicted by both the S-T and the F-A models. In the case of the S-T model the front velocities were calculated by the derivative of the expression obtained from the fit of Figure 12 (dashed line). For the F-A model the front velocities were calculated using equation (17). In both models after the front velocities were known, the pressures were calculated from equation (13). The difference between values predicted for each model is of several orders with the S-T model predicting the higher values. The pressure plot shows that for both gas backgrounds the F-A model predicts with good agreement the position at which the plasma exerted pressure equals that of the gas surrounding it. If the value for the maximum expansion distance is correct, it is to be expected that from this point on the plasma expands similarly to the spreading of a cloud of ions by diffusion through a gas.

The ionic number density n at a radius r from the release point and time t can be defined as follows [Mason, et al., 1988]:

$$n = \frac{N_0}{\left(4\pi D(t - t_0)\right)^{3/2}} e^{-\frac{r^2}{4D(t - t_0)}} \qquad (20)$$

where N_0 is the initial number of ions at the point where the diffusion starts and D is the diffusion coefficient. The parameter t_0 was introduced for fitting purposes to allow for an offset in the initial time.

Figure 15 shows the results of the fit performed using equation (20) to the ion current measured at 4 cm in an oxygen background of 0.4 mbar. The best values obtained in the fit were r = 1.01 cm and t_0 = 8.66 μs, both in good agreement with the previous models. Note that at this pressure the maximum expansion distance, predicted by both equation (19) and the

F-A model, was 3.1 cm and that the front arrival time to this distance was 8.2 μs (Figure 11). The lack of experimental data in diffusion of zinc ions prevents a comparison with the value obtained for the diffusion coefficient but longitudinal diffusion coefficients of the same order have been reported for ions of elements as heavy as Rb [Mason, et al., 1988].

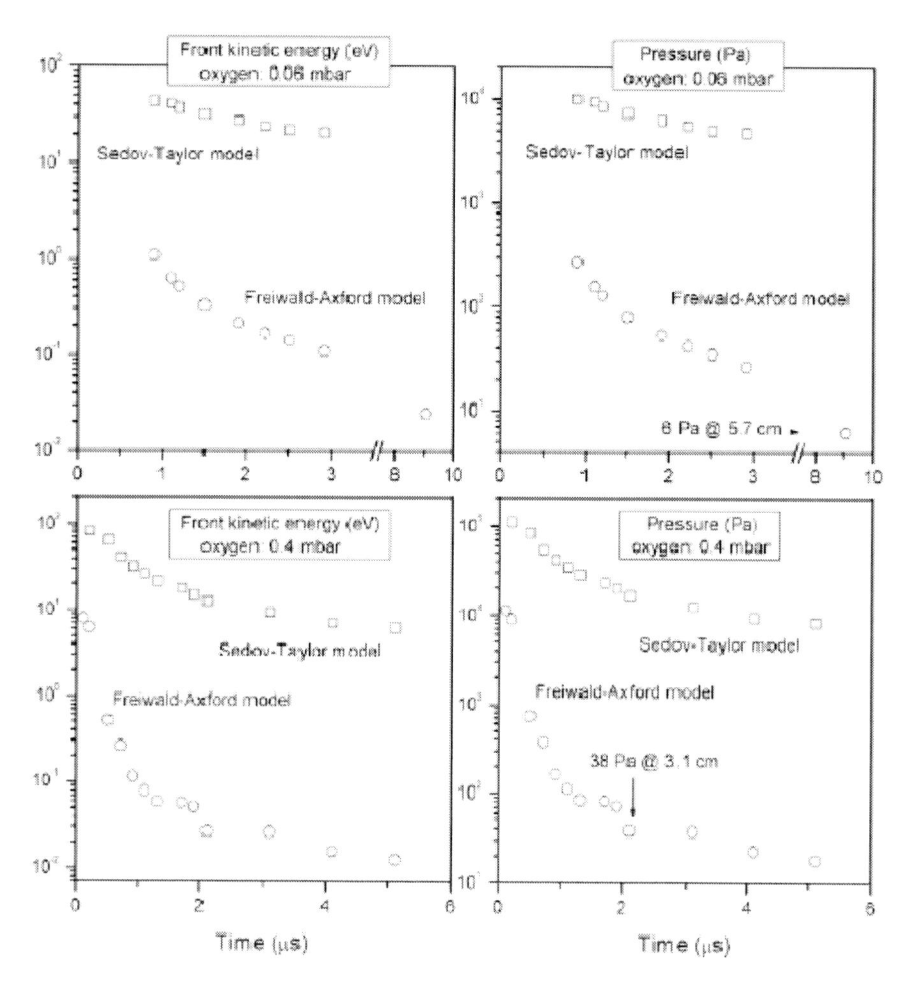

Figure 14. Values of pressure and front kinetic energy as predicted by both the S-T and F-A models. Left column presents the kinetic energy in eV and the right column presents the pressure in Pa. Top row for a gas background pressure on 0.06 mbar and bottom 0.4 mbar.

In Sedov-Tylor model subsection inconsistencies were found among results from using the S-T to fit data belonging to the front of the ion signal or its peak values (Figure 11). The same was observed when comparing results from using ICCD images front edge data (Figure 12) and any of the two options in ion signals. In Freiwald-Axford model subsection (Figure 13) the F-A model described successfully both ICCD images front edge data and ion signal front arrival times but only at 0.06 mbar oxygen pressure. Although no ICCD front edge data was available to compare with, the F-A model was used to model the front expansion as monitored using the Langmuir probe but after increasing the laser fluence to 2.5 J/ cm^2. For this laser fluence the mass loss measurements gave ~ 4 μg per pulse.

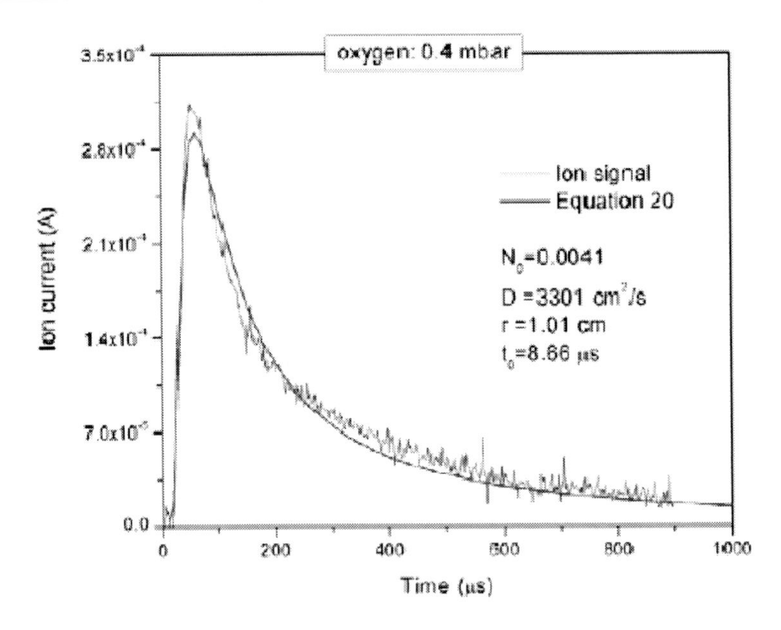

Figure 15. Shows the ion current signal measured at 4 cm from the target and in 0.4 mbar of oxygen background. Dotted line represents the fit obtained by using equation (20).

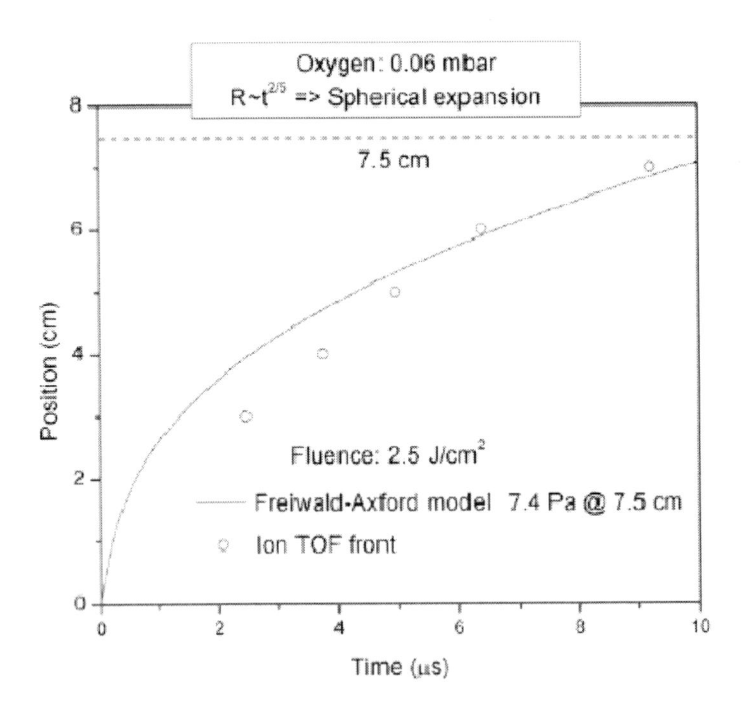

Figure 16. Result from using the F-A model (solid line) to predict plasma expansion in 0.06 mbar of oxygen background but after increasing the laser fluence to 2.5 J/cm^2. Circles represent the front arrival time at each position as measured using a Langmuir probe.

The in vacuum plasma expansion velocity was measured to be the same as for the previous used laser fluence so γ = 1.1 and an initial plasma energy E_p = 0.12 J was used as input data. This value of energy comes after taking into account the energy released during the vaporization of the above mentioned mass of material, using the same procedure as explained at the end of Expansion in vacuum section.

Figure 16 shows the results of the numerical calculation (solid line) together with the plot of the arrival time at each position of the front in the ion signals. The dashed line represents the maximum travel distance from equation (19). A reasonable agreement can be observed for the plasma front positions and the predicted position at which pressure equals that of the background gas. Again deviations appeared when applied to the 0.4 mbar case.

Although not commented upon, a similar discrepancy can be observed in the data presented by Geohegan et al [Geohegan, 1992b], when applying these two techniques in studying the expansion of laser produced plasmas in presence of a gas background. An explanation for such behaviour is not clear but the coincidence of both techniques at background pressures lower than 0.1 mbar might be related to the strong overlap that still exists between the first and second peak of the plume splitting for the distances and laser fluences presented here. On the other hand at this low pressure neutralization of ions in the contact front might not be as efficient as for higher pressures.

Thin Film Growth

Thin films of ZnO have been produced by wide range of techniques, including chemical vapour deposition (CVD) [Calli, et al., 1970], metallo-organic chemical vapour deposition (MOCVD) [Lau, et al., 1980] and pulsed laser deposition (PLD) [Sankur, et al., 1983; Prekumar, et al., 2009; Late, et al., 2009; Ayouchi et al., 2009; Jang, et al., 2010; Bruncko, et al., 2010; Fasio, et al., 2011]. These films have many potential applications in solar cells, light emitting devices, spin electronics, surface acoustic wave devices etc. ZnO is a wide bandgap (3.3 eV) II-VI semiconductor, which shows an efficient band-edge emission. The near UV interband emission, and its relation to the structural, stoichiometric and surface properties of the ZnO thin films, has been discussed in several papers in the recent years [Tang, et al., 1998; Wu, et al., 2001]. Studies of the influence of the film thickness [Shim, et al., 2002], oxygen pressure [Im, et al., 2000] and growth time [Vasco, et al., 2001] on the optical properties have been reported, and suggest that the oxygen stoichiometry is critical to obtaining good quality films. This chapter presents the characterization of the obtained thin films. Structure, morphology, stoichiometry and optical properties of the films will be analysed. Using the results obtained from the previous sections conclusions on how the plasma conditions are related to the properties of the final films will be obtained.

ZnO Thin Films

Figure 17 shows XRD patterns of four ZnO samples obtained for different growth conditions.

The strong (002) and (004) peaks show that there is preferential growth in the direction [00*l*]. There is relatively weak diffraction due to the presence of small quantity of (100) oriented material. The intensity and position of the (002) peak is similar for all the four samples, including the sample that was post-annealed in the chamber for 20 minutes (Figure 17 (d)).

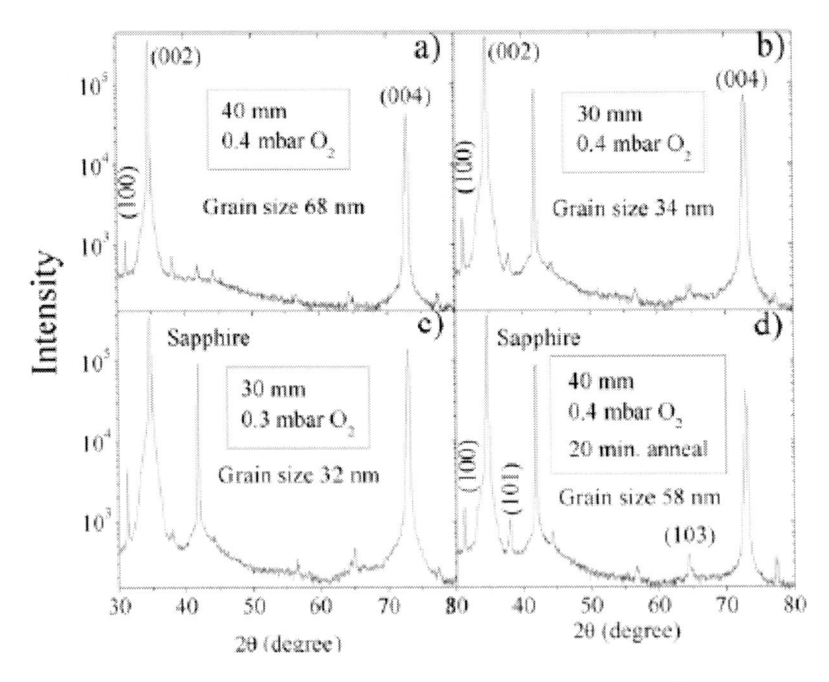

Figure 17. XRD pattern of c-oriented ZnO films and the corresponding grain size, as calculated using Scherrer formula.

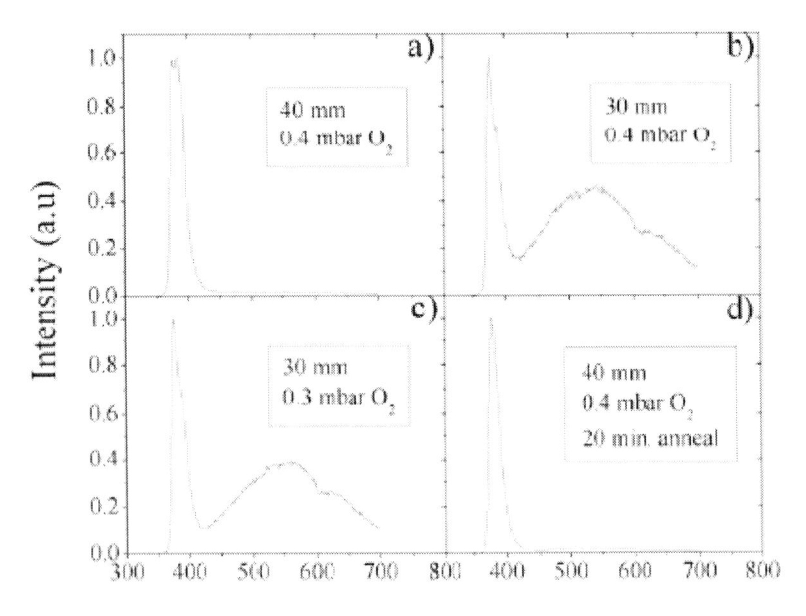

Figure 18. The PL spectra of all the ZnO films show the expected blue band-edge transition. The films in (b) and (c) also show yellow-green and orange bands.

All four samples were grown at the same temperature, but the two (Figure 17 (b) and 17 (c)), which were deposited closer to the target at 30 mm, have a smaller grain size, as revealed by the broader diffraction lines and calculated according to the Scherrer formula. These two samples also show a higher level of diffraction from (100) planes. Comparing the XRD for the samples in Figure 17 (a) and 17 (d), it would seem that post-annealing does not significantly affect the crystallinity of films deposited at 40 mm.

Figure 18 shows the Photoluminescence (PL) spectra of the same films as above.

All show the expected blue band associated with the band-edge transition (3.3 eV) characteristic of ZnO. However, samples grown at 30 mm Figure 18 (b) and 18 (c) also show strong yellow-green and orange bands, which are understood to be associated with oxygen vacancies [Im, et al., 2000]. Figure 19 shows an example of the typical surface morphology obtained in the grown thin films.

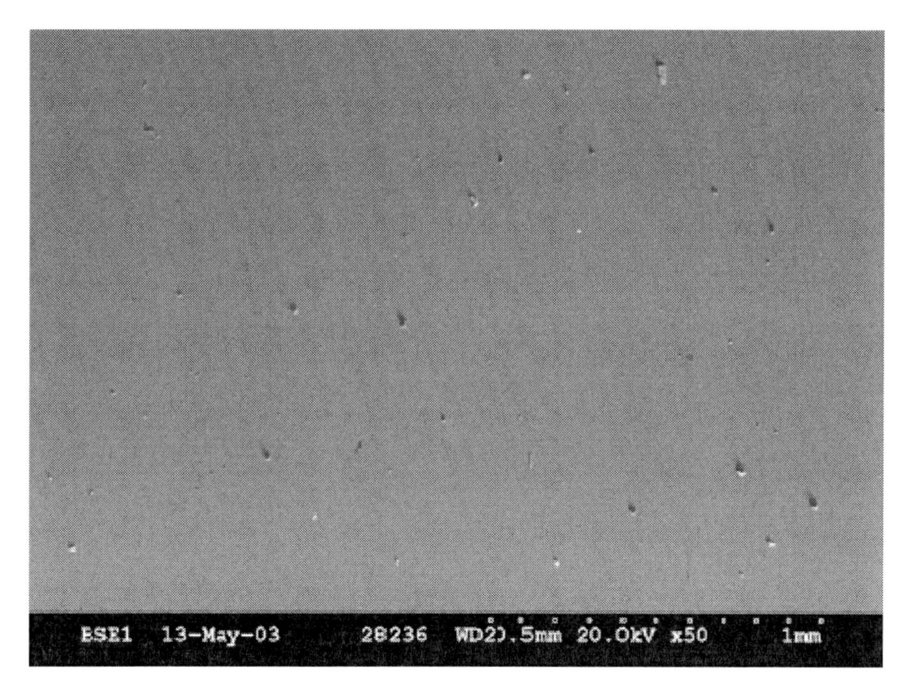

Figure 19. Example of the typical surface morphology of the grown thin film. The photograph was acquired using SEM with the specifications marked in the bottom of the figure. The distance from the first to the last square in the bottom of the figure is equivalent to 1 mm.

For all films the characteristic droplets were detected. Apart from this the films showed a reasonable smoothness. Several mechanisms have been discussed to be responsible for the droplet generation in PLD [Chrisey, et al., 1994]. In the case presented here seems to be due to the direct transport of target grains onto the substrate. This can be understood by looking at the SEM photograph of the target (Figure 20).

In general the results can be summarised as follow. The films were grown with similar growth parameters, yet important differences in their final optical properties were obtained. Besides, annealing of a sample with good as-grown quality, did not improve it. From the PL study can be concluded that such a quality difference is related to the oxygen stoichiometry in the films.

In explaining the above results the first step will be to discuss the possible mechanisms of oxygen incorporation.

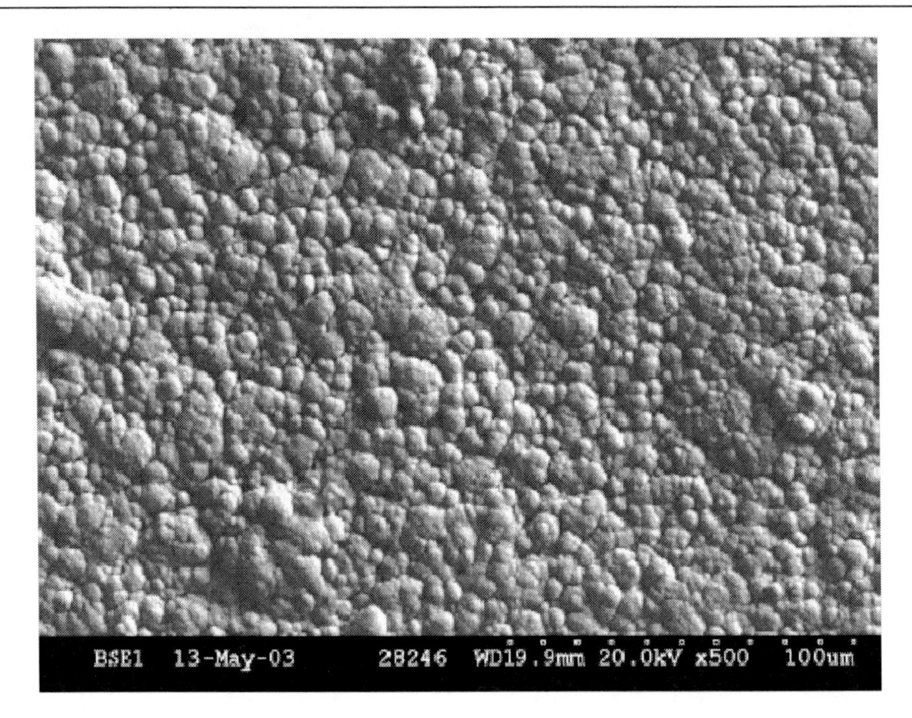

Figure 20. Surface morphology of the target used for the thin film growth. The photograph was acquired using SEM with the specifications marked in the bottom of the figure. The distance from the first to the last square in the bottom of the figure is equivalent to 100 μm.

During the plasma expansion the following reactions can take place:

$$\begin{aligned} &Zn^{+} + O_2 \rightarrow ZnO^{+} + O \\ &Zn^{*} + O_2 \rightarrow ZnO + O \\ &Zn + O_2 + e^{-} \rightarrow ZnO + O + e^{-} \end{aligned} \tag{21}$$

The third reaction could be obtained as well with an energetic atom instead of the electron. The first reaction in equation (20) has been reported [Fisher, et al., 1990] to have a threshold of about 3 eV and a maximum cross section at 6 eV. Although no data was found for the energy requirement of the second reaction, it can be expected that the excitation of the atomic zinc should, at least, equal the dissociation energy of the oxygen molecule (5.12 eV). Similar energy requirement should see the electron participating in the third reaction. Reactions with atomic oxygen need the participation of a third body.

It was shown before that for the pressures at which the films were grown a clear blast wave is developed as early as 500 ns after the laser pulse arrival (Figure 13). Besides, such a blast wave was decelerated to kinetic energies lower than 1 eV (Figure 14). From this the possibility for the first reaction of taking place, in the contact front, can be ruled out. Similar behaviour would be expected for the electrons in the contact front of the plasma, ruling out the third reaction.

It can be expected that the strong deceleration of the produced plasmas induces a quick cool down of plasma species temperature, this can be corroborated by optical spectroscopy but it is out of the scope of this chapter, so reaction inside the plasma plume will have a low probability. Furthermore, incorporation of oxygen due to reactions involving charged particles (Zn^+, e^-) would be negligible compared to the reaction with neutral species. This is due to the low ionization ratio of produced plasmas for the used laser fluence.

The probability of occurrence for the second reaction could be favoured for the predominance of atomic zinc in the plasma plume. Yet, should the second reaction be taking place little or no differences would be expected for an increase from 0.3 mbar to 0.4 mbar which is the difference in the pressures used for the growth of the films under analysis.

It seems from above that determining processes of oxygen incorporation take place at the substrate surface. If this is the case the oxidation of the film could be divided into two steps. The first step takes place during the material arrival onto the substrate; the second step will take place in the period of time between pulses.

The material during a PLD process does not arrive to the substrate in a continuum basis but as "packs" of material over a period of time defined by the time over which the plasma plume remains in contact with the substrate.

Figure 21 shows the normalised ion signals acquired with the same conditions as the films under discussion were grown. The ion signals show that for 40 mm the residence time of the plasma in the vicinity of the substrate deposition time is about 90 μs, whereas at 30 mm it is only about 40 μs. It means that at 40 mm material have more than twice the time to react.

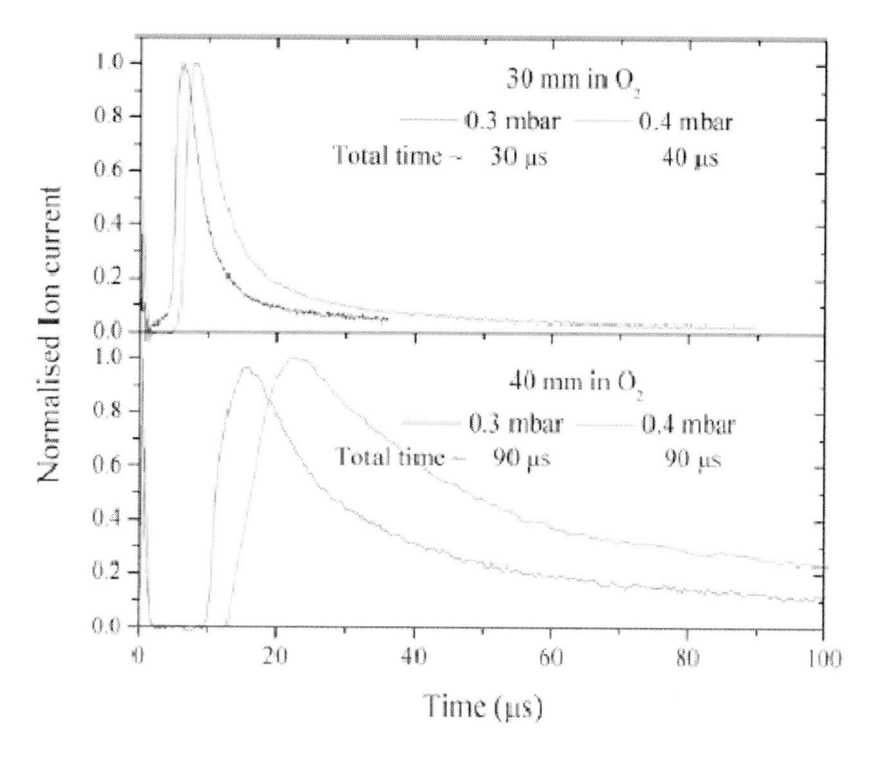

Figure 21. Normalised ion currents acquired with the same conditions used for the growth of the ZnO films under discussion.

During the second step, which takes place after the plasma plume moved away the substrate position, the time depends only on the laser frequency, being the same for all growth should make no difference. Furthermore, at this stage the material will be, at least, partially oxidized. It means that oxygen incorporation will be mainly due to the diffusion into the formed ZnO [Robin, et al., 1973]. This is a process which involves interaction with defects and hole migration. The diffusion coefficient has been measured to vary from 4.9×10^{-16} to 3.2×10^{-15} cm^2/s within the temperature range from 940 to 1141 °C and at 700 torr of oxygen. Diffusion data for lower pressures and temperatures are not available, but it is to be expected that there will be a decrease in the coefficient values. In general the depth of the diffusion will at most be of the order of Ångströms, which is only noticeable at earlier times of the growth when the film thickness is of this order.

CONCLUSIONS

The chapter was divided in three main sections, two of them were devoted to the study and analysis of the laser produced plasma expansion, both in vacuum and in a background gas. Finally in the third section the results obtained in the first two were used as the basis for the explanation and discussion of the results obtained from the film growth.

In vacuum all the analysis was supported on the Anisimov model. Using ion signals, acquired with a Langmuir probe, it was corroborated that this model describes successfully the expansion of laser produced plasma.

In the case of plasma expansion in a background gas, firstly, a full characterization of the expansion was performed using ICCD photography and ion signals. The Sedov-Taylor model was applied and shown that whilst it describes the general features of the expansion, i.e. front expansion $\sim t^{2/5}$, it failed in a full description of the plasma properties.

It was proposed, that, the Freiwald-Axford model may be more appropriate for the description of the plasma expansion in a background gas. Though, this model was not applied in its original form. Based on some assumptions, which were used by the authors during the mathematical formulation of the model, it was proposed the idea of describing the plasma contact front instead of the commonly described shock front.

The above proposition allowed us at the same time to use the plasma specific heats ratio as obtained from Anisimov model instead of approximated values of the same physical magnitude but for the gas background. A detailed comparison between the two models and the correctness of the plasma expansion velocities, positions and pressures corroborated that such approach was correct.

As result of the model comparison it was shown that the use of ion signals as extracted using the Langmuir probe is not correct for describing the plasma front. On the other hand the use of ICCD photography is correct.

It was corroborated that due to the interaction with the background gas the plasma reaches a maximum in its expansion which is well predicted by the Predtechensky model. Beyond this distance the plasma enters in a diffusive regime which is well described by a 3-D diffusion model.

In general was shown that plasma properties such as initial energy and specific heats ratio obtained from the Anisimov model can be used as input for the initial conditions of the plasma expansion in a background gas.

In the last section results obtained from the thin film depositions were presented. Structure, surface smoothness and optical properties of the ZnO thin films were studied and shown to present good quality.

It was shown that oxygen incorporation to the growing thin film occurs, preferentially, during the material condensation on the substrate.

Finally, the following question was posed in the introduction of the chapter; the problem that arises is whether the optimum conditions for the growth of good quality thin films can be predicted from models or at least bounded between some limits?

The answer is, yes. In order to produce a thin film of a given oxide material the following steps will lead to a faster and easier growth process:

To search and identify the mechanisms which direct a normal oxidation of the evaporated elements, i.e. optimum pressure and temperature. At the same time look for the cross sections for chemical collision.

From an ablation in vacuum monitored with Langmuir probe and emission spectroscopy the plasma would be characterised. That is, after calculating the mass loss the initial energy of the plasma can be extracted which gives a rather accurate idea of the amount of energy couple to the plasma from the laser pulse. The specific heats ratio is determined.

Using these parameters and the Freiwald-Axford model the behaviour of the plasma expansion in a given gas background pressure can be obtained. Possibility of reaction in the plasma front can be assessed from the calculated expansion kinetic energy. Maximum expansion distance can be obtained as well from this model and corroborated with the Predtechensky model.

In this way it can be estimated whether chemical reactions are possible during the plasma expansion or if the oxidation will take place on the substrate. Depending of all the results it can be decided whether to increase the laser energy or the pressure of the gas background or decide if the substrate will be placed in region is which the plasma is still driving a blast wave or is diffusing.

References

Anisimov, S. I.; Luk'yanchuk, B. S.; Luches, A. *Appl. Surf. Sci.* 1996, 96-98, 24.

Ayouchi, R.; Bentes, L.; Casteleiro, C.; Conde, O.; Marques, C.P.; Alves, E.; Bobin, J. L.; Durand, Y. A.; Langer, Ph. P.; Tonon, G. *J. Appl. Phys.* 1968, 39, 4184.

Bruncko, J.; Vincze, A.; Netrvalova, M.; Uherek, F.; Sutta, P. *Appl. Phys. A-Mater.* 2010, 101, 665.

Bulgakov, A. V.; Bulgakova, N. M. *J. Phys. D Appl. Phys.* 1995, 28, 1710.

Calli, G.; Coker, J. E. *Appl. Phys. Lett.* 1970, 16, 439.

Chrisey, D. B.; Hubler, G. K. *Pulsed Laser Deposition of Thin Films*; Wiley, New York, NY, 1994;

Cowin, J. P.; Auerbach, D. J.; Becker, C.; Wharton, L. *Surf. Sci.* 1978, 78, 545.

Dinescu, M.; Verardi, P. *Appl. Surf. Sci.* 1996, 106, 149.

Dyer, P. E.; Issa, A; Key, P. H. *Appl. Phys. Lett.* 1990, 57, 186.

Fazio, E.; Mezzasalma, A. M.; Mondio, G.; Serafino, T.; Barreca, F.; Caridi, F.; *Appl. Surf. Sci.* 2011, 257, 2298.

Freiwald, D. A.; Axford, R. A. *J. Appl. Phys.* 1975, 46, 1171.

Geohegan, D. B. *Appl. Phys. Lett.* 1992a, 60, 2732.

Geohegan, D. B. Laser ablation of Electronic Materials: Basic Mechanisms and Applications; North-Holland, 1992b; Gonzalo, J.; Afonso, C. N.; Madariaga, I. J. Appl. Phys. 1997, 81, 951.

Gupta, A.; Braren, B.; Casey, K. G.; Hussey, B. W.; Kelly, R. Appl. Phys. Lett. 1991, 59, 1302.

Hansen, T. N., Schou, J.; Lunney, J. G. Appl. Phys. A-Mater.1999, A69, 601.

Harilal, S. S., Bindhu, C. V.; Tillack, M.S.; Najmabadi, F.; Gaeris, A. C. J. Appl. Phys. 2003, 93, 2380.

Im, S.; Jim, B. J.; Yi, S. J. Appl. Phys. 2000, 9, 4558.

Jang, Y. R.; Yoo, K. H.; Park, S. M. J. Mater. Sci. Technol. 2010, 26, 973.

Jeong, S. H.; Greif, R.; Russo, R. E. Appl. Surf. Sci. 1998a, 127-129, 177.

Jeong, S. H.; Greif, R.; Russo, R. E. Appl. Surf. Sci. 1998b, 127-129, 1029.

Kapitan, D.; Coutts, D. W. Europhys. Lett. 2002, 57, 205.

Kelly, R., P. Soc. Photo- Opt. Inst. 1989, 1056, 258.

Kelly, R.; Dreyfus, R. W. Nucl. Instrum. Meth. B 1988, B32, 341.

Koopman, D. W. Phys. Fluids 1971, 14, 1707.

Langmuir, I.; Mott-Smith, H. M. Phys. Rev. 1926, 28, 727.

Late, D. J.; Misra, P.; Singh, B. N.; Kukreja, L. M.; Joag, D. S.; More, M. A. Appl. Phys. A-Mater. 2009, 95, 613.

Lau, C. K.; Tiku, S. K.; Lakin, K. M. J. Electrochem. Soc. 1980, 127, 1843.

Leboeuf, J. N.; Chen, K. R.; Donato, J. M.; Geohegan, D. B.; Liu, C. L.; Puretzky, A. A.; Wood, R. F. Appl. Surf. Sci. 1996, 96-98, 14.

Linlor, V. I. Phys. Rev. Lett. 1963, 12, 383.

Mason, E. A.; McDaniel, E. W. Transport properties of ions in gases; Wiley, NY, 1988.

Misra, A.; Mitra, A.; Thareja, R. K. Appl. Phys. Lett. 1999, 74, 929.

Moutinho, A.M.C.; Marques, H.P.; Teodoro, O.; Schwarz, R. Appl. Surf. Sci. 2009, 255, 5917.

Namiki, A.; Kawai, T.; Yasuda, Y.; Nakamura, T. Jpn. J. Appl. Phys. 1985, 24, 270.

Ohtomo, A.; Kawasaki, M.; Sakurai, Y.; Koinuma, H.; Yu, P.; Tang, Z. K.; Wong, G. K. L.; Segawa, Y. Mat. Sci. Eng. B- Solid 1998, B54, 24.

Predtechensky, M. R.; Mayorov, A. P. Appl. Supercond. 1993, 1, 2011.

Prekumar, T.; Manoravi, P.; Panigrahi, B. K.; Baskar, K. Appl. Surf. Sci. 2009, 255, 6819.

Robin, R.; Cooper, A. R.; Heuer, A. H. J. Appl. Phys. 1973, 44, 3770.

Ryu, Y. R., Zhu, S.; Budai, J. D.; Chandrasekhar, H. R.; Miceli, P. F.; White, H. W. J. Appl. Phys. 2000, 88, 201.

Sankur, H.; Cheung, J. T. J. Vac. Sci. Technol. A 1983, 1, 1806.

Sankur, H.; Cheung, J. T. J. Vac. Sci. Technol. A. 1983, 1, 1806.

Searcy, A.; Ragone, D. V.; Columbo, U. Chemical and Mechanical Behavior of Inorganic Materials; John Wiley and Sons Inc: New York, NY, 1970; pp 150-152 .

Sedov, L. I., Similarity and dimensional methods in mechanics; New York : Academic Press, NY, 1959; .

Shim, S., Kang, H. S.; Kang, J. S.; Kim, J. H.; Lee, S. Y. Appl. Surf. Sci. 2002, 186, 474.

Shim, S.; Kang, H. S.; Kang, J. S.; Kim, J. H.; Lee, S. Y. Appl. Surf. Sci. 2002, 186, 474.

Sushmita, R. F.; Thareja, R. K. Appl. Surf. Sci. 2001, 177, 15.

Tang, Z. K., Wong, G. K. L.; Yu, P.; Kawasaki, M.; Ohtomo, A.; Koinuma, H.; Segawa, Y. Appl. Phys. Lett. 1998, 72, 3270.

Tang, Z. K.; Wong, G. K. L.; Yu, P.; Kawasaki, M.; Ohtomo, A.; Koinuma, H.; Segawa, Y. *Appl. Phys. Lett.* 1998, 72, 3270.

Vasco, E.; Zaldo, C; Vazquez, L. *J. Phys.-Condens. Matt.* 2001, 13, 663.

Wood, R. F.; Leboeuf, J. N.; Geohegan, D. B.; Puretzky, A. A.; Chen, K. R. *Phys. Rev. B* 1998, 58, 1533.

Wu, X. L., Siu, G. G.; Fu, C. L.; Ong, H. C. *Appl. Phys. Lett.* 2001, 78, 2285.

Wu, X. L.; Siu, G. G.; Fu, C. L.; Ong, H.C. *Appl. Phys. Lett.* 2001, 78, 2285.

Zel'dovich, Y. B.; Raizer, Y. P. *Physics of Shock Waves and High Temperature Hydrodynamic Phenomena;* Academic Press: New York, NY, 1966; Vol. 2, pp 465-681.

In: Laser-Induced Plasmas
Editor: Ethan J. Hemsworth, pp. 35-54
ISBN 978-1-61324-851-5

Chapter 2

CLUSTER-CONTAINING PLASMA PLUMES: THE ATTRACTIVE MEDIA FOR HIGH-ORDER HARMONIC GENERATION OF LASER RADIATION

R. A. Ganeev*
Institute of Electronics, Akademgorodok, 33 Dormon Yoli street,
Tashkent 100125, Uzbekistan

ABSTRACT

A review of recent studies and applications of cluster-containing plasmas as the nonlinear optical media for frequency conversion of ultrashort laser pulses towards the extreme ultraviolet range is presented. The studies of conditions when the plasma producing on the surface of laser-ablated targets contains nanoparticles, clusters, and nanotubes are presented. The results are presented which show that nanoparticle formation in the plasma plumes can be accomplished using relatively long laser pulses under tight focusing condition. The nanoparticles were produced by the interaction of 300 ps, 20 mJ laser pulses with bulk Ag target at an intensity of $\sim 1\times10^{13}$ W cm^{-2}. The laser ablation of chromium in vacuum at the tight and weak focusing conditions of a Ti:sapphire laser radiation allowed to synthesis the 60 nm particles. The studies show that nanoparticle formation during ablation of metals by laser pulses strongly depends on the concentration of surrounding gas. At moderate laser intensity of the 300 ps pulses on the surface of nanoparticle-containing materials ($<5\times10^{9}$ W cm^{-2}), the deposited material remains approximately the same as the initial nanoparticles. The ablation of carbon nanotubes and C_{60} powders has been proved the effective technique to create the plasma plumes containing these species. The integrity of carbon nanotubes and C_{60} within the plasma plume was confirmed by structural studies of plasma debris. These cluster-containing plasma plumes proved to be the effective media for high-order harmonic generation of femtosecond laser pulses. The advantages of clustered plasmas are shown allowing the enhancement of low-order harmonic yield compared with atomic/ionic plasmas.

---

* e-mail: rashid_ganeev@mail.ru

INTRODUCTION

The nonlinear optical properties of nanoparticles took much attention due to their potential applications in optoelectronics, mode-locking technologies, optical limiting, etc. Currently, nanoparticle formation during laser ablation of solid-state targets using the ultrashort (femtosecond) laser pulses is a widely accepted technique. Together with formation of the ripples of wavelength range sizes, this technique gives the opportunity of creating the exotic structures with variable physical and chemical properties. The increase of surface area of the nanostructured materials allows for enhancement of the velocity of catalytic reaction, provides the opportunity of application of such structures for the information writing, manufacturing of lubricants, semiconductor technologies, etc. The structural, optical, and nonlinear optical parameters of nanoparticles are known to differ from those of the bulk materials due to the quantum confinement effect. Silver [1], copper [2], and gold [3] are among the most useful metals suited for the nanoparticles preparation for optoelectronics and nonlinear optics. Further search of prospective materials in clustered form, their preparation and application are of considerable importance nowadays. In particular, the formation of cluster-containing laser plumes using relatively long (nanosecond and picosecond) laser pulses seems to be an attractive area of studies due to availability of such lasers in many laboratories. The knowledge of the conditions when clusters are presented in the plasma plumes opens the door for the study of various properties of these media, in particular their high-order nonlinear optical properties, which allow the creation of the coherent sources of efficient high-order harmonic generation (HHG) of laser radiation [4].

The nanoparticle formation during laser ablation of targets has previously been described as a process of short excitation of electronic gas and transfer of this energy to the atomic cell with further aggregation processes, which continue during evaporation of the material [5,6]. In the case of bulk target ablation, the attention is taken for creation of the conditions when laser energy is accumulated for a short period at a small area to maintain the conditions of non-equilibrium heating. In that case, the extremely heterogeneous conditions help creating the nanoparticles in the small areas of heated samples. The ablation-induced nanoparticle formation in laser plumes was carefully documented in multiple experiments using femtosecond laser pulses [7,8]. One can maintain the conditions when the aggregated atoms do not disintegrate during evaporation from the surface. The analysis of the aggregation state of evaporated particles was carried out by different techniques. Among them the time resolved emission spectroscopy, CCD camera imaging of plasma plume, Rayleigh scattering, laser-induced fluorescence, etc, has shown the ability of defining the presence of nanoparticles in the laser plumes.

Currently, the nanoparticles, fullerene clusters, and carbon nanotubes (CNTs) with different sizes and shapes are commercially available from various manufacturers and can be attached to the surfaces and then evaporated using the laser ablation technique. In that case one has to carefully define the optimum laser intensity, pulse duration, and focusing conditions for heating of the nanoparticles until they evaporate from the targets. The history of ablated nanoparticles in that case can be difficult to analyze using the above techniques due to some restrictions in identification of the clusters in plasma plumes at moderate heating of the targets. In that case the comparison of the size characteristics of initial nanoparticles and deposited debris becomes a versatile approach for definition of the changes of nanoparticle morphology during laser ablation. Another indirect method is the analysis of the harmonic spectra generating in the plasmas containing nanoparticles and atoms/ions of the same origin.

It is well accepted that, when a solid target is ablated by the laser radiation, the ablating material consists on atoms and clusters. These atoms and clusters tend to aggregate during laser pulse or soon afterwards, leading to the formation of larger clusters. The reported results (see for example Ref. [9]) also indicate that the ablation processes in the picosecond and femtosecond time scales considerably differ compared to the nanosecond one. In addition to earlier experimental observations, several theoretical studies have suggested that rapid expansion and cooling of the solid-density matter heated by short laser pulse may result in nanoparticle synthesis via different mechanisms. Heterogeneous decomposition, liquid phase ejection and fragmentation, homogeneous nucleation and decomposition, and photomechanical ejection are among those processes that can lead to nanoparticles production [10-12]. Short pulses, contrary to the nanosecond pulses, do not interact with the ejected material thus avoiding complicated secondary laser interactions. Further, short pulses heat a solid to higher temperature and pressure than do longer pulses of comparable fluence, since the energy is delivered before significant thermal conduction can take place.

Pulsed laser deposition techniques in rare gas ambiences have been used for nanometer-size particle preparation, multi-component thin film deposition, and carbon nanotube syntheses [13]. As a result of frequent collisions of ablated particles with gas atoms in plume, the particles cool down and form nanoparticles [14]. The development of the nanoparticle synthesis at gas conditions and the characterization techniques that make possible the control of nanoparticle features within a few nm have attracted renewed interest to the production of metal nanoparticles, as this opens the possibility of taking advantage of their special properties for the development of applications such as new catalysts, tunnel resonance resistors, or optical devices [15]. The search of optimal gas pressure for nanoparticles formation during laser ablation of solids thus remains an open issue.

It is known that metal ablation in air is less efficient than that in vacuum due to re-deposition of the ablated material. The influence of surrounding gas on the conditions of cluster formation during laser ablation of the metals by short laser pulses has previously been analyzed only at two conditions, when the target was placed in vacuum or ambient air. The ablation rates in vacuum can be calculated using a thermal model, as well as from basic optical and thermal properties [16]. It is of interest to analyze the influence of the concentration of surrounding gas on cluster formation. One can study this process using noble gases of different atomic numbers at the pressures varying between 10^{-2} torr (i.e. vacuum conditions) and 760 torr (i.e. atmospheric pressure). It would be interesting to analyze whether there is a threshold in gas pressure scale above which the conditions of nanoparticle formation get spoiled.

The use of nanoparticles for efficient conversion of the wavelength of ultrashort laser toward the deep UV spectral range through harmonic generation seems an attractive application of cluster-containing plasmas. Note that earlier observations of HHG in nanoparticles were limited by using the exotic gas clusters formed during fast cooling of atomic flow from the gas jets [17-20]. One can assume the difficulties in definition of the structure of such clusters and the ratio between nanoparticles and atoms/ions in the gas flow. The characterization of gas phase cluster production was currently improved using the sophisticated techniques (e.g., a control of nanoparticle mass and spatial distribution, see the review [21]). In the meantime, the plasma nanoparticle HHG has demonstrated some advantages over gas cluster HHG [22]. The application of commercially available nanopowders allowed for precisely defining the sizes and structure of these clusters in the plume. The laser ablation technique made possible the predictable manipulation of plasma consistence, which led to the creation of laser plumes containing mainly nanoparticles with known spatial structure. The latter allows the application of such plumes in nonlinear optics, x-ray emission of clusters, deposition of nanoparticles with fixed parameters on the substrates for semiconductor industry, production of nanostructured and nanocomposite films, etc.

Other nanostructures, which attract the attention during last time, are the fullerenes and carbon nanotubes. Recently, the application of laser ablation technique allowed the creation of relatively dense C_{60}-rich plasma ($\sim 5\times10^{16}$ cm^{-3}), in a stark contrast with the heat oven-based methods ($\sim 10^{14}$ cm^{-3}). It was demonstrated the efficient broadband HHG in C_{60}-rich plasmas [23]. The changes in fundamental wave characteristics allowed the dramatic manipulation of the harmonic spectrum and intensity at well-defined conditions of C_{60}-containing plasma. It is also proposed the application of CNT-containing plasma for efficient harmonic generation. Carbon nanotubes have remarkable electronic and optical properties due to their particular structure combining one-dimensional solid-state characteristics with molecular dimensions. While their structure has extensively been studied by means of transmission electron microscopy and scanning tunneling microscopy, only few experimental studies have been reported on their nonlinear optical properties. Moreover, their high-order nonlinearities can be analyzed solely by probe experiments in the media containing sufficient amount of these species and transparent in the extreme ultraviolet (XUV) range. For these purposes, one has to create the CNT-containing plasmas, where the presence of nanotubes should be proven by indirect methods.

In this review, we analyze the studies of conditions when the plasma producing on the surface of targets contains the nanoparticles, clusters, and nanotubes. These studies show that nanoparticle formation in the plasma plumes can be accomplished using relatively long (subnanosecond) laser pulses under tight focusing condition. Laser ablation of various metals in vacuum at tight and weak focusing conditions of a Ti:sapphire laser radiation leads to synthesis of the 60 nm nanoparticles. Nanoparticle formation during laser ablation of metals strongly depends on the concentration of surrounding gas. The studies show that destruction of nanoparticle formation was attributed to the negative influence of surrounding gas particles on ablated particles aggregation. We discuss the morphology of ablated nanoparticles after their laser-induced deposition on various substrates thus confirming presence of nanoparticles in the plasmas at appropriate ablation conditions. At moderate laser intensity of subnanosecond pulses on the surface of nanoparticle-containing materials ($<5\times10^{9}$ W cm^{-2}), the deposited material remains approximately the same as the initial nanoparticles. We also analyze the ablation of carbon nanotubes and C_{60} powders to create the plasma plumes containing these species. The integrity of CNT and C_{60} within the plasma plume was confirmed by structural studies of plasma debris. The morphologies of the fullerene and nanotube targets as well as their ablated debris are analyzed. The cluster-containing plasma plumes proved to be the effective media for high-order harmonic generation of femtosecond laser pulses. We describe the HHG in various plasmas containing nanoparticles, clusters, and nanotubes.

1. Experimental Arrangements for Cluster-Containing Plasma Formation and High-Order Harmonic Generation

The studies describing below were carried out using the Ti:sapphire lasers. To create the ablation, part of uncompressed radiation from the Ti:sapphire laser ($\lambda = 800$ nm, $t = 210$ - 300 ps) was focused on a target placed in the vacuum chamber. The spot size of this radiation on the target surface was maintained in the range of 0.5 – 0.8 mm. The intensity of this radiation (I) on the target surface was varied between 2×10^9 and 5×10^{10} W cm^{-2}, depending on the target material. The laser fluence during ablation was in the range of 0.4 - 1 J cm^{-2}. The chamber was maintained at the vacuum pressure of 8×10^{-4} torr. The debris from plasma plume were deposited on nearby placed semiconductor wafers, glass plates, and metal foils. The distance between the target and substrates was 40 - 70 mm. The structure of deposited material was analyzed by the scanning electron microscope (SEM, JEOL JSN-5600) and transmission electron microscope (TEM, JEM-201OF). The absorption spectra of deposited debris were analyzed using the fiber spectrometer (USB2000, Ocean Optics, USA).

At the first set of experiments, the bulk targets, such as silver, were ablated to create the conditions for nanoparticles formation in the laser plumes. Further, the targets containing nanoparticles were ablated as well. The various commercially available nanoparticles (Alfa Aesar, USA) were glued on the glass substrates by mixing with the drop of superglue. The bulk materials of the same origin as nanoparticle powders were used to compare the ablation properties of these targets. The targets containing silver, gold, platinum, ruthenium, and palladium nanoparticles (Wako Industries, Japan) were also used in these studies for creation of clustered plumes. The latter nanoparticles were purchased in the form of suspensions and were dried on the surfaces of glass substrates prior to laser ablation. Other ablated materials were the C_{60} and CNT powders. These powders were also glued on different substrates. More details on targets preparation are presented below in the corresponding subsections.

The cluster-containing plumes were used for the HHG of femtosecond laser pulses. The harmonic generation in nanoparticle-containing laser plumes was carried out using the femtosecond pulses ($\lambda = 800$ nm, $t_{fp} = 35$ - 130 fs), which were focused, after some delay (6 – 74 ns) with regard to the beginning of laser ablation, on the plasma from the orthogonal direction (Figure 1).

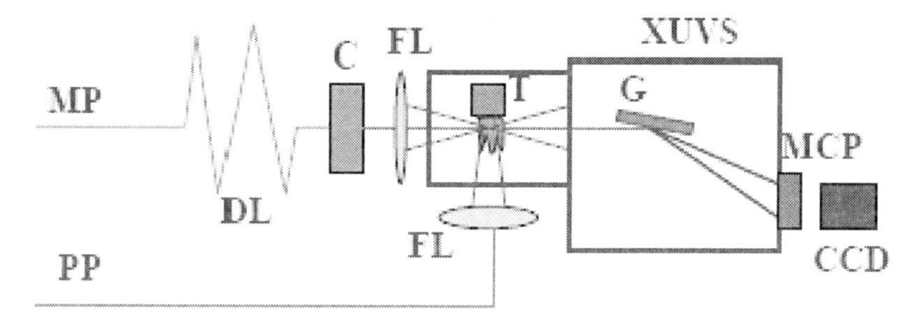

Figure 1. Experimental setup for high-order harmonic generation in laser plasma. MP: main femtosecond pulse; PP: heating picosecond pulse; DL: delay line; C: grating compressor; FL: focusing lenses; T: target; G: gold-coated grating; XUVS: extreme ultraviolet spectrometer; MCP: micro-channel plate; CCD: charge-coupled device.

The harmonics were spectrally dispersed by an XUV spectrometer, detected by a micro-channel plate, and recorded using a CCD camera. The details of harmonic generation set-up can be found elsewhere [24].

2. Ablation of Metal Nanoparticles

Initially, the size distribution of the clusters used in these studies was defined. The initial sizes of metal powder nanoparticles were varied in the range of 30 – 300 nm. Figure 2 presents the SEM images of some nanoparticles before and after ablation. The experimental conditions are presented on the figure captions. The presence of nanoparticles in the plumes was confirmed by analyzing the morphology of the ablated material deposited on the substrates placed at the distance of 40 mm from the targets. It was shown that, at optimal ablation conditions, the nanoparticles remain almost intact in the plasma plume.

Figure 2. SEM images of Ag (left column) and Cu (right column) nanoparticle powders obtained (a) before ablation and (b) on the deposited substrates. The length of white lines is 500 nm. The powders were glued by superglue on the glass substrates. The ablation was accomplished at vacuum conditions during 50 laser shots using the 300-ps heating pulses at the intensity of $I = 7\times10^9$ W cm^{-2}. The ablated material was deposited on the glass substrates placed at the distance of 40 mm from the target surface.

The material directly surrounding nanoparticles is the glue, which has a lower ablation threshold than the nanoparticles. Therefore the glue starts to ablate carrying the nanoparticles at relatively low intensities of laser radiation. This feature allowed for easier creation of the optimum plasma conditions, which resulted in the presence of the nanoparticles with defined size characteristics in the laser plumes [22,25-33].

The laser ablation was carried out at different laser intensities at the surface of nanoparticle-containing targets. The laser intensity was maintained at conditions when the size characteristics of deposited material remained intact with regard to the initial morphology of nanoparticles. The intensity of subnanosecond (t=300 ps) pulse at which these conditions were satisfied was in the range between 3×10^9 and 1×10^{10} W cm^{-2}. At these conditions, the sizes of the nanoparticles deposited on the substrates during laser ablation (Figure 2b) were close to those glued on the surface of targets (Figure 2a). The increase of laser intensity above certain level ($I \sim 3\times10^{10}$ W cm^{-2}, laser fluence 4 J cm^{-2}) led to considerable disintegration or aggregation of nanoparticles on the target surface. This followed by the appearance of chaotic drops of aggregates on the substrate surface.

The SEM images of deposited nanoparticles in most cases, when the optimal ablation of nanoparticle-containing targets was maintained, revealed that they remain approximately same as the initial powders. At the same time broader wings of size distribution observed in the histograms point out the appearance of both small and large nanoparticles due to partial aggregation and disintegration of some nanoparticles. The sizes of deposited Ag clusters were in the range of 30 - 150 nm. The same can be said about the Au (40 – 180 nm), Cu (30 - 80 nm) and other deposited clusters.

Analogous results were obtained in the case of dried nanoparticle suspensions (Ag, Au, Pd, Pt, Ru). The initial sizes of nanoparticles (Ag: 6 nm, Au: 12 nm, Pt and Ru: 3 nm, Pd: 5 nm) did not change considerably during drying of the suspensions, since they were protected against aggregation. Laser ablation and deposition of these nanoparticle-containing films on the nearby substrates at moderate laser intensities ($I = (3\text{-}8)\times10^9$ W cm^{-2}) led to appearance of the deposited nanoparticles possessing analogous morphology.

The presence of original nanoparticles on the deposited substrates confirms their presence in the laser plumes. The absorption spectra of deposited materials also confirmed the presence of nanoparticles, since these spectra demonstrated the surface plasmon resonance (SPR) - induced absorption bands (Figure 3a).

3. Ablation of Bulk Metals

Here we present two sets of laser ablation experiments using different laser sources, when nanoparticle formation during ablation of bulk targets was confirmed by both SEM and optical absorption analysis of deposited materials [34,35].

In the first set of these studies, the ablation of various bulk materials was carried out in vacuum using the 210 ps pulses. Uncompressed pulses from the Ti:sapphire laser were focused on a bulk target at two regimes of focusing. In the first case (referred to as tight focusing), the intensity of laser radiation was in the range of 5×10^{12} W cm^{-2}, and in the second case (referred to as weak focusing), the intensity was considerably lower (9×10^{10} W cm^{-2}). Silver and chromium were used as the targets. Float glass and GaAs wafer were used as the substrates, and were placed at a distance of 50 mm from the targets.

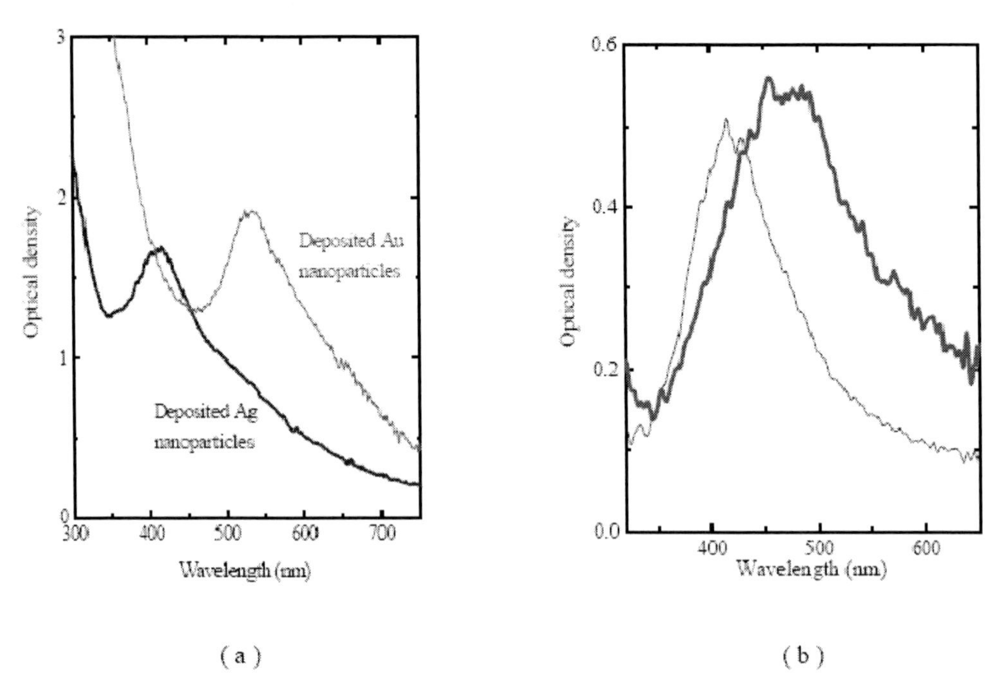

Figure 3. (a) Absorption spectra of deposited Ag (thick curve) and Au (thin curve) nanoparticles during ablation of nanoparticle powder-containing tagets. The experimental conditions were analogous to those described in the caption to Figure 2. (b) Absorption spectra of the silver films deposited at different intensities of the heating pulse (thin curve: $I = 1\times10^{12}$ W cm^{-2}, thick curve: $I = 6\times10^{12}$ W cm^{-2}) during ablation of the bulk Ag target. In these experiments, which were carried out at vacuum conditions, the 800-nm, 210-ps laser pulses irradiated the bulk target during 180 s at 10 Hz pulse repetition rate. The ablated material was deposited on the glass substrates placed 50 mm away from the target.

Figure 3b presents the absorption spectra of the silver films deposited on float glass substrates at the tight and weak focusing conditions. It was observed a variation of the position of the absorption peak of Ag deposition, which depended on the conditions of focusing of laser pulses on the bulk target. In particular, in these cases, the peaks of SPR were centered in the range of 440 nm ($I = 1\times10^{12}$ W cm^{-2}) and 490 nm ($I = 6\times10^{12}$ W cm^{-2}). In the case of the deposition of silver film at smaller excitation of bulk target ($I < 5\times10^{15}$ W cm^{-2}), no absorption peaks were observed in this region, indicating the absence of nanoparticles.

The SEM studies of the structural properties of deposited films showed that, in the tight focusing condition, these films contain a lot of nanoparticles with variable sizes (Figure 4a). In the weaker focusing condition, the concentration of nanoparticles was considerably smaller compared to the tight focusing condition (Figure 4b). The mean size of particles producing during ablation of bulk chromium target was estimated to be 70 nm. The same behavior was observed in the case of other targets. These studies showed that the material of the target does not play much significant role in the formation of nanoparticles in the case of laser ablation using the 210 ps laser pulses at tight focusing conditions.

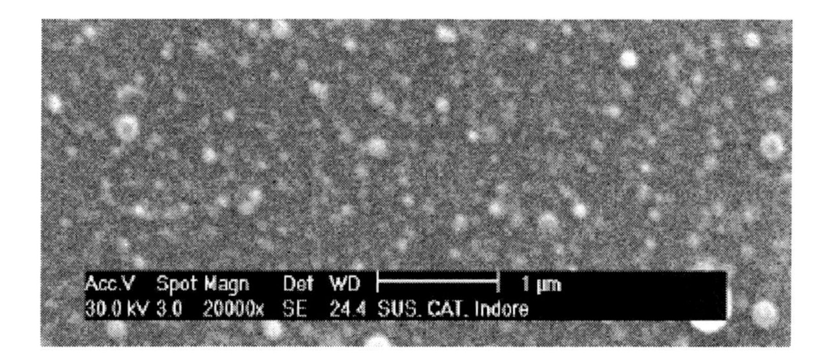

(a)

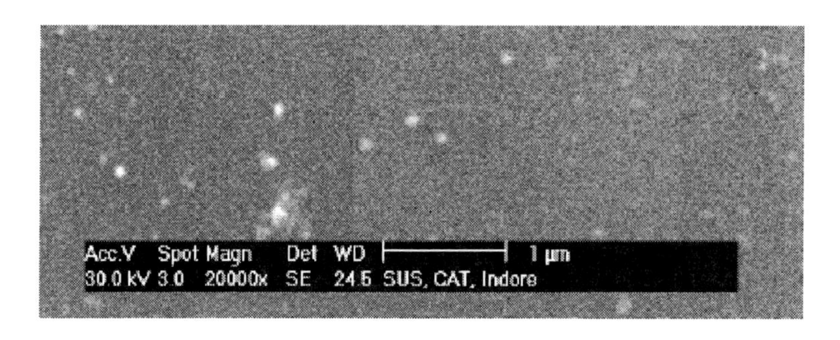

(b)

Figure 4. SEM images of Cr nanoparticles deposited on the GaAs wafer placing at the distance of 50 mm from the ablation area at (a) $I = 3\times10^{12}$ W cm^{-2} and (b) $I = 7\times10^{11}$ W cm^{-2} during ablation of bulk chromium target. The experiments were carried out at vacuum conditions. 210-ps laser pulses irradiated the chromium target during 200 s at 2 Hz pulse repetition rate.

Further studies on the characteristics of nanosized structures of the deposited materials were carried out using the atomic force microscopy (AFM). AFM measurements were carried out in non-contact mode under ambient environment. The mean size of silver nanoparticles deposited at tight focusing conditions was 60 nm. In contrast to this, the AFM images obtained from the deposited films prepared at weak focusing conditions showed considerably smaller number of nanoparticles. These images showed the presence of very few nanoparticles. The same difference in AFM pictures was observed in the case of indium deposition under the two focusing conditions.

In the second set of experiments, it was demonstrate that nanoparticle formation during laser ablation of metals strongly depends on the concentration of surrounding gas [36]. While, at vacuum conditions, nanoparticle formation shows very "sharp" AFM images of aggregated clusters, following with clear appearance of the SPR on the absorption spectra of deposited films, an addition of gas particles during laser ablation starts to decrease the probability of cluster formation. The destruction of nanoparticle formation is attributed to the negative influence of surrounding gas particles on ablated particles aggregation. The same feature was reported in Refs [37,38].

The experiments were carried out using the Nd:YAG laser. To create the ablation, a single pulse from the pulse train of oscillator was amplified (λ = 1064 nm, t = 38 ps, E = 10 mJ, 2 Hz pulse repetition rate) and focused on a target placed in the vacuum chamber. The laser pulses were focused on a bulk target (Ag or Cu plates). The chamber was maintained at the vacuum pressure of $p = 8\times10^{-4}$ torr. The pressure was varied by adding different noble gases up to p = 300 torr. The debris from the plasma plume was deposited on a nearby-placed BK7 glass plates.

The analysis of the films deposited at various ambient conditions has shown a considerable difference in their morphology. The histograms of nanoparticle size distribution corresponding to the AFM images of these films produced at vacuum conditions showed that mean size of nanoparticles was 100 nm (in the case of Ag ablation) and 80 nm (in the case of Cu ablation). The sharp images of nanoparticles indicated that concentration of the deposited atomic layer containing atoms is insignificant. The analysis of AFM images has clearly indicated a difference in the "sharpness" of nanoparticle images in the case of vacuum ($p < 10^{-2}$ torr) and gas deposition. Figure 5 shows the AFM images of the deposited Ag and Cu films produced during laser ablation at air conditions (p = 15 torr).

A considerable decrease of nanoparticle concentration was observed at these conditions compared to the case of the ablation at vacuum conditions. The analysis of nanoparticle formation at different pressures of surrounding air showed a decrease of the "sharpness" of nanoparticle images at the pressures of up to 20 – 30 torr. Further growth of surrounding pressure led to disappearance of nanoparticles on the deposited films.

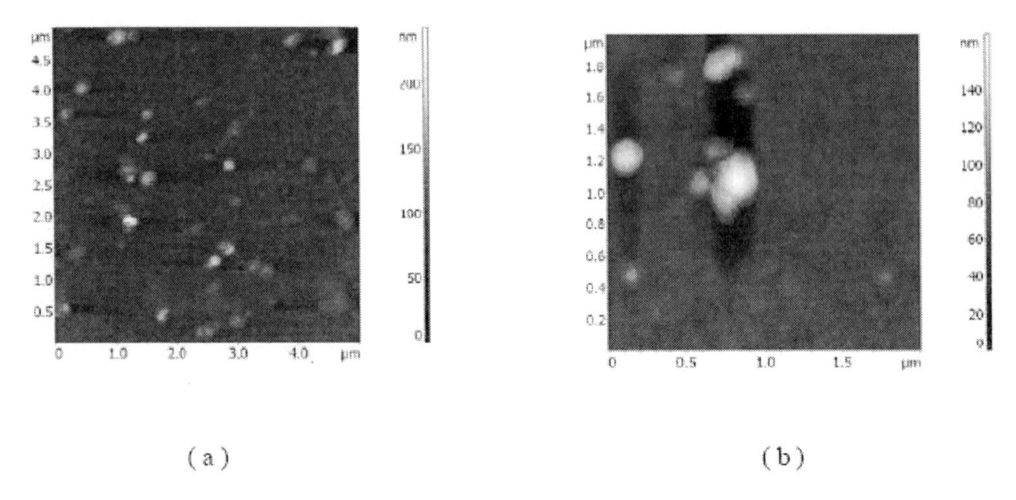

Figure 5. AFM images of the (a) Ag and (b) Cu films deposited during laser ablation in air. These experiments were accomplished at the air pressure of 15 torr. Solid silver and copper targets were ablated at $I = 2\times10^{12}$ W cm^{-2} using 1064-nm, 38-ps laser operating at 2 Hz pulse repetition rate. The ablated material was deposited on the BK7 glass plates placed at the distance of 30 mm from the ablated target.

The absorbance of the films deposited at different conditions (vacuum, air) also showed a stark difference with each other. The absorption curves in the case of vacuum deposition clearly showed the strong absorption bands related with presence of Ag and Cu nanoparticles, which caused the appearance of SPRs at 470 and 590 nm respectively. In contrast, in the case of laser ablation at air conditions (p = 760 torr), no resonance absorption bands appeared in the absorption spectra of deposited materials.

The light (He, Z = 4) and heavy (Xe, Z = 131) noble gases were used to define the influence of weight characteristics of the surrounding gas particles on the nanoparticle formation during laser ablation of silver target at different gas pressures. The analysis of the absorption spectra of deposited Ag films in the case of different pressures of xenon shows that SPR appears up to 12 torr. Above this pressure, only monotonic growth of absorption towards blue side was observed. The same measurements in the case of helium showed analogous tendency, when the SPR of silver nanoparticles disappeared from the absorption spectra of the deposited films obtaining at gas pressure above 33 torr. A shift of SPR towards the longer wavelengths with increase of gas pressure was observed [36].

4. Ablation of Fullerene- and Carbon Nanotube-Containing Targets

Below, the experimental studies of the structural modifications of ablated material during laser heating of the C_{60}- and CNT-containing targets are analyzed. The morphology of initial material and deposited films was studied by analyzing the debris deposited on nearby substrates. This technique allowed the optimization of laser ablation parameters for maintaining the fullerenes and nanotubes in the laser plumes [39-41].

The target used in the fullerene ablation experiments was a C_{60} powder glued on glass slides. To create ablation, the laser beam was focused on a target placed in the vacuum chamber. The spot size of this beam on the target surface was maintained in the range of 0.5 – 0.8 mm. The laser energy density during ablation was kept at 0.4 - 1 J cm^{-2}. The chamber was maintained at a pressure of 8×10^{-4} torr. The debris was deposited from plasma plume on Si substrate and copper grids with carbon films placed nearby, which were analyzed using the transmission electron microscopy [23].

The high-resolution TEM (HRTEM) images of the edge of an initial C_{60} aggregated cluster powder showed a regular spacing of the lattice planes in the ranges of 0.6 and 0.8 nm for the C_{60} clusters, which is consistent with the lattice spacing of these face-centered cubic structures. Analogous features remained in the case of the HRTEM of deposited debris of the fullerene powder after laser ablation at moderate intensities of the heating pulse radiation ($\leq 7\times10^{9}$ W cm^{-2}). Another pattern of HRTEM of the debris of ablated fullerene powder appeared at the pulse intensities above 1×10^{10} W cm^{-2}. In that case, different spacing of the lattice planes created on the surface of substrates and grids after laser ablation was observed. The regular spacing of these lattice planes was 0.36 nm, which is consistent with the inter-planar lattice spacing for graphitic layers, i.e., 0.34 nm. HRTEM images pointed to the microstructure typical of carbon black: an intermediate structure between amorphous and fully graphitized carbon. These studies revealed the range of laser intensities, which could be useful for maintaining the fullerenes in the laser plumes after laser ablation of C_{60}-containing targets. For C_{60} plasma, the fullerene density was estimated to be no less than 5×10^{16} cm^{-3}.

Single-walled CNT powder glued on glass substrate was used as a target for laser ablation. Laser ablation technique was used to produce a plasma plume containing CNTs. The analysis of the morphology of the CNT powder and the deposited CNT debris was carried out using a transmission electron microscope. The morphological characteristics of the CNT prior to laser ablation were analyzed and compared with the ablated material debris deposited on a copper grid and carbon film. The diameter of initial CNTs was 3 - 6 nm, with length varying from 0.3 to 10 μm. The debris of plasma plumes was studied at various pump pulse intensities. At a pump pulse intensity in the range of 3×10^9 - 2×10^{10} W cm^{-2}, CNTs were observed to be deposited. Figure 6 shows the TEM image of the deposited CNTs. The ability of CNTs to survive an appropriate excitation was confirmed in the aforementioned range of laser intensities. Higher heating intensities led to disintegration of the CNTs.

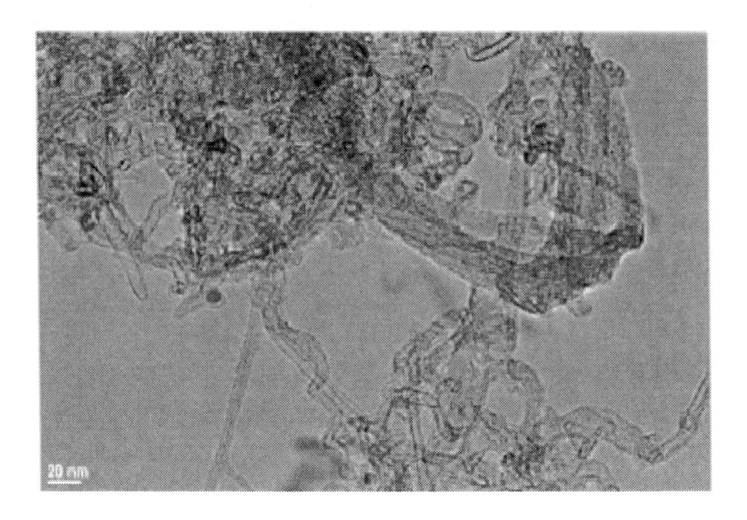

Figure 6. TEM image of the deposited CNTs. The length of marked line is 20 nm. The CNT powder was glued by superglue on the glass substrates. The ablation of the CNT powder was carried out during 10 s at the intensity of $I = 7\times10^9$ W cm^{-2} using the 800-nm, 300-ps laser pulses at 2 Hz pulse repetition rate. The deposited debris were collected on the copper grids and carbon films and analyzed using the transmission electron microscope.

5. Application of Cluster-Containing Plasma for Efficient High-Order Harmonic Generation of Ultrashort Radiation

A few examples of application of this deposition analysis technique are presented below, which allowed defining the presence of clusters in the plumes, for the studies of the high-order nonlinear optical properties of nanosized species in the laser plasmas. The plumes containing various metal nanoparticles, fullerenes, and carbon nanotubes were used for the high-order harmonic generation of the femtosecond radiation propagating through the laser plasmas. The harmonics were generated effectively at the conditions when the presence of nanoparticles in the plumes was confirmed by morphological analysis of the debris. The high-order harmonics were observed in all of studied nanoparticle-containing laser plasmas. The focusing of femtosecond radiation in front of or after the laser plume optimized the harmonic yield. The focusing inside the plasma area led to a decrease of harmonic efficiency due to over-ionization of nanoparticles by femtosecond pulses and appearance of additional free electrons. The latter led to the phase mismatch of the HHG, which has previously been reported in the case of over-excitation of bulk targets [4].

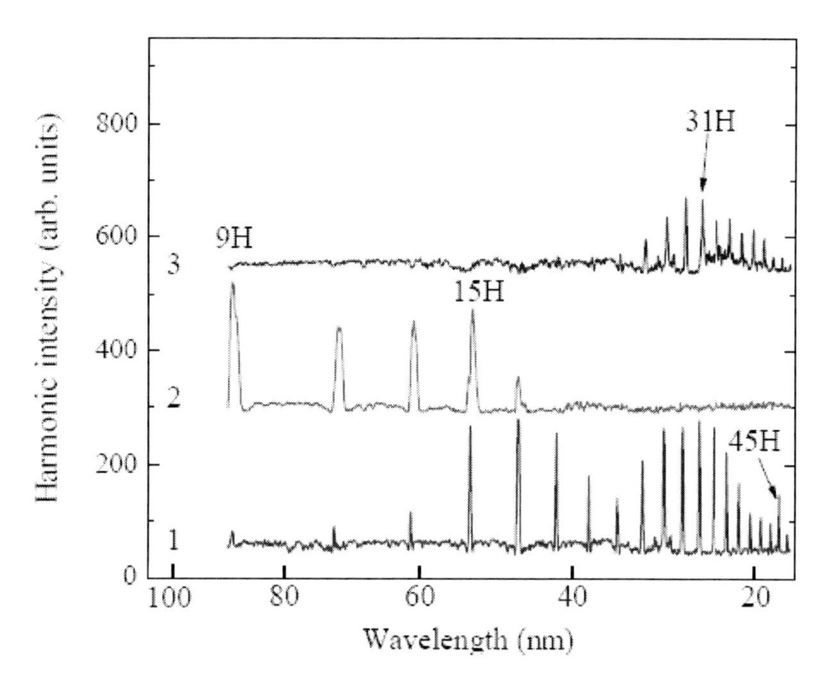

Figure 7. Harmonic spectra obtained from the plasmas produced on the surfaces of (1) bulk silver, and (2,3) silver nanoparticle-containing target. The laser intensities of heating subnanosecond pulses on the target surfaces were (1) 3×10^{10} W cm^{-2}, (2) 7×10^{9} W cm^{-2}, and (3) 2×10^{10} W cm^{-2}. The intensity of femtosecond pulse in the plasma plume was $I_{fp} = 4\times10^{14}$ W cm^{-2}. The delay between the subnanosecond and femtosecond pulses was maintained at 20 ns. The femtosecond pulse propagated through the plasma 100 μm above the target surface.

The details of the HHG from the ablated nanoparticles can be found elsewhere [22,25-30]. Here we show the characteristic peculiarity of this process — the enhancement of low-order harmonic yield from 100-nm silver nanoparticles compared with atoms and ions of the same origin (Figure 7). One can see the six- to twenty-fold enhancement of low-order (9th, 11th, and 13th) harmonic generation efficiency in the Ag nanoparticles-containing plasma plumes (Figure 7, curve 1) compared with the same process in the plasmas produced on the surface of bulk silver (Figure 7, curve 2). The increase of heating pulse intensity from 7×10^{9} W cm^{-2} (Figure 7 curve 2) to 2×10^{10} W cm^{-2} (Figure 7, curve 3) led to disintegration of nanoparticles and corresponding disappearance of strong low-order harmonics, while the higher-order harmonics appearance confirmed the involvement of atomic/ionic particles in the laser frequency conversion.

One can assume that, in the case of small nanoparticles, the ejected electron, after returning back to the parent particle, can recombine with any of atoms in the nanoparticle due to enhanced cross section of the recombination with parent particle compared with single atom, which considerably increases the probability of harmonic emission in the former case. Thus the enhanced cross section of the recombination of accelerated electron with parent particle compared with single atom can be a reason of observed enhancement of the HHG yield from the nanoparticle-containing plume compared with atoms/ions-containing plasma.

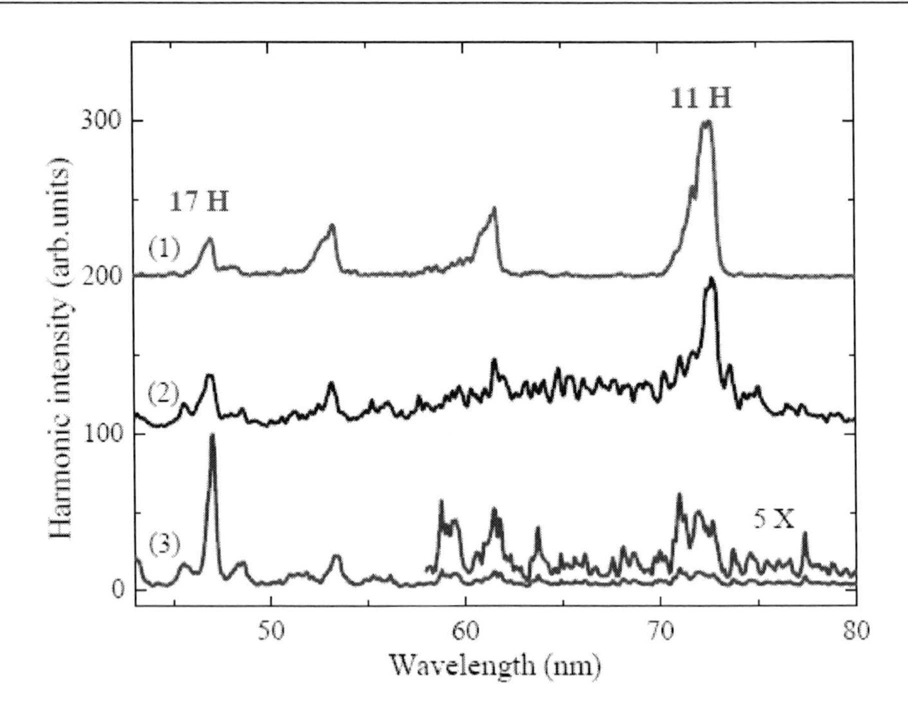

Figure 8. Comparison of the HHG spectra from (1) silver nanoparticles glued on the target, (2) *in-situ* produced silver nanoparticles, and (3) silver plasma without the nanoparticles.

The generation of high-order harmonics from the interaction of *in-situ* produced nanoparticles with intense ultrashort laser pulse has also been studied. The interaction of the sub-nanosecond prepulse with the silver target at intensity of $\sim 1\times10^{13}$ W cm^{-2} generates the nanoparticles. The mean size of silver nanoparticles was 30 nm. High-order harmonics were generated through the interaction of *in-situ* produced nanoparticles with ultrashort laser pulses. The spectrum of HHG from *in-situ*-produced nanoparticles was compared with the HHG spectrum from bulk Ag plumes and 9-nm Ag nanoparticles glued on the target (Figure 8). The intensities of 9th-15th harmonics were less compared to the 17th harmonic in the case of HHG from atoms/ions-containing silver plumes. In the meantime, the intensity of harmonics from the plumes created on the target coating by silver nanoparticles decreases slowly from 9th harmonic to 17th harmonic. Comparison of the HHG spectral characteristics from *in-situ* produced nanoparticles with that from bulk silver and glued silver nanoparticles indicates that HHG in that case is from Ag nanoparticles rather than from Ag atoms and ions. As the intensity of pump pulse on the bulk silver surface was increased from $\sim10^{10}$ W cm^{-2} to $\sim8\times10^{11}$ W cm^{-2}, the HHG spectra gradually reduced and completely vanished. Then, at 8×10^{12} - 1×10^{13} W cm^{-2}, HHG suddenly reappears, as we focused the radiation at tight focusing conditions. The intensity of HHG radiation from *in-situ* produced nanoparticles is lower compared to that of from plasma produced on the Ag nanoparticle-coated surface. However, the intensity of the HHG could be further enhanced by improving the methods for nanoparticle formation [33].

The systematic studies of the HHG in fullerene-containing plumes have been reported in Refs. [23,39-41]. Those studies also revealed the advantages of the use of the C_{60}-containing plasma as a nonlinear optical medium for harmonic generation. Here some examples of harmonic spectra from C_{60}-containing plasmas are presented (Figure 9). Harmonics up to the

33rd order were observed in these studies. The harmonic yield from fullerene-rich plasma was a few times larger compared with those produced from a bulk carbon target. Odd and even harmonics were generated using two-color pump (800 nm + 400 nm). The comparative studies of harmonic spectra from CNTs and fullerenes have shown better HHG efficiency in the latter case.

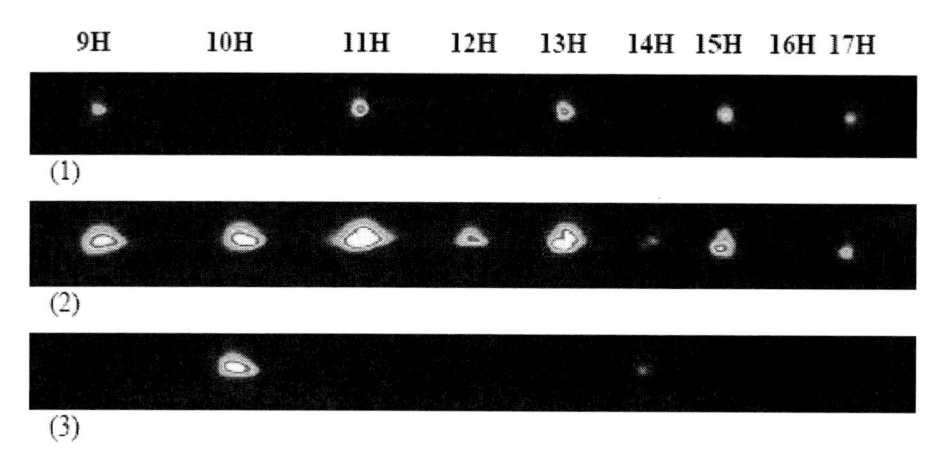

Figure 9. CCD images of the harmonic spectra generated in C_{60} plasma in the cases of: (1) single-color fundamental pump (800 nm), (2) two-color pump (800 nm + 400 nm), and (3) single-color second harmonic pump (400 nm). The data were collected under similar experimental conditions.

Discussion and Conclusions

The analysis of the plasma obtained by the subnanosecond laser-target interaction showed the typical features already found in femtosecond ablation [42]. Optical emission spectroscopy revealed the presence of a primary emission of material, showing the presence of neutral and ionized particles. Aforementioned studies showed the presence of melted material on the target and in the expanding plume, so the ablation mechanism should involve the emission of both gaseous material and melted particles. To characterize the ablation process, the temporal and spectral characteristics of the light emitted by the plume were also studied. The oscilloscope traces showed a considerable increase in the duration of the plasma emission in the case of tight focusing that could be expected considering the excitation conditions. A combined analysis of the spectra and oscilloscope traces in the cases of two different regimes of the excitation of surface plasma revealed that the structureless continuum appearing in the spectrum in the case of tight focusing is due to the emission from hot nanoparticles produced during laser ablation. Such hot nanoparticles behave like black-body radiators emitting for a longer time till they get cooled down.

The fullerene medium has previously demonstrated both direct and delayed ionization and fragmentation processes during interaction with femtosecond pulses. The ionization and fragmentation of fullerene plasma in moderate laser fields occur predominantly via multiphoton excitation of the 20 eV plasmon resonance of C_{60}. The experimentally observed survival of fullerene at rather intense laser fields can be ascribed to the very large number of internal degrees of freedom of the C_{60} molecule, which leads to the fast dissipation of the excitation energy followed by inefficient redistribution into ionizing and fragmenting modes.

These peculiarities of fullerenes can be effectively used for creation of the plasma medium containing considerable amount of C_{60} molecules for their further studies in laser-matter interaction. The same can be said about the ablation and maintenance of the CNTs in plasma plumes.

It was found that HHG from the plume containing Ag multiatomic particles became more efficient at considerably smaller pulse intensities (4×10^9 W cm^{-2}) compared to the case of the plasma appearing on a bulk silver target (2×10^{10} W cm^{-2}). This tendency, which was also observed in the case of other nanoparticles-containing targets, was attributed to the lower ablation threshold and lower cohesive energy of the host substrate that contains nanoparticles compared to those of the bulk material. The material directly surrounding the nanoparticles is a polymer (epoxy glue), which has a considerably lower ablation threshold than the metallic materials. Therefore, the polymer-carrying nanoparticles begin to ablate at relatively low intensities resulting in the lower laser intensity required for the preparation of the appropriate nonlinear medium for the HHG. This feature allowed for the easier creation of the optimum plasma conditions, which resulted in a better HHG conversion efficiency from the nanoparticle-containing plume compared to the plume from the bulk target.

In conclusion, the studies of the plasma formations containing various clusters using laser radiation have been reviewed. It is shown that, at moderate laser intensity on the surface of cluster-contained materials, the deposited material demonstrates the structure, which was close to the initial structure of ablated nanoparticles. The reviewed studies confirmed the presence of nanoparticles in the laser plumes after the optimal excitation of cluster-contained targets. It was demonstrated that the nanoparticle formation on the surface of target during laser ablation of metals strongly depends on the concentration of surrounding gas. While, at vacuum conditions, nanoparticle formation shows very "sharp" AFM images of aggregated clusters, following with appearance of plasmon resonance on the absorption spectra of deposited films, an addition of gas particles starts to decrease the probability of cluster formation. The destruction of nanoparticle formation was attributed to the negative influence of surrounding gas particles on the ablated particles aggregation.

The morphology of ablated fullerenes before and after their laser-induced deposition on various substrates has been analyzed. It is observed that, at moderate laser intensity of the 210 ps pulses on the surface of C_{60}-rich materials (up to 7×10^9 W cm^{-2}), the deposited material remains approximately the same as the initial clusters, while, at higher pulse intensity, the disintegration of fullerenes leads to the formation of graphite on the substrate placed near the ablation area. The morphology of ablated CNTs before and after laser-induced deposition on substrates has also been analyzed. At $I \leq 7\times10^9$ W cm^{-2}, the deposited material remained approximately the same as the initial nanotubes, while, at higher pump pulse intensity, the disintegration of CNTs takes place.

It was assumed, throughout the description of reviewed studies, that there are no decomposition processes in the plume, or nucleation on the substrates at optimal excitation of the targets. Arguments supporting these assumptions are based on the analysis of the deposited debris. Both structural studies of deposited nanoparticles and absorption studies of the deposited films confirmed the presence of the nanoparticles of the same morphology as initial nanoparticles.This means that they exist during laser plasma formation.

The reviewed studies also demonstrated the effective high-order harmonic generation of the femtosecond radiation propagating through these cluster-contained plumes. The comparison of the intensities of the harmonics generating in two plumes (with and without nanoparticles) at equal experimental conditions has shown that the enhancement factor in the case of cluster-containing plumes was in the range of 6 to 20. Summarizing, we have analyzed the method for defining the presence of nanoparticles in the plasmas and it's use for the optimization of high-order harmonic generation in cluster-containing laser plumes.

ACKNOWLEDGMENTS

The author thanks H. Kuroda, T. Ozaki, P. D. Gupta, P. A. Naik, M. Suzuki, M. Baba, L. B. Elouga Bom, and H. Singhal for the fruitful discussions and support at various stages of these studies.

REFERENCES

[1] R. A. Ganeev, A. I. Ryasnyansky, A. L. Stepanov, C. Marques, R. C. da Silva, E. Alves, "Application of RZ-scan technique for investigation of nonlinear optical characteristics of sapphire doped with Ag, Cu, and Au nanoparticles," *Opt. Commun.* 253, 205-13 (2005).
[2] M. Falconieri, G. Salvetti, E. Cattaruza, F. Gonella, G. Mattei, P. Mazzoldi, M. Piovesan, G. Battaglin, R. Polloni, "Large third-order optical nonlinearity of nanocluster-doped glass formed by ion implantation of copper and nickel in silica," *Appl. Phys. Lett.* 73, 288-290 (1998).
[3] S. Debrus, J. Lafait, M. May, N. Pinçon, D. Prot, C. Sella, J. Venturini, "Z-scan determination of the third-order optical nonlinearity of gold:silica nanocomposites," *J. Appl. Phys.* 88, 4469-4475 (2000).
[4] R. A. Ganeev, "High-order harmonic generation in laser plasma: a review of recent achievements", *J. Phys. B: At. Mol. Opt. Phys.* 40, R213-R253 (2007).
[5] S Amoruso, G. Ausanio, R. Bruzzese, M. Vitiello, X. Wang, "Femtosecond laser pulse irradiation of solid targets as a general route to nanoparticle formation in a vacuum," *Phys. Rev.* B 71, 033406 (2005).
[6] D. Perez, L. J. Lewis, "Molecular-dynamics study of ablation of solids under femtosecond laser pulses," *Phys. Rev.* B 67, 184102 (2003).
[7] D. Scuderi, O. Albert, D. Moreau, P. P. Pronko, J. Etchepare, "Interaction of a laser-produced plume with a second time delayed femtosecond pulse," *Appl. Phys. Lett.* 86, 071502 (2005).
[8] J. Perrière, C. Boulmer-Leborgne, R. Benzerga, S. Tricot, "Nanoparticle formation by femtosecond laser ablation," *J. Phys. D: Appl. Phys.* 40, 7069-7075 (2007).
[9] R. Teghil, L. D.`Alessio, A. Santagata, M. Zaccagnino, D. Ferro, D. J. Sordelet, "Picosecond and femtosecond pulsed laser ablation and deposition of quasicrystals," *Appl. Surf. Sci.* 210, 307-317 (2003).
[10] T. E. Glover, "Hydrodynamics of particle formation following femtosecond laser ablation," *J. Opt. Soc. Am. B* 20, 125-131 (2003).

[11] H. O. Jeschke, M. E. Garsia, K. H. Bennemann, "Theory for the ultrafast ablation of graphite films," *Phys. Rev. Lett.* 87, 015003 (2001).

[12] V. Kabashin, M. Meunier, "Synthesis of colloidal nanoparticles during femtosecond laser ablation of gold in water, " *J. Appl. Phys.* 94, 7941-7943 (2003).

[13] Y.Suda, T.Nishimura, T.Ono, M.Akazawa, Y.Sakai, N. Homma, "Deposition of fine carbon particles using pulsed ArF laser ablation assisted by inductively coupled plasma," *Thin Solid Films* 374, 287-290 (2000).

[14] T.Makimura, Y.Kunii, K.Murakami, "Light emission from nanometer-sized silicon particles fabricated by the laser ablation method," *Jpn. J. Appl. Phys.* 35, 4780-4784 (1996).

[15] J. Gonzalo, A. Perea, D. Babonneau, C. N. Afonso, N. Beer, J.-P. Barnes, A. K. Petford-Long, D. E. Hole, P. D. Townsend, "Competing processes during the production of metal nanoparticles by pulsed laser deposition," *Phys. Rev. B* 71, 125420 (2005).

[16] S. I. Anisimov, Y. A. Imas, G. S. Romanov, Y. V. Khodyko, "*High power radiation effect in metals*," Moscow, Nauka (1970).

[17] T. D. Donnelly, T. Ditmire, K. Neuman, M. D. Pery, R. W. Falcone, "High-order harmonic generation in atom clusters," *Phys. Rev. Lett.* 76, 2472-2475 (1996).

[18] J. W. G. Tisch, T. Ditmire, D. J. Fraser, N. Hay, M. B. Mason, E. Springate, J. P. Marangos, M., H. R. Hutchinson, "Investigation of high-harmonic generation from xenon atom clusters," *J. Phys.* B 30, L709-L714 (1997).

[19] C. Vozzi, M. Nisoli, J.-P. Caumes, G. Sansone, S. Stagira, S. De Silvestri, M. Vecchiocattivi, D. Bassi, M. Pascolini, L. Poletto, P. Villoresi, G. Tondello, "Cluster effects in high-order harmonics generated by ultrashort light pulses,"*Appl. Phys. Lett.* 86, 111121 (2005).

[20] C.-H. Pai, C. C. Kuo, M.-W. Lin, J. Wang, S.-Y. Chen, J.-Y. Lin, "Tomography of high harmonic generation in a cluster jet," *Opt. Lett.* 31, 984-986 (2006).

[21] K. Wegner, P. Piseri, H. V. Tafreshi, P. Milani, "Cluster beam deposition: a tool for nanoscale science and technology," *J. Phys. D: Appl. Phys.* 39, R439-R460 (2006).

[22] R. A. Ganeev, "High-order harmonic generation in nanoparticle-containing laser-produced plasmas," *Laser Phys.* 18, 1009-1015 (2008).

[23] R. A. Ganeev, L. B. Elouga Bom, J. Abdul-Hadi, M. C. H. Wong, J. P. Brichta, V. R. Bhardwaj, T. Ozaki, "High-order harmonic generation from fullerene using the plasma harmonic method," *Phys. Rev. Lett.* **102**, 013903 (2009).

[24] R. A. Ganeev, "Generation of high-order harmonics of high-power lasers in plasmas produced under irradiation of solid target surfaces by a prepulse," *Phys. – Usp.* 52, 55-77 (2009).

[25] R. A. Ganeev, M. Suzuki, P. V. Redkin, M. Baba, H. Kuroda, "Variable pattern of high harmonic spectra from a laser-produced plasma by using the chirped pulses of narrow-bandwidth radiation," Phys. Rev. A 76, 023832 (2007).

[26] R. A. Ganeev, M. Suzuki, M. Baba, M. Ichihara, H. Kuroda, "Low- and high-order nonlinear optical properties of BaTiO3 and SrTiO3 nanoparticles," J. Opt. Soc. Am. B 25, 325-333 (2008).

[27] R. A. Ganeev, M. Suzuki, M. Baba, M. Ichihara, H. Kuroda, "High-order harmonic generation in Ag nanoparticle-contained plasma," J. Phys. B: At. Mol. Opt. Phys. 41, 045603 (2008).
[28] R. A. Ganeev M. Suzuki, M. Baba, M. Ichihara, H. Kuroda, "Low- and high-order nonlinear optical properties of Au, Pt, Pd, and Ru nanoparticles," J. Appl. Phys. 103, 063102 (2008).
[29] R. A. Ganeev, L. B. Elouga Bom, T. Ozaki, "Comparison of high-order harmonic generation from various cluster- and ion-contained laser plasmas," J. Phys. B: At. Mol. Opt. Phys. 42, 055402 (2009).
[30] L. B. Elouga Bom, R. A. Ganeev, J. Abdul-Hadi, F. Vidal, T. Ozaki, "Intense multi-microjoule high-order harmonics generated from neutral atoms of In2O3 nanoparticles," Appl. Phys. Lett. 94, 111108 (2009).
[31] R. A. Ganeev, L. B. Elouga Bom, T. Ozaki, "Deposition of nanoparticles during laser ablation of nanoparticle-containing targets," Appl. Phys. B 96, 491 – 498 (2009).
[32] R. A. Ganeev, L. B. Elouga Bom, T. Ozaki, "Application of nanoparticle-containing laser plasmas for harmonic generation," J. Appl. Phys. 106, 023104 (2009).
[33] H. Singhal, R. A. Ganeev, P. A. Naik, A. K. Srivastava, A. Singh, R. Chari, R. A. Khan, J. A. Chakera, P. D. Gupta, "Study of high-order harmonic generation from nanoparticles," J. Phys. B: At. Mol. Opt. Phys. 43, 025603 (2010).
[34] R. A. Ganeev, A. I. Ryasnyanskiy, U. Chakravarty, P. A. Naik, H. Srivastava, M. K. Tiwari, P. D. Gupta, " Structural, optical, and nonlinear optical properties of indium nanoparticles prepared by laser ablation," Appl. Phys. B 86, 337-341 (2007).
[35] R. A. Ganeev, U. Chakravarty, P. A. Naik, H. Srivastava, C. Mukherjee, M. K. Tiwari, R. V. Nandedkar, P. D. Gupta, "Pulsed laser deposition of metal films and nanoparticles in vacuum using subnanosecond laser pulses," Appl. Opt. 46, 1205-1210 (2007).
[36] R. A. Ganeev, G. S. Boltaev, R. I. Tugushev, T. Usmanov, "Nanoparticle formation during laser ablation of metals at different pressures of surrounding noble gases," Appl. Phys. A 100, 119-123 (2010).
[37] T. Seto, T. Orii, M. Hirasawa, N. Aya, "Fabrication of silicon nanostructured films by deposition of size-selected nanoparticles generated by pulsed laser ablation," Thin Solid Films 437, 230-234 (2003).
[38] J.J. Lin, L.S. Loh, P. Lee, T.L. Tan, S.V. Springham, R.S. Rawat, "Effects of target–substrate geometry and ambient gas pressure on FePt nanoparticles synthesized by pulsed laser deposition," Appl. Surf. Sci. 255, 4372-4377 (2009).
[39] R. A. Ganeev, L. B. Elouga Bom, M. C. H. Wong, J.-P. Brichta, V. R. Bhardwaj, P. V. Redkin, T. Ozaki, "High-order harmonic generation from C60-rich plasma," Phys. Rev. A 80, 043808 (2009).
[40] R. A. Ganeev, H. Singhal, P. A. Naik, J. A. Chakera, A. K. Srivastava, T. S. Dhami, M. P. Joshi, P. D. Gupta, "Influence of C60 morphology on high-order harmonic generation enhancement in fullerene-containing plasma," J. Appl. Phys. 106, 103103 (2009).
[41] R. A. Ganeev, H. Singhal, P. A. Naik, J. A. Chakera, A. K. Srivastava, T. S. Dhami, M. P. Joshi, P. D. Gupta, "Enhanced harmonic generation in C60-containing plasma plumes," Appl. Phys. B 100, 581-585 (2010).

[42] O. Albert, S. Roger, Y. Glinec, J. C. Loulergue, J. Etchepare, C. Boulmer-Leborgne, J. Perriere, E. Millon, "Time-resolved spectroscopy measurements of a titanium plasma induced by nanosecond and femtosecond lasers," Appl. Phys. A 76, 319-323 (2003).

In: Laser-Induced Plasmas
Editor: Ethan J. Hemsworth, pp. 55-77
ISBN 978-1-61324-851-5

Chapter 3

PULSED LASER DEPOSITION OF V_2O_5 AND TIME-OF-FLIGHT ANALYSIS OF THE LASER-INDUCED PLASMA

Szabolcs Beke
Laser Technology Laboratory, RIKEN – Advanced Science Institute,
Hirosawa 2-1, Wako, Saitama 351-0198, Japan

ABSTRACT

This chapter presents results of pulsed laser deposition of nanocrystalline V_2O_5 thin films along with a time-of-flight (TOF) analysis of the laser-induced plasma. Additionally, a brief historical overview of this deposition method is provided. In this study, the deposited films were characterized by X-ray diffraction (XRD), X-ray photoelectron spectroscopy (XPS), high-resolution transmission electron microscopy (HRTEM), and by optical transmission spectroscopy. The films were deposited on amorphous glass substrates while keeping the O_2 partial pressure at 13.33 Pa and the substrate temperature at 220 °C. The characteristics of the films were changed by varying the laser fluence and repetition rate. XRD revealed that films are nanocrystalline with an orthorhombic structure. XPS shows the substoichiometry of the films that generally relies on the fact that during the formation process of V_2O_5 films lower valence oxides are also created. From the HRTEM images, the size evolution and distribution characteristics of the clusters as a function of the laser fluence were observed. From the spectral transmittance, the absorption edge using the Tauc plot was determined. Time-of-flight (TOF) optical spectroscopy was used to analyze the elemental composition, and temporal- and spatial distributions of vanadium atoms and ions in the laser-induced V_2O_5 plasma. Neutral atoms and singly charged V ions were detected in the TOF spectra. The time evolution of emission lines is discussed along with the velocity of the ablated species. The data are discussed and conclusions are drawn for the elemental analysis by laser ablation of solid vanadia samples.

Keywords: vanadium oxide thin films, ablation, pulsed laser deposition, X-ray diffraction, transmission electron microscopy, X-ray photoelectron spectroscopy, absorption edge, optical transmission, Bohr radius, time-of-flight.

1. INTRODUCTION

1.1. The Vanadium Oxides

The principal oxides of vanadium occur as single valence in oxidation states from V^{2+} to V^{5+}, in the form of VO, V_2O_3, VO_2, and V_2O_5. However, the vanadium–oxygen phase diagram also includes mixed valence oxides containing two oxidation states, such as V_6O_{13} with V^{5+} and V^{4+} and a series of oxides between VO_2 and V_2O_3 (e.g., V_8O_{15}, V_7O_{13}, V_6O_{11}, etc.) which contain V^{4+} and V^{3+} species.

The mixed valence oxides are formed by introducing oxygen vacancy defects. If the number of oxygen vacancies exceeds a certain value, the vacancies tend to correlate and form crystallographic shear planes (i.e., the vacancies associate along a lattice plane) and are subsequently eliminated by reorganization of V-O coordination units. The result is a series of oxides with related stoichiometries, such as the Magnéli phases with V_nO_{2n-1} and the Wadsley phases with $V_{2n}O_{5n-2}$ formulas [1].

V_2O_5 undergoes a semiconductor/metal phase transition near 257 °C [2]. A large change in electrical properties accompanies this transition. Thus, thermally activated electrical and/or optical switches can be fabricated. Since optical and electrical behaviors are correlated, V_2O_5 is especially interesting in thin film form due to the possibility of integration into microelectronics circuits. V_2O_5 has been suggested for use as a variable transmittance electrochromic device for controlling sunlight through windows [3, 4].

1.2. The Crystal Structure and Usability of V_2O_5

V_2O_5 is thermodynamically the most stable phase in the V-O system. It crystallizes with an orthorhombic unit cell structure belonging to the P_{mnm} space group with lattice parameters a = 11.510 Å, b = 3.563 Å, and c = 4.369 Å. It has a layer-like structure and it is composed of distorted trigonal bipyramidal coordination polyhedra of O atoms around V atoms. The polyhedra share edges to form $(V_2O_4)_n$ zigzag double chains along the (001) direction and are cross linked along (100) through shared corners as shown in Figure 1. [5]. The distorted polyhedra have a short (1.58 Å) vanadyl bond, $[VO]^{2+}$, and four O atoms located in the basal plane at distances ranging from 1.78 to 2.02 Å. The sixth O atom in the coordination polyhedron lies along the vertical axis opposite to the V-O bond at a distance of 2.79 Å.

From a chemical point of view, V_2O_5 is an excellent catalyst [6] due to its rich and diverse chemistry that is based on two factors: the large variety of vanadium oxidation states, ranging from 2^+ to 5^+, and the variability of oxygen coordination geometries. This structural richness is the source for the existence of differently coordinated oxygen ions which provide an important ingredient for controlling physical and chemical surface properties. Due to its layered structure, V_2O_5 is a promising material for energy storage systems and has a high ionic storage capacity.

Among transition metal oxides (e.g., WO_3, MoO_3), V_2O_5 has drawn significant interest in the past decades due to industrial applications such as electrochromic devices [7], optical switching devices [8], and reversible cathode materials for Li batteries [9-12]. V_2O_5 thin films have been prepared by various methods, on different substrates, including electron beam evaporation [10-12, 13-17], magnetron sputtering [18-24], pulsed laser deposition [7, 9, 38-41], chemical vapor deposition [42-61], spray pyrolysis [62-64], electrospinning [65-70], sol-gel [71-83], spin coating [84-88], and film growth in the field of a CW IR laser beam [89-105]. These deposition techniques are described and summarized with the corresponding application areas and optical band gap values in ref. [106].

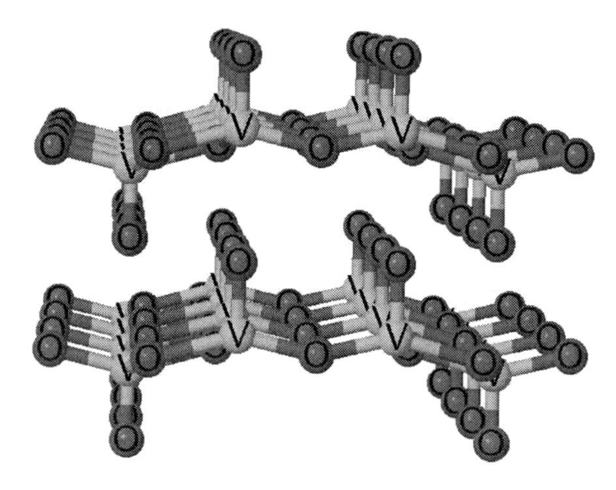

Figure 1. Perspective view of two layers of V_2O_5. V atoms are represented as grey balls, O atoms as red balls. Weak van der Waals bonds are omitted for clarity.

In general, vanadium oxide thin films are amorphous when deposited at a relatively low substrate temperature (< 300 °C). The temperature of crystallization depends on the deposition technique as well as the structure (crystalline or amorphous) of the substrate material. For example, in the case of pulsed laser deposition (PLD), when ejected species in the laser induced plasma plume condense on the substrate surface with high average kinetic energies, the crystallization process can take place at a lower substrate temperature due to enhanced adatom mobilities.

2. A Brief Historical Overview of Pulsed Laser Deposition [7-9, 25-41]

Pulsed laser deposition is one of the primary deposition techniques of V_2O_5 thin films and is an excellent tool for the fabrication of vanadium oxide clusters in the nanoscale size range. PLD is a physical vapor deposition technique in which the vapor flux is formed by the ablation of material from a target due to a series of high-energetic laser pulses. Due to its versatility and flexibility, it is widely used for deposition of complex oxide films. The deposition environment (gas type and pressure) can be used to control the energy and extent of the expanding laser plume (which includes activated species) and subsequent film development.

Laser-assisted film growth began in 1962, soon after the technical realization of the first solid-state ruby laser in 1960 by Maiman [107]. The initial ablation experiments were carried out in 1962 by Breech and Cross who studied the laser-vaporization and excitation of atoms from solid surfaces using a ruby laser [26]. However, the deposited films were inferior to those obtained by other techniques such as chemical vapor deposition and molecular beam epitaxy. Three years later, in 1965, Smith and Turner [27] utilized a ruby laser to deposit the first thin films on semiconductors, dielectrics, and metal-organic surfaces. They found that many materials (Sb_2S_3, As_2S_3, Te, Se, ZnTe, MoO_3, PbTe, Ge) can be vaporized in vacuum by a directed laser beam and condensed on substrates as thin films.

Until the end of the 1980s, few reports had been published on PLD; however, some research groups achieved remarkable results in controlling thin film structures utilizing PLD [28,29]. The breakthrough of this technique came in 1987 when Dijkkamp et al. [30] laser-deposited thin films of $YBa_2Cu_3O_7$, a high temperature superconductive material, with outstanding properties. In the same year, Cheung and Madden [31] deposited HgCdTe epitaxial layers by pulsed-laser induced evaporation.

Following these earlier achievements, many research groups began to utilize PLD for the fabrication of high quality crystalline films. By 1991, the number of PLD-related publications reached 128 [32] and this number increased to 180 in 1994 [33]. The deposition of ceramic oxides, nitrides, metallic multilayers, and superlattices has been demonstrated [34-37]. In the 1990s, the development of UV excimer lasers (high repetition rate with short pulse durations) made PLD a very competitive tool for the growth of well-controlled thin films with complex stoichiometries.

The first pulsed laser ablated vanadium oxide deposition was reported in 1993 by Borek et al. [38]. A KrF excimer laser was utilized to ablate a pure vanadium target in an O_2/Ar atmosphere. The partial pressure of O_2 in the chamber was critical in stabilizing the VO_2 phase and 13 different phases ranging from V_4O to V_2O_5 were reported.

3. Experimental Details

High-purity (99.99 %, Sigma-Aldrich) V_2O_3 powder was subjected to a pressure of $2x10^9$ Pa to form pellets of 13 mm in diameter and 4 mm in thickness. Before the laser irradiation, the pellets were annealed at 550 °C in air for 24 hours. It is worth noting the coloration of the pellets from dark blue to orange, which exhibits the transition of $V_2O_3 \rightarrow V_2O_5$.

PLD was performed in a stainless steel vacuum chamber evacuated by a turbomolecular pump. The experimental setup can be seen in Figure 2. A fused silica glass substrate was placed in front of the target equipped with a heater allowing the variation of the substrate temperature in the range of 200-400 °C. Target – substrate distance was about 5 cm. A continuous flux of oxygen was introduced into the chamber after pumping down the pressure to about 1.33×10^{-5} Pa.

The vanadia targets were ablated by a pulsed ArF excimer laser (λ=193 nm, pulse duration 15 ns) at a fluence level of 0.9-2.7 J/cm^2. Repetition rate was 2 Hz. The laser beam was focused by a lens (f=160 mm) onto the target at an incident angle of 45°. The target was continuously rotated by an electric motor to avoid the depletion at any spot.

The structural characterization of films was studied by XRD using a Philips PW 1830 powder diffractometer using Ni-filtered Cu-K_α radiation (λ = 0.15406 nm wavelength). The mean size (D) of crystallite was calculated from full-width at half-maxima (FWHM) of corresponding X-ray diffraction peaks using Scherrer's formula $D = K\lambda/(\beta\cos\theta)$, where K=0.9 (K is a constant depending on the particle shape), λ is the wavelength of the X-ray radiation, β is the FWHM and θ is the angle of reflection [108].

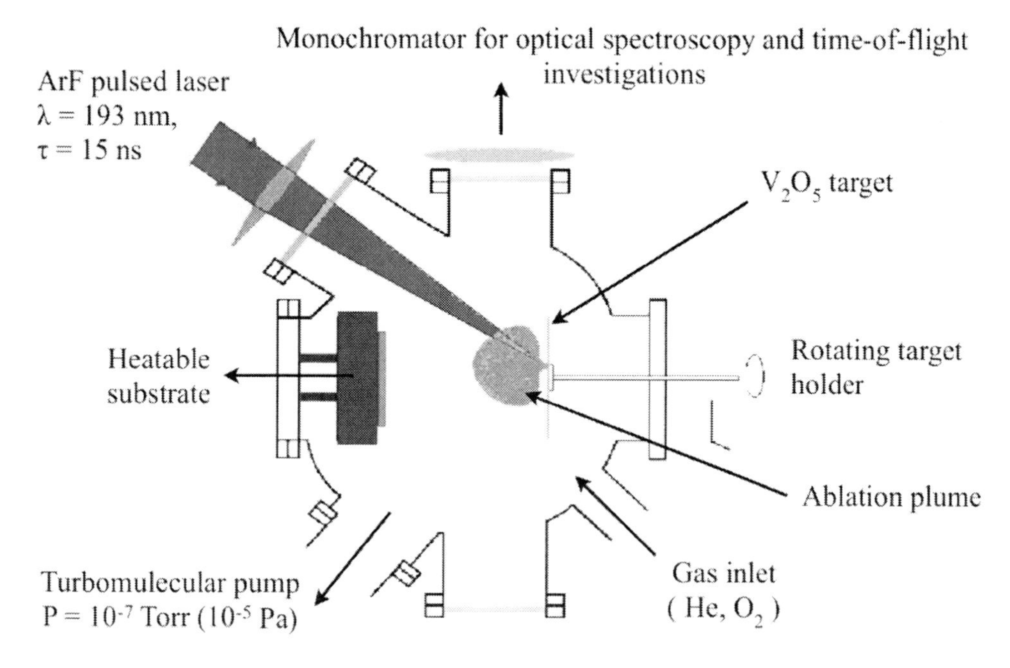

Figure 2. Experimental scheme for PLD and optical investigations of V_2O_5.

Optical transmission measurements were taken by StellarNet EPP2000 spectrometer with a wavelength resolution of ± 0.75 nm.

XP spectra were taken with a SPECS instrument equipped with a PHOIBOS 150 MCD 9 hemispherical electron energy analyzer operated in the FAT mode. The excitation source was the non-monochromatic K_α radiation of a magnesium anode (hν = 1253.6 eV). The X-ray gun was operated at 210 W electrical power (14 kV, 15mA). The incident angle of the X-ray beam was 45° with respect to the surface normal. The pass energy was set to 20 eV, the step size was 25 meV, and the collection time in one channel was 100 ms. Typically five scans were added to acquire a single spectrum. The C 1s binding energy of adventitious carbon was used as energy reference: it was taken as 285.1 eV. For data acquisition and evaluation both manufacturer's (SpecsLab2) and commercial (CasaXPS or Origin) software were used.

As X-ray diffraction is unable to indicate the size of small clusters, the samples were investigated by Transmission Electron Microscopy (TEM) with a Jeol 3010 electron microscope working at 300 kV. Small pieces of the surface were collected by the microcleavage technique.

4. Results and Discussion

4.1. Characterization of the Deposited Films

4.1.1. X-Ray Diffraction Measurement

Figure 3a shows a set of XRD patterns of V_2O_5 films deposited in O_2 partial pressure of 13.33 Pa with varying the fluence. These patterns reveal the polycrystalline structure of the films deposited at a substrate temperature as low as T_S = 220 °C. Moreover, it shows the predominant (001) peak of orthorhombic V_2O_5 phase. However, the fact that V_2O_5 is polycrystalline at this deposition temperature is one of the main advantages of PLD compared to other physical vapor deposition methods, such as electron-beam evaporation and sputtering, where the reported deposition temperature were in the range 300-500 °C.

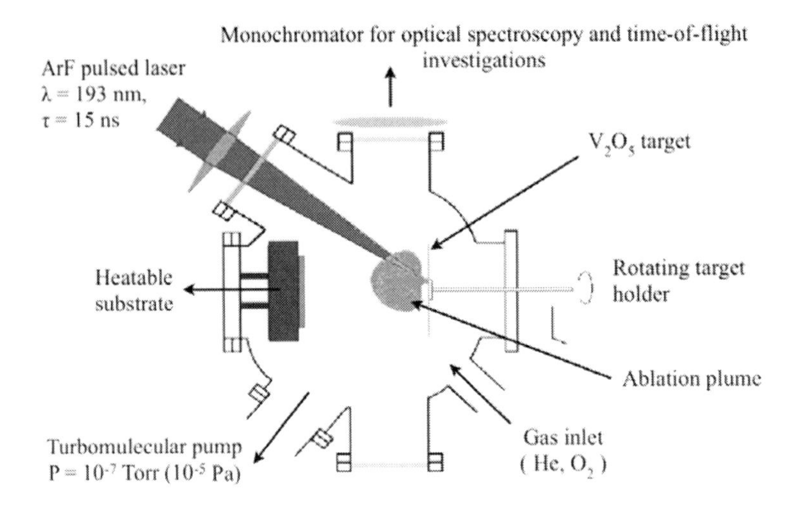

Figure 3.a. X-ray diffraction patterns of PLD V_2O_5 films deposited at various fluences in O_2 partial pressure of 100 mTorr (13,33 Pa) at T_s=220 °C (6000 pulses).

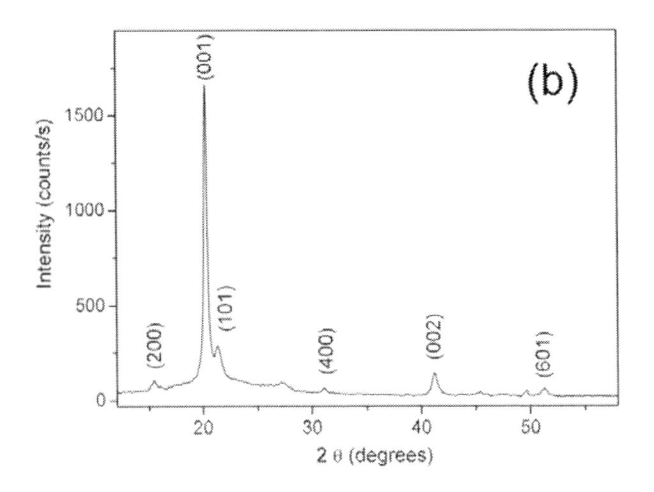

Figure 3.b. XRD of V_2O_5 film produced by 15000 laser pulses and F=2.7 J/cm^2 (T_s= 220 °C).

For example, Kumar et al. reported amorphous structure at 300 °C using vacuum evaporation technique [109]. This lower temperature of crystallization in case of PLD can be explained by the high kinetic energy (>1 eV) of the ejected species in the laser-produced plasma. The condensed particles have high energies after being deposited. The ad-atom mobility is able to produce highly oriented nano-crystalline films even at low temperature.

Table 1. Cluster sizes as a function of fluence and number of pulses

Name of the sample	Fluence (J/cm^2)	Intensity (W/cm^2)	Number of pulses	Mean crystallite size calculated from XRD, D (nm)
VOa	0.9	60×10^6	6000	-
VOb	1.2	80×10^6	6000	18.7
VOc	2.0	133×10^6	6000	21.5
VOd	2.7	180×10^6	15000	21.8

From XRD measurements it can be concluded that each film has nanocrystalline structure. The peaks located at $2\theta \approx 20.2°$, 21.5°, 41° represent the orthorhombic V_2O_5 phase. They are assigned to the (001), (101) and (002) reflections, respectively. It is observed that peaks become narrower with higher intensities when increasing the laser energy density. Consequently, using higher fluences result in larger crystallite sizes. At the same time we can draw the conclusion that in case of low fluences, films become more amorphous. As (002) peak is not present in case of the "VOa" sample (see Table 1), one can use the TEM images to measure the crystallite size.

Figure 3b shows an XRD pattern when using a fluence of 2.7 J/cm^2 and 15000 subsequent laser pulses. It can be concluded that higher energy densities and the use of more laser pulses yield a film of higher crystallinity [25].

The calculated crystallite sizes are shown in Table.1. The diameters of the V_2O_5 clusters were around 20 nm. It should be noted that these values correspond to an average crystallite size on the spot where the X-ray passed through.

4.1.2. Optical Transmission and Band Gap Calculations

The variation in spectral transmittance (T) as a function of the fluence is shown in Figure 4. It is clearly visible that using higher fluences, thicker films can be obtained, i.e., T globally decreases with the fluence. The peak around 330 nm wavelength indicates that the films are non-homogenous and lower oxidation states of vanadium are also present. Thus, it can be concluded that the films exhibit a multiphase composition as it was also reported in other works in for vanadium oxide films thinner than 400 nm [114].

Determining the fundamental absorption edge from T measurements, the following expression can be used:

$$\alpha(h\nu) = B(h\nu - E_g)^n, \quad (1)$$

where α is the optical absorption coefficient, hν is the incident photon energy, B is a constant called edge width parameter, E_g is the optical band gap of the material and n is the exponent, that determines the type of electronic transition causing the absorption. It can take values of (1/2) for direct allowed, (3/2) for direct forbidden, 2 for indirect allowed and 3 for indirect forbidden transitions [25]. The absorption coefficient α can be calculated by using the relation:

$$\alpha = - \{1 / t\} \ln \{ T / (1-R)^2 \} , \tag{2}$$

where T is the transmittance, R is the reflectance and t is the thickness of the film. In case of thin and highly transparent films, R can be considered to be 0.

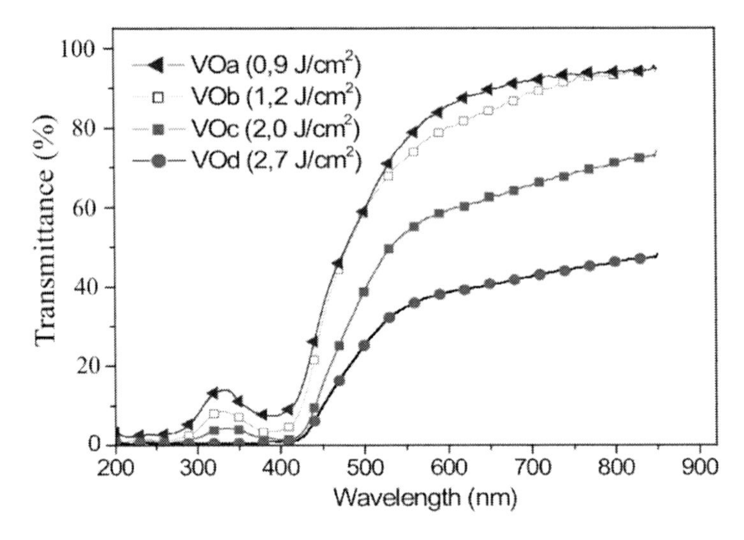

Figure 4. Optical transmission characteristics of V_2O_5 thin films formed at 220 °C as a function of different fluences in case of 6000 pulses. VOd shows the transmission in case of 15000 pulses at a fluence of 2.7 J/cm^2.

Despite that the film thickness (t) is unknown in this case, one can deduce the value of the absorption edge by using the Tauc plot since the thickness depends only on laser fluence and the number of pulses. Consequently, we can plot $\alpha t h\nu$ vs. $h\nu$ (Figure 5), where t means a constant multiplier that does not influence the evaluation of the optical band gap. The optical band gap can be evaluated by extrapolating the linear region of the plot to zero (α = 0).

As it is shown in Figure 5, both the 1/2 (indirect allowed) and the 2/3 (direct forbidden) least-squares-fits give similar result for the absorption edge value, but 2/3 is preferred as it gives a better fit. The calculated optical band gaps for the VOb and VOc samples are 2.53 eV and 2.52 eV, respectively.

Both types of transitions have already been reported for vanadia [110,111]. For example, in case of single crystals of V_2O_5, Bodó and Hevesi [112] and Kenny et al. [113] reported that α varies linearly with $(h\nu)^{3/2}$ which is characteristic of direct forbidden transition across the band gap of a crystalline solid.

The calculated optical band gaps were around 2.5 eV for V_2O_5, which is in good agreement with the vanadium oxide thin films (t < 400 nm) reported in [114].

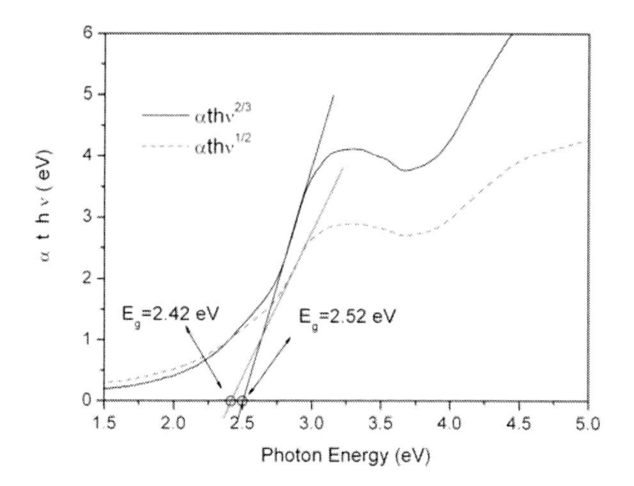

Figure 5. $\alpha t h\nu$ vs. $h\nu$ (Tauc-) plots for a V_2O_5 thin film grown by PLD.

4.1.3. X-Ray Photoelectron Spectroscopy

Vanadium oxide film compositions can be determined using X-ray photoelectron spectroscopy (XPS). Figure 6 shows representative vanadium peaks as a function of oxidation state: 2p core level XP spectra of the principal vanadium oxides from V_2O_5 to V_2O_3 and of several mixed valent compounds [115].

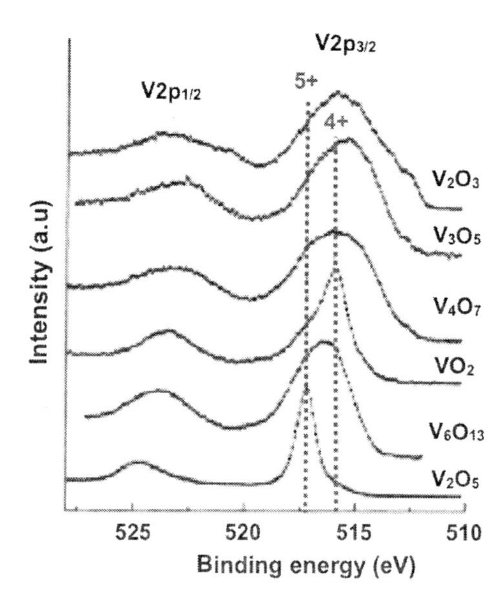

Figure 6. V_{2p} core level spectra for the vanadium oxides V_2O_5, V_6O_{13}, V_4O_7, V_3O_5, and V_2O_3. Data adapted from ref. [106, 115].

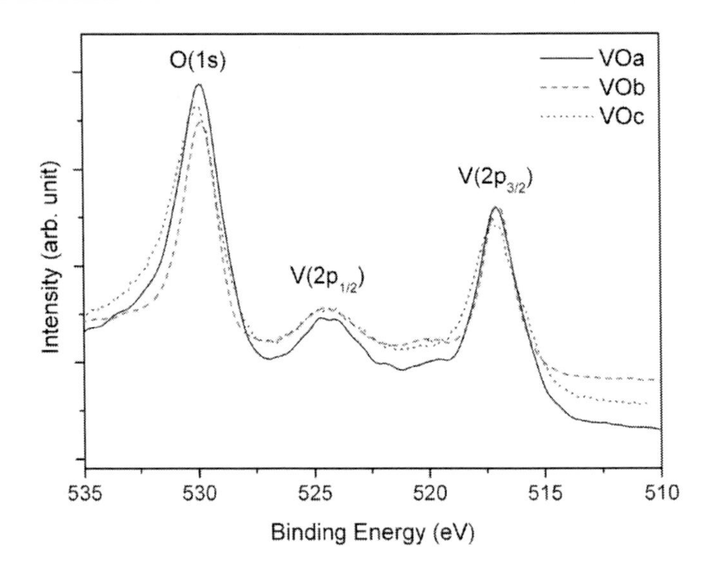

Figure 7. X-ray photoelectron spectra of different V_2O_5 films prepared using 6000 pulses with fluence varying from 0.9 to 2.0 J/cm^2 [25].

In this study, high resolution XP spectra were acquired in the binding energy range of 510-540 eV (Figure 7.). The recorded XP spectra show three intensive peaks corresponding to the core level binding energies of O (1s), V ($2p_{1/2}$) and V ($2p_{3/2}$), respectively. The main signal of the O (1s) spectrum has a binding energy at about 530 eV.

The core level binding energy (BE) peaks V (2p) and O (1s) are used to characterize the chemical state of vanadium in V_2O_5 films. There is a relationship between the core level BE and the charge state associated with an individual atom. An increase in the core level BE indicates an increase in the positive charge of the atom. The core level BE of the V ($2p_{3/2}$) peak of vanadium metal is 512.4 eV. For polycrystalline stoichiometric V_2O_5, the V ($2p_{3/2}$) peak position is reported at 517.0 eV with a chemical shift of 4.6 eV characterizing the highest oxidation state of vanadium [25]. The full-width at half maximum (FWHM) of V ($2p_{1/2}$) is much broader than that of V ($2p_{3/2}$) due to Coster-Kronig transitions [116].

The sample stoichiometry ratio $S_{i,j}$ can be estimated from XP spectra using the following relationship [25]:

$$S_{i,j} = \frac{C_i}{C_j} = \frac{I_i / ASF_i}{I_j / ASF_j} \tag{3}$$

where C_i and C_j are the concentrations of the elements i and j, I_i and I_j the background corrected intensities of the photoelectron emission lines, and ASF_i and ASF_j the atomic sensitivity factors for photoionization of the ith and j^{th} elements. This formula (3) is only valid for samples with homogenous elemental distributions. For V_2O_5, the typical spin-orbit splitting between V ($2p_{3/2}$) and V ($2p_{1/2}$) peaks for all the films deposited at different fluences is about 7.2 eV. This value is observed in the present work and is in good agreement with the previous works [117]. Binding energies, their chemical shifts and FWHMs are used for the analysis. The calculated values and comparison with other works are listed in Table 2. They are in agreement with those reported for orthorhombic bulk V_2O_5 [110].

Table 2. Comparison of binding energies, FWHMs, and core level area ratios of V_2O_5 samples

Sample name and other works	V ($2p_{3/2}$)		O (1s)		O/V
	BE (eV)	FWHM(eV)	BE (eV)	FWHM(eV)	
Bulk [118]	517.0	-	529.8	-	-
Films [119]	516.8	2.1	529.6	2.3	2.41±0.05
Pellet [119]	517.0	2.3	529.7	2.6	2.46±0.04
Films [120]	516.9	2.0	529.8	2.2	2.5±0.2
This work (Films): VOa (F=0.9 J/cm^2)	517.1	1.7	529.9	2.1	2.41
VOb (F=1.2 J/cm^2)	517.0	1.5	529.8	1.5	2.38
VOc (F=2.0 J/cm^2)	517.2	2.1	530.1	2.2	2.50

4.1.4. Transmission Electron Microscopy

Figures 8a-b correspond to sample VOc. Collections of nanoclusters (NCs) assembled in a continuous layer are observed and one can see that diameters of the NCs vary between 20 and 60 nm.

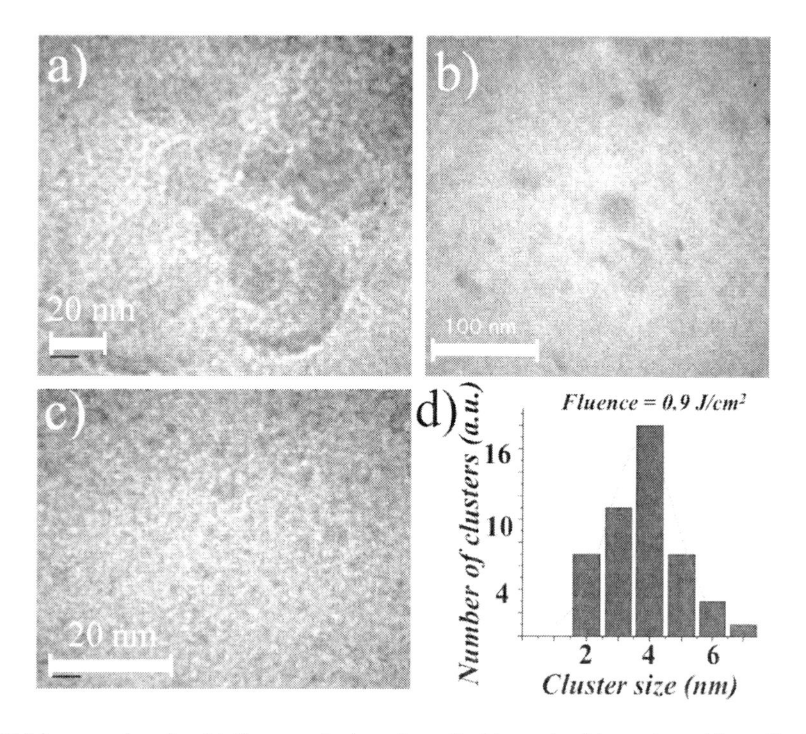

Figure 8. TEM images showing V_2O_5 nanoclusters deposited by pulsed laser deposition after being subjected to (8a-b) 2 J/cm^2 and (8c) 0.9 J/cm^2 laser irradiation (ArF excimer laser, λ=193 nm, pulse duration 15 ns at FWHM, repetition rate 2 Hz, 6000 pulses). Figure 8(d) presents a nanocluster size histogram corresponding to sample 8c showing the size distribution obtained with a laser fluence of 0.9 J/cm^2 [106].

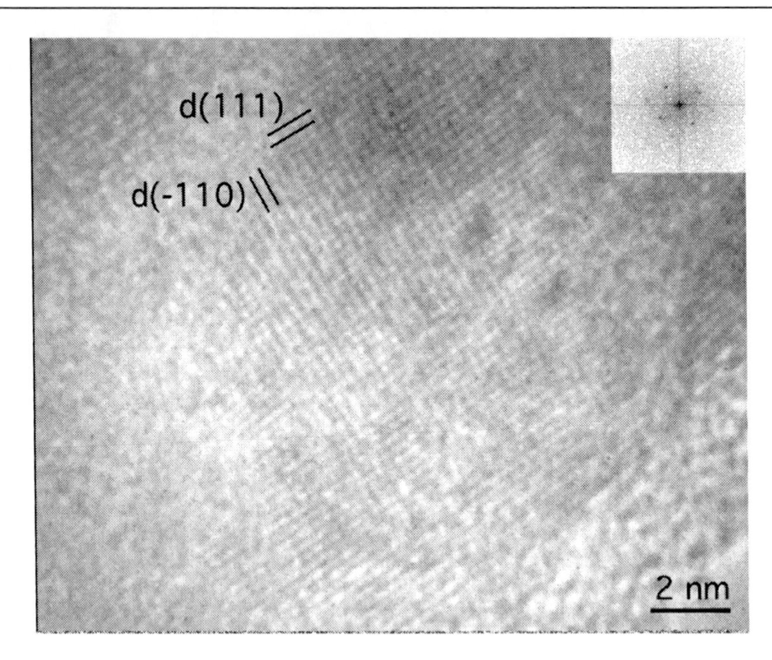

Figure 9. HRTEM image of a crystalline domain with the V_2O_3 stoichiometry observed on VOb. The inset shows the corresponding electron diffraction pattern.

In contrast, as it is shown in Figure 8c., when using lower fluence (0.9 J/cm^2), most of the NCs are smaller than 5 nm. The size histogram corresponding to this sample (VOa) is shown in Figure 8d. The average crystallite diameter is about 4 nm.

Figure 9 shows a large V_2O_3 domain observed along the [110] direction in the VOb sample. The electron diffraction pattern is given in the inset. As it was mentioned earlier, during the formation of the vanadium oxide, lower oxides can be created. However, the evolution of V_2O_5 to V_2O_3 during the exposure in the electron beam of the microscope cannot be excluded.

HRTEM applied to small pieces of the oxide surfaces is not a statistic method to determine the structure and orientation of the deposit. However, the TEM observations give more precise measurements of the cluster size according to the fluence and a better description of the polycrystalline layer. The V_2O_3 structure was not detected by X-rays, thus it seems reasonable to believe that the quantity is very low in the sample and it is also produced upon the interaction with the electron beam.

4.1.4.1. Effect of Cluster Size on the Absorption Edge

Much of the excitement surrounding nanochemistry resides in the promise that materials fabricated in the nm size range exhibit unique structural properties and chemical reactivity. It is known that the reduced size of V_2O_5 nanoparticles yields more efficient catalysis due to the fact that the surface-to-volume ratio is increased and surface atoms have lower bond coordination. The higher the catalyst active surface area, the greater is the reaction efficiency. This also results in a change of physical properties opening another field of exploiting nanoclusters. In addition, when the nanocluster is sufficiently small -- typically 10 nanometers or less -- the physical properties change due to quantum confinement that results from electrons and holes being confined in a dimension that approaches the exciton Bohr radius [25, 106, 122].

4.1.4.2. The V_2O_5 Bohr Radius

From transmission electron microscopy (TEM) images of V_2O_5 layers grown by pulsed laser deposition (Figure 8c), it is clearly visible that nanoclusters are present with sizes in the range of 2-5 nanometers. It is well known that due to the quantum confinement effect, semiconducting particles with a diameter below the Bohr radius have a larger band gap. The Bohr radius for V_2O_5 can be calculated using the following equation [25,106]:

$$a_B = 4\pi\,\varepsilon_0\hbar^2\varepsilon\,/e^2 m_0\,\mu \quad (4)$$

where $\mu = m_e m_h /(m_e + m_h)$ is the reduced exciton mass and m_e (0.24) and m_h (0.5) are the electron and the hole effective masses [25,106,121], ε_0 and ε are the dielectric constants of vacuum and V_2O_5, respectively, and e is the elementary charge.

As the calculated Bohr radius from (4) is 4.52 nm and the dimensions of some clusters in Figure 8c are below this value, these small-sized nanoclusters have an increased energy gap yielding a blue shift in the absorption edge [25].

5. Time of Flight Analysis (TOF) of the Laser-Induced Plasma

Figure 10 shows the schematic of the experimental apparatus used for optical emission spectroscopy observations. The same ArF excimer laser (λ = 193 nm, FWHM = 15 ns) was used as for the PLD experiments. The laser beam was focused onto a rotating target at an angle of incidence of 45° with the target surface. Single laser pulses were used for the analysis. The pressure in the vacuum chamber during the optical emission measurements was 5 x 10^{-7} Torr.

For the TOF investigations, the image of the plasma plume was projected on the slit of the monochromator (Instruments SA, Inc) using a convex lens (f=8 cm) as shown in Figure 10. The setup allows moving the slit (0.1 mm width) of the monochromator for spatial- and time resolved plasma plume investigations.

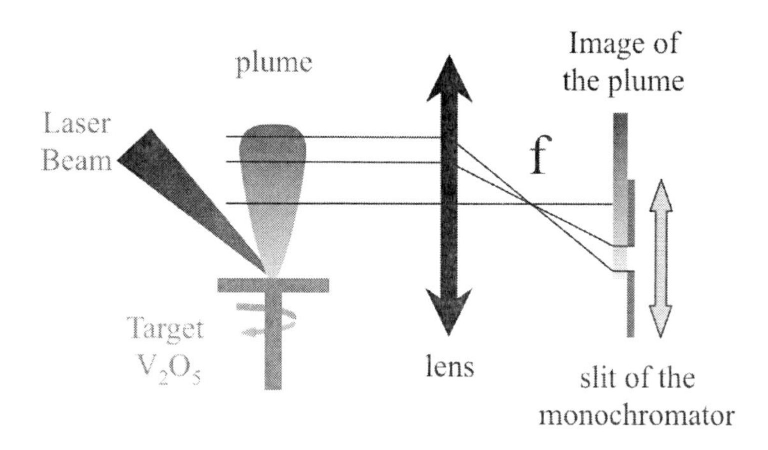

Figure 10. Experimental setup for the optical emission spectroscopic experiments.

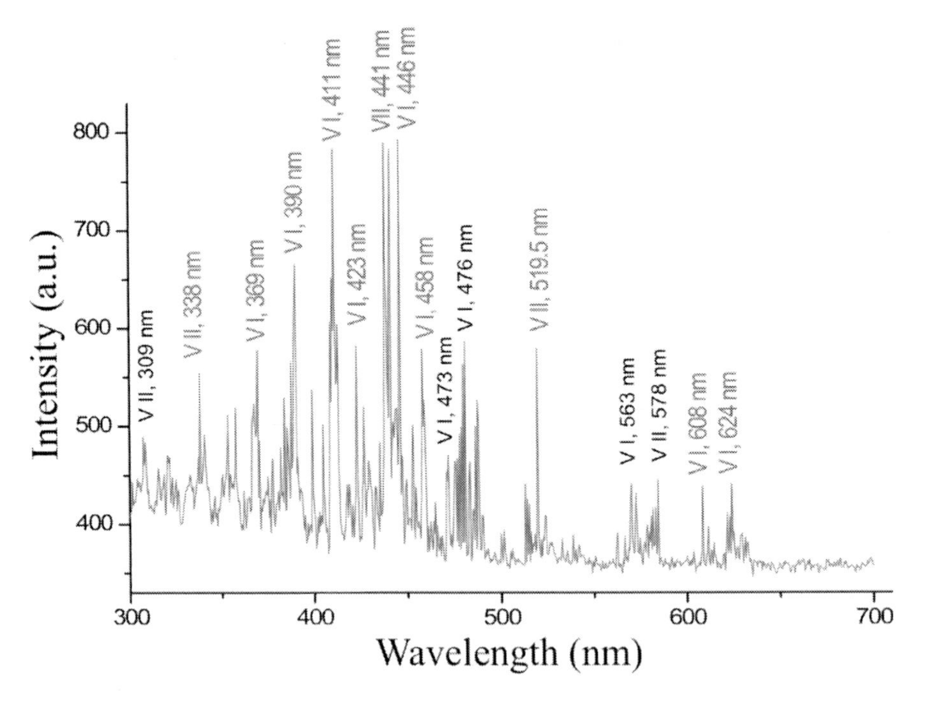

Figure 11. The optical emission spectrum of V_2O_5 generated by the 193 nm excimer laser showing predominantly the spectral lines originating from transitions in V atoms (VI) and the singly ionized vanadium (VII).

Figure 11 shows a typical TOF spectrum, recorded in the visible optical range upon the ablation of V_2O_5 target by a single ns-laser pulse irradiation at an average fluence of 2.0 J/cm^2. Two main species are observed: neutral vanadium atoms (VI) and singly charged vanadium ions (VII). Some of the most intense peaks are assigned.

Table 3. Optical emission lines with corresponding electron transitions in V atoms (VI)

Wavelength (nm)	Transition
369	$^6D_{1/2} - {}^6P^0{}_{3/2}$
390	$^4F_{9/2} - {}^4F^0{}_{9/2}$
411	$^6D_{9/2} - {}^6D^0{}_{9/2}$
423	$^6D_{9/2} - {}^2G^0{}_{7/2}$
446	$^6D_{9/2} - {}^6P^0{}_{7/2}$
458	$^4F_{5/2} - {}^4G^0{}_{7/2}$
473	$^6D_{5/2} - {}^4F^0{}_{7/2}$
476	$^6D_{5/2} - {}^4F^0{}_{5/2}$
563	$^4F_{9/2} - {}^6D^0{}_{7/2}$
608	$^4D_{3/2} - {}^6F^0{}_{5/2}$
624	$^4F_{9/2} - {}^6G^0{}_{7/2}$

The intense peaks at 369 nm, 390 nm, 411 nm, 423 nm, 446 nm, 458 nm, 473 nm, 476 nm, 563 nm, 608 nm, and 624 nm correspond to transitions between atomic energy levels (also referred as „term values") in the neutral vanadium atoms. Neutral V atom has 468 energy levels with 11219 emission lines in the spectral range 260 nm – 762 nm, while singly charged V ion has 323 energy levels with 9767 emission lines in the spectral range 206 nm – 885 nm [123]. Some of the detected emission lines are summarized in Tables 3-4. The peak at 309 nm can be associated with both V atoms and V ions, however, the relative intensity of V II is much higher than V I, so the peak is assumed to originate from V ions.

Table 4. Optical emission lines with corresponding electron transitions in singly charged V^+ ions (VII)

Wavelength (nm)	Transition
309	$^5F_5 - {}^5G^0_6$
338	$^1G_4 - {}^5D^0_4$
441	$^3P_2 - {}^5G^0_2$
519.5	$^1G_4 - {}^5F^0_4$
578	$^1G_4 - {}^5G^0_5$

Figure 12 shows two emission spectra taken 5 mm and 15 mm away from the target surface. As it can be seen, the signal of V ions (V II at 519.5 nm) looses relatively more intensity compared with V atoms as a function of the target distance. This can be explained by the recombination phenomenon of V ions while moving away from the sample surface.

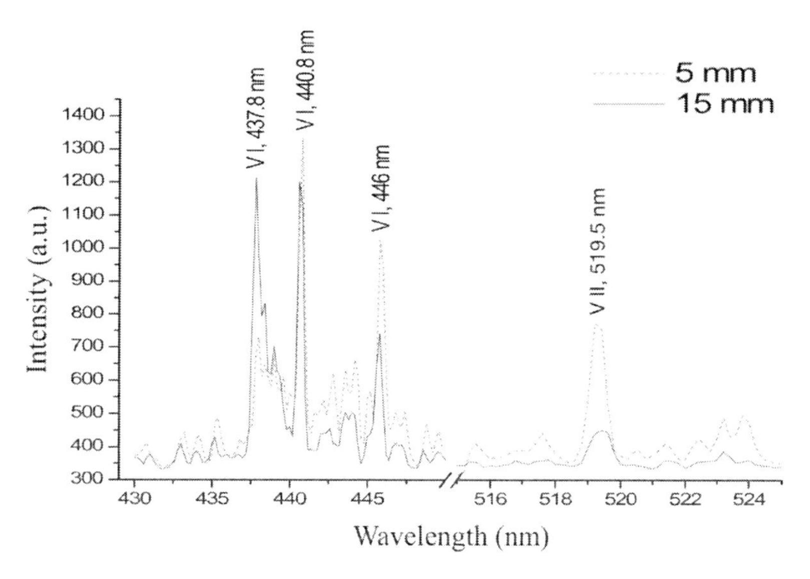

Figure 12. Comparison of two emission spectra taken at 5 mm and 15 mm distance from the target surface.

To study the dynamics of a selected species, its propagation can be observed by moving the slit of the monochromator. In this case, the sensitivity of the monochromator is adjusted to the corresponding wavelength of the emission peak of the observed species. The mean velocity of species can be obtained from the temporal change in the position of the emission peak in a time-resolved representation. Figure 13 shows the temporal evolution of the strongest V I emission line at 411 nm for different distances (5 – 25 mm) from the target A decrease of intensity peaks as a function of the target distance can be seen.

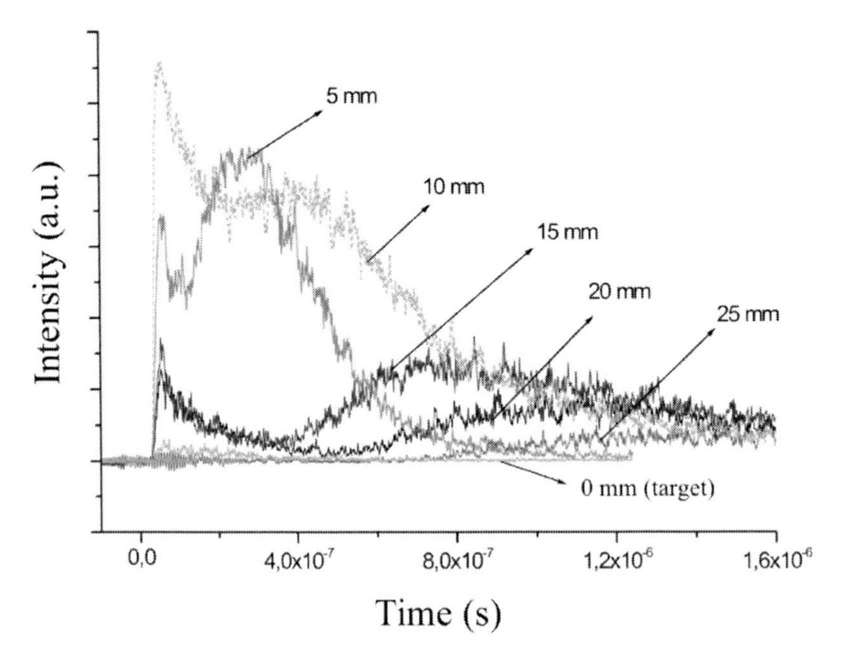

Figure 13. Intensity vs. time plot for the V I emission line observation (411 nm) at 5 different distances from the target.

By representing the intensity maximums on a target distance vs. time graph, the mean velocity of the selected specie can be obtained, as demonstrated in Figure 14. The linear fit represents the mean velocity of the V atom and it was calculated to be around 15.7 km/s after the first 25 mm from the target.

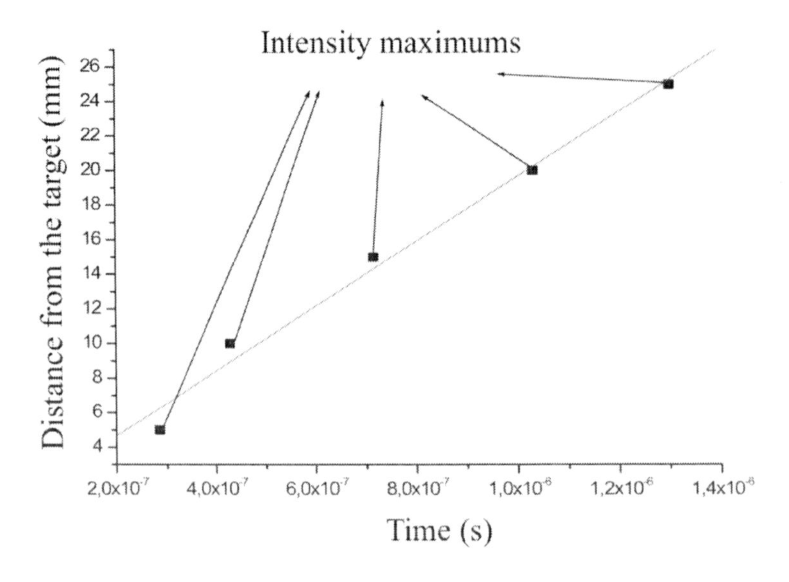

Figure 14. Distance-time plot from emission peak position of neutral V I species.

CONCLUSIONS

Pulsed laser deposited, highly oriented nanocrystalline V_2O_5 films using different fluences and pulse numbers were synthesized and investigated. From TEM images we can conclude that the size of NCs were in the range of 2-60 nm. From experimental results, it is evident that NC size depends on the laser fluence used during deposition. Consequently, NC size can be controlled by adjusting the laser fluence.

An estimation of the optical band gap was carried out and found to be around 2.52 eV, and the Bohr radius to be 4.52 for V_2O_5. The shift of the absorption edge with respect to the bulk value was observed. It was assumed that the shift of the edge relies on two facts. Firstly, a part of NCs with diameters smaller than the Bohr radius, can increase the energy gap. Secondly, during the formation of V_2O_5 thin films, lower oxidation states of vanadium are also created that give rise to the energy gap. These two factors together influence the band gap and yield a relatively higher value.

A plasma investigation in vacuum was additionally presented and has proven the existence of two species in the laser-induced plasma: neutral V atoms and singly charged V ions. Mean velocity of neutral V atoms was estimated using the temporal- and spatial dynamics of neutral V species in the plasma.

ACKNOWLEDGMENTS

S. Beke is grateful for J. Bonse and L. Nánai for useful discussions. L. Kőrösi, S. Papp and A. Oszkó are acknowledged for the XRD measurements, SEM images, and XPS measurements, respectively.

REFERENCES

[1] S. Surnev, M. G. Ramsey and F. P. Netzer, Surf. Sci. 73 (2003) 117.
[2] A. Kamper, I. Hahndorf and M. Baerns, Top. Catal. 11 (2000) 77.
[3] M. Weckhuysen and D.E. Keller, Catal. Today 78 (2003) 25.
[4] D. Wruck, S. Ramamurthi and M. Rubin, Thin Solid Films 182 (1989) 79.
[5] J. B. Goodenough, Prog. in Solid State Chem. 5 (1971) 145.
[6] A. Suli, M. I. Torok and I. Hevesi, Thin Solid Films 139 (1986) 233.
[7] J. G. Zhang, J. M. McGraw, J. Turner and D. Ginley, J. Electrochem. Soc. 144 (1997) 1630.
[8] G. Guzman, B. Yebka, J. Livage and C. Julien, Solid State Ion. 4 (1996) 407.
[9] C. Julien, E. Haro-Poniatowski, M. A. Camacho-López, L. Escobar-Alarcón and J. Jímenez-Jarquín, J. Mater. Sci. Eng. 65 (1999) 170.
[10] A. Kumar, P. Singh, N. Kulkarni and D. Kaur, Thin Solid Films 516 (2008) 912.
[11] C. V. Ramana, O. M. Hussain, B. S. Naidu and P. J. Reddy, Thin Solid Films 305 (1997) 219.
[12] C. V. Ramana, O. M. Hussain, B. S. Naidu, C. Julien and M. Balkansy, Mater. Sci. Eng. B 52 (1998) 32.

[13] C.V. Ramana, O. M. Hussain, S. Uthanna, et al. Disordered Mater. – Current developments 223 (1996) 449.
[14] C.V. Ramana, O.M. Hussain, B.S. Naidu, et al. Vacuum 48 (1997) 431.
[15] C.V. Ramana, O.M. Hussain, Adv. Mater. Opt. Electr. 7 (1997) 225.
[16] C. V. Ramana, O. M. Hussain, S. Uthanna and B. S. Naidu, Opt. Mater. 10 (1998) 101.
[17] A. Lourenco, A. Gorenstein, S. Passerini, W. H. Smyrl, M. C. A. Fantini and M.H. Tabacniks, J. Electrochem. Soc. 45 (1998) 706.
[18] Aita C.R., Liu Y.L., Kao M.L., Hansen S.D., J. Appl. Phys. 60 (1986) 749.
[19] L. J. Meng, M. Andritschky and M. P. Dos Santos, Thin Solid Films 223 (1993) 242.
[20] L. J. Meng, M. Andritschky and M. P. Dos Santos, Vacuum 45 (1994) 19.
[21] L. J. Meng and M. P. Dos Santos, Thin Solid Films 237 (1994) 112.
[22] L. J. Meng, R. A. Silva, H. N. Cui, V. Teixeira, M. P. Dos Santos and Z. Xu, Thin Solid Films 515 (2006) 195.
[23] I. Petrov, P.B. Barna, L. Hultman, and J.E. Greene, "Microstructural Evolution during Film Growth," J. Vac. Sci. Technol. 21 (2003) S117.
[24] J.E. Greene, "Thin Film Nucleation, Growth, and Microstructural Evolution: an Atomic Scale View," in Handbook of Deposition Technologies for Thin Films and Coatings, Third Edition, ed. by P. Martin, William Andrew Publications (Elsevier), Burlington, MA (2010).
[25] S. Beke, S. Giorgio, L. Korosi, L. Nanai, W. Marine, Thin Solid Films 516 (2008) 4659.
[26] F. Breech, L. Cross: Appl. Spect. 16 (1962) 59.
[27] H. M. Smith, A. F. Turner: Appl. Opt. 4 (1965) 147
[28] S. V. Gaponov, A. A. Gudkov, V. I. Luskin, I. Salaschenko: Sov. Tech. Phys. Lett. 5 (1979) 210.
[29] H. Oesterreicher, H. H. Bittner, B. Kothari: J. Solid State Chem. 26 (1978) 97.
[30] D. Dijkkamp, T. Venkatesan, X. D. Wu, S. A. Shaheen, N. Jisrawi, Y. H. Min-Lee, W. L. McLean, M. Croft, Appl. Phys. Lett. 51 (1987) 619.
[31] J. T. Cheung, J. Madden: J. Vac. Sci. Tech A1(3) (1987) 1604.
[32] F. Beech, I. W. Boyd in Photochemical Processing of Electronic Materials (Academic Press, New York), pp.387-429.
[33] K. L. Saenger: "Bibliography of films deposited by pulsed laser deposition" in Pulsed Laser Deposition of Thin Films (Wiley, New York 1994), pp. 581-613.
[34] C.M. Cotell, Appl. Surf. Sci. 69 (1993) 140.
[35] X.D Wu, S.R Foltyn, R.C Dye, A.R Garcia, N.S Nogar, R.E Muenchausen, Thin Solid Films 218 (1992) 310.
[36] V. Craciun, D. Cracium, I.W. Boyd, Mater. Sci. Eng. B 18 (1993) 178.
[37] H.-U. Habermeier, Appl. Surf. Sci. 69 (1993) 204.
[38] M. Borek, F. Qian, V. Nagabushnam et al., Appl. Phys. Lett. 63 (1993) 3288.
[39] G.C. Tyrrell, T. York, N. Cherief, D. Givord, J.G. Lunney, M. Buckley and I.W. Boyd Microelectr. Eng. 25 (1994) 247.
[40] D. Lubben, S.A. Barnett, K. Suzuki, and J.E. Greene, "Primary and Secondary Ion Deposition of Epitaxial Semiconductor Films From Laser□Induced Plasmas," in Laser Controlled Chemical Processing of Surfaces, ed. by A.W. Johnson, North Holland Publishing Co., New York, (1984).

[41] D. Lubben, S.A. Barnett, K. Suzuki, S. Gorbatkin, and J.E. Greene, "Laser Induced Plasmas for Primary Ion Deposition of Epitaxial Ge and Si Films," J. Vac. Sci. Technol. B3 (1985) 968.
[42] T.N. Kennedy, R. Hakim, J.D. Mackenzie, Mater. Res. Bull. 2 (1967) 193.
[43] Chang et al., Thin Solid Films 517 (2004) 2653.
[44] H. Groult, E. Balnois, A. Mantoux, K. Le Van and D. Lincot, Appl. Surf. Sci. 252 (2006) 5917.
[45] M. Schuisky, A. Harsta, A. Aidla, K. Kukli, A. A. Kiisler and J. Aarik, J. Electrochem. Soc. 147 (2000) 3319.
[46] M. Vehkamaki, T. Hatanpaa, T. Hanninen, M. Ritala and M. Leskela, Electrochem. Solid-State Lett. 2 (1999) 504.
[47] P. Tagtstrom, P. Martensson, U. Jansson and J. O. Carlsson, J. Electrochem. Soc. 146 (1999) 3139.
[48] M. Cassir, F. Goubin, C. Bernay, P. Vernoux and D. Lincot, Appl. Surf. Sci. 193 (2002) 120.
[49] C. Bernay, A. Ringuede, P. Colomban, D. Lincot and M. Cassir, J. Phys. Chem. Solids 64 (2003) 1761.
[50] E.B. Yousfi, J. Fouache and D. Lincot, Appl. Surf. Sci. 153 (2000) 223.
[51] J. J. Ganem, I. Trimaille, J. C. Vickridge, D. Blin and F. Martin, Nucl. Instr. Met. Phys. Res. B 219-220 (2004) 856.
[52] J. C. Badot, S. Ribes, E.B. Yousfi, V. Vivier, J.-P. Pereira-Ramos, N. Baffier and D. Lincot, Electrochem. Solid-State Lett. 3 (2000) 485.
[53] R. Baddour-Hadjean, V. Golabkan, J.-P. Pereira-Ramos, A. Mantoux and D. Lincot, J. Phys. IV France, 11 (2001) Pr11–Pr85.
[54] A. Mantoux, J.-C. Badot, N. Baffier, J. Farcy, J.-P. Pereira-Ramos, D. Lincot and H. Groult, J. Phys. IV France 12 (2002) Pr2–Pr111.
[55] R. Baddour-Hadjean, V. Golabkan, J.P. Pereira-Ramos, et al. J. Raman Spectr. 33 (2002) 631.
[56] H. Groult, E. Balnois, A. Mantoux et al., Appl. Surf. Sci. 252 (2006) 5917.
[57] M.G. Willinger, G. Neri, E. Rauwel et al., Nano Lett. 8 (2008) 4201.
[58] H. Kim, Han-Bo-Ram Lee, W.-J. Maeng, Thin Solid Films 517 (2009) 2563.
[59] J.-C. Badot, S. Ribes, E.B. Yousfi, V. Vivier, J.-P. Pereira-Ramos, N. Baffier and D. Lincot, Electrochem. Solid-State Lett. 3 (2000) 485.
[60] R. Baddour-Hadjean, V. Golabkan, J.-P. Pereira-Ramos, A. Mantoux and D. Lincot, J. Phys. IV France 11 (2001) Pr11–Pr85.
[61] A. Mantoux, J.-C. Badot, N. Baffier, J. Farcy, J.-P. Pereira-Ramos, D. Lincot and H. Groult, J. Phys. IV France 12 (2002) Pr2–Pr111.
[62] A. Bouzidi, N. Benramdane, A. Nakrela, C. Mathieu, B. Khelifa, R. Desfeux and A. Da Costa, Mater. Sci. Eng. B- Solid State Mater. Adv. Techn. 95 (2002) 141.
[63] L. Boudaoud, N. Benramdane, R. Desfeux, B. Khelifa and C. Mathieu, Catal. Today 113 (2006) 230.
[64] B. Wang, K. Konstantinov, D. Wexler, H. Liu and G. Wang, Electrochim. Acta 54 (2009) 1420.
[65] J. G. Lu; P. Chang and Z. Fan, Mater. Sci. Eng. R: Rep., 52 (2006) 49.
[66] I. S. Chronakis, J. Mater. Proc. Techn. 167 (2005) 283.

[67] S. V. N. T. Kuchibhatla, A. S. Karakoti, D. Bera and S. Seal S, Prog. Mater. Sci. 52 (2007) 699.
[68] P. Viswanathamurthi, N. Bhattarai, H. Y. Kim, D. R. Lee, Scr. Mater. 49 (2003) 577.
[69] C. Ban and M. S. Whittingham, Solid State Ion. 179 (2008) 1721.
[70] C.M. Ban, N.A. Chernova, M.S. Whittingham, Electrochem. Comm. 11 (2009) 522.
[71] Z. S. El Mandouh and M. S. Selim, Thin Solid Films 371 (2000) 259.
[72] B. Samuneva, Y. Dimitriev, V. Dimitrov, E. Kashchieva and G. Encheva, J. Sol-Gel Sci. Techn. 13 (1998) 969.
[73] Y. Dimitriev, R. Iordanova, M. Mancheva and D. Klissurski, Chem. for Sust. Dev. 13 (2005) 185.
[74] M. Mancheva, R. Iordanova, Y. Dimitriev and D. Klissurski, Nanosci. Nanotechn. 5 (2005) 123.
[75] J. Livage, Chem. Mater. 3 (1991) 578.
[76] A. C. R. Ditte, Acad. Sci. Park 101 (1885) 698.
[77] W. Biltz, Ber. Dtsch. Chem. Ges. 37 (1904) 3036.
[78] W. Ostermann, Wiss. 2nd. Hamburg 1 (1922) 17.
[79] E. Z. Müller, Chem. 2nd. Kolloide 8 (1911) 302.
[80] N. Gharbi, C. Rkha, D. Ballutaud, M. Michaud, J. Livage, J. P. Audiere and G. Schiffmacher, J. Non-Cryst. Solids 46 (1981) 247.
[81] Aiping Jin, Wen Chen, Quanyao Zhu, Ying Yang, V.L. Volkov, G.S. Zakharova Thin Solid Films 517 (2009) 2023.
[82] M. Kakihana, J. Sol-Gel Sci. Tech., 6 (1996) 7.
[83] S. Beke, L. Kőrösi, S. Papp, L. Nánai, A. Oszkó, J.G. Kiss, V. Safarov, Appl. Surf. Sci. 254 (2007) 1363-1368.
[84] Y. Shimizu, K. Nagase, N. Miura and N. Yamazoe, Solid State Ion. 53-6 (1992) 490.
[85] M. B. Sahana, C. Sudakar, C. Thapa, G. Lawes, V. M. Naik, R. J. Baird, G. W. Auner, R. Naik and K. R. Padmanabhan, Mater. Sci. Eng. B – Solid State Mater. Adv. Techn. 143 (2007) 42.
[86] A. Z. Moshfegh, A. Ignatiev, Thin Solid Films 198 (1991) 251.
[87] S. Passerini, D. Chang, X. Chu, D. Ba Le, W. Smyrl, Chem. Mater. 7 (1995) 780.
[88] N. Ozer, Thin Solid films 305 (1997) 80.
[89] Nánai L. Laser induced synthesis and optical strength of vanadium pentoxide, PhD Thesis, Moscow, 1982.
[90] A. T. Fromhold "Theory of Metal Oxidation", North Holland Publishing Company, Amsterdam (1976).
[91] A. Atkinson, Rev. Mod. Phys. 57 (2) (1985) 437.
[92] R. Vajtai, C. Beleznai, L. Nánai, Z. Gingl, T.F.George, Appl. Surf. Sci. 106 (1996) 247.
[93] Bert M. Weckhuysen, Daphne E. Keller, Catal. Today 78, (2003) 25.
[94] K.H. Kim, D.K. Roh, I.K. Song et al., J. Solid State Electrochem. 14 (2010) 1801.
[95] A.M. Glushenkov, M. F. Hassan, V. I. Stukachev et al., J. Solid State Electrochem 14 (2010) 1841.
[96] H.S. Kim, B.W. Cho, Bull. Kor. Chem. Soc. 31 (2010) 1267.
[97] T.Y. Zhai, H.M Liu, H.Q. Li et al., Adv. Mater. 22 (2010) 2547.
[98] J.H. Song, H.J. Park, K.J. Kim et al., J. Pow. Sources 195 (2010) 6157.
[99] N.S, Choi, J.S. Kim, R.Z.Yin et al., Mater. Chem. Phys. 116 (2009) 603.
[100] R.A. Timm, M.P.H Falla, M.F.G Huila et al., Sens. Actuat. B - Chem. 146 (2010) 61.

[101] E. M. Guerra, M. Mulato, J. Sol-gel Sci. Techn. 52 (2009) 315.
[102] A.V. Grigorieva, A.B Tarasov, E. A. Goodilin et al., Mengeleev Comm. 18 (2008) 6.
[103] Y.B. Li, Z. W. Huang, S. Q. Rong, Sens. Mater. 18 (2006) 241.
[104] G Micocci, A. Serra, A. Tepore et al. J Vac. Sci. Techn. A – Vac. Surf. Films 15 (1997) 34.
[105] H.L. Zhang, L.C. Zhang, J. Hu et al. Talanta 82 (2010) 733.
[106] S. Beke, Thin Solid Films, 519 (2011) 1761.
[107] Maiman, T. H. Nature 187 (1960) 493.
[108] E.F. Kaelbe, Handbook of X-Rays, McGraw-Hill, New York, 1967.
[109] R.T. Rajendra Kumar, B. Karunagaran, V. Senthil Kumar, Y.L. Jeyachandran, D. Mangalaraj, Sa. K. Narayandass, Mat. Sci. in Semicond. Proc. 6 (2003) 543.
[110] C.V. Ramana, O.M. Hussain, S. Uthanna, B. Srinivasulu Naidu, Opt. Mat. 10 (1998) 101.
[111] M. Ghanashyam Krishna, A. K. Bhattacharya, Mat. Sci. Eng. B 86 (2001) 41.
[112] Z. Bodó, I. Hevesi, Phys. Stat. Sol. 20 (1967) K45.
[113] N. Kenny, C.R. Kannewurf, D. H. Whitmore, J. Phys. Chem. Solids 27 (1966) 1237.
[114] M. Ghanashyam Krishna, Y. Debauge, A.K. Bhattacharya, Thin Solid Films 312 (1998) 116.
[115] M. Demeter, M. Neumann and W. Reichelt, Surf. Sci. 454 (2000) 41.
[116] E. Antonides, E. C. Janse and G. A. Sawatzky, Phys. Rev. B. 15 (1977) 4596.
[117] K.Wandelt, Surf. Sci. Rep. 2 (1982) 1.
[118] R.J. Colton, A.M. Guzman, J.W. Rabalais, J. Appl. Phys. 49 (1978) 409.
[119] A.Z. Moshfegh, A. Ignatiev, Thin Solid Films 198 (1991) 251.
[120] C. V Ramana, O. M. Hussain, B. Srinivasulu Naidu, C. Julien and M. Balkanski, Mat. Sci. Eng. B 52 (1998) 32.
[121] G.V. Samsonov, The Oxide Handbook, Institute of Problems in Materials Science, Academy of Sciences of the Ukrainian SSR, Kiev, 1973.
[122] S. Beke, L. Kőrösi, S. Papp, A. Oszkó, L. Nánai, Appl. Surf. Sci. 255 (2009) 9779.
[123] R. Payling and P. Larkins, Optical emission lines of the elements, John Wiley and Sons Ltd., England, 2000.

In: Laser-Induced Plasmas
Editor: Ethan J. Hemsworth, pp. 77-96
ISBN 978-1-61324-851-5

Chapter 4

NANOSECOND AND FEMTOSECOND LASER ABLATION OF TeO_2 CRYSTALS: SURFACE CHARACTERIZATION AND PLASMA ANALYSIS

S. Beke[1*], K. Sugioka[1], K. Midorikawa[1] and J. Bonse[2]
[1]Laser Technology Laboratory, RIKEN – Advanced Science Institute,
Hirosawa 2-1, Saitama 351-0198, Japan
[2]BAM Federal Institute for Materials Research and Testing,
Unter den Eichen 87, 12205 Berlin, Germany

ABSTRACT

Near-IR femtosecond (fs) (pulse duration = 150 fs, wavelength = 775 nm, repetition rate 1 kHz) and VUV nanosecond (ns) (pulse duration = 20 ns, wavelength = 157 nm, repetition rate 1 to 5 Hz) laser pulse ablation of single-crystalline TeO_2 (c-TeO_2) surfaces was performed in air using the direct focusing technique. A multi-method characterization using optical microscopy, atomic force microscopy and scanning electron microscopy revealed the surface morphology of the ablated craters. This allowed us at each irradiation site to characterize precisely the lateral and vertical dimensions of the laser-ablated craters for different laser pulse energies and number of laser pulses per spot. Based on the obtained information, we quantitatively determined the ablation threshold fluence for the fs laser irradiation when different pulse numbers were applied to the same spot using two independent extrapolation techniques. We found that in the case of NIR fs laser pulse irradiation, the ablation threshold significantly depends on the number of laser pulses applied to the same spot indicating that incubation effects play an important role in this material. In the case of VUV ns laser pulses, the ablation rate is significantly higher due to the high photon energy and the predominantly linear absorption in the material. These results are discussed on the basis of recent models of the interaction of laser pulses with dielectrics. In the second part of this chapter, we use time-of-flight mass spectrometry (TOFMS) to analyze the elemental composition of the ablation products generated upon laser irradiation of c-TeO_2 with single fs- (pulse duration ~ 200 fs, wavelength 398 nm) and ns-pulses (pulse duration 4 ns, wavelength 355 nm). Due to the

* Email: beke@scientist.com; phone +81-48462-1111 ext.8540; fax +81-48462-4682

three order of magnitude different peak intensities of the ns- and fs laser pulses, significant differences were observed regarding the laser-induced species in the plasma plume. Positive singly, doubly and triply charged Te ions (Te^+, Te_2^+, Te_3^+) in the form of many different isotopes were observed in case of both irradiations. In the case of the ns-laser ablation, the TeO^+ formation was negligible compared to the fs case and there was no Te trimer (Te_3^+) formation observed. It was found that the amplitude of Te ion signals strongly depends on the applied laser pulse energy. Singly charged oxygen ions (O^+) are always present as a byproduct in both kinds of laser ablation.

Keywords: Tellurium dioxide crystals, femtosecond laser ablation, VUV nanosecond laser ablation, multiphoton absorption, time-of-flight mass spectroscopy, incubation, optical properties, Scanning Electron Microscopy, Atomic Force Microscopy, isotopes.

1. Introduction

Tellurium dioxide crystals (c-TeO_2) play an important role in acousto-optical devices (deflectors, light modulators, tuneable optical filters) due to their beneficial photo-elastic properties, high optical homogeneity, low light absorption and scattering and high optical damage resistance [1,2]. TeO_2 has high refractive indices and it transmits in the mid-infrared part of the electromagnetic spectrum; therefore, it is of technological interest for optical devices. The optical band gap of TeO_2 is <4.05 eV. TeO_2 has also been shown to exhibit Raman gain up to 30 times that of silica; therefore, it is useful in optical fiber amplification [3]. Crystalline TeO_2 exists in two different forms, the yellow orthorhombic mineral tellurite, β-TeO_2, and the synthetic, colorless tetragonal (paratellurite) α-TeO_2 phase. Most of the information regarding the reaction chemistry has been obtained in studies involving paratellurite, α-TeO_2 [4,5]. Thermal and photo-induced (He–Cd laser irradiation) surface damage treatments in paratellurite have been already reported [6]. Another group used pulsed laser (second harmonic of Nd:YAG laser irradiation) ablation to produce tellurium oxide species (Te_2O_2, TeO, TeO_2, Te_2O and Te_2O_4) in argon and nitrogen matrices [7].

In this work, single and multiple-pulse ablation of single-crystalline TeO_2 by near infrared (IR) femtosecond (fs) and vacuum ultraviolet (VUV) nanosecond (ns) laser microfabrication is demonstrated as well as time-of-flight mass spectroscopy (TOFMS) of TeO_2 crystals. The combination of several characterization techniques allows us to reveal the precise shapes of the ablated craters and the resulting surface morphologies and to quantify the effects of incubation (i.e., the decrease in the damage threshold fluence with increasing number of laser pulses per spot).

2. Experimental Details

2.1. Crystal Growth Procedure and Sample Preparation

Single-crystals of paratellurite (TeO_2) were grown by using the balance controlled Czochralski method. The description of the growth apparatus is given in ref. [8]. For the crystal growth, a resistance furnace was employed in order to maintain low thermal gradients and minimize the eventual temperature fluctuation during the growth and post-growth processes. The crystals were grown from 6N pure TeO_2. Single-crystals were pulled along the <110> direction in sizes up to 50 mm in diameter and 50 mm in length. The seed rotation rate varied between 15 and 20 rounds per minute and the pulling rate was 0.8-1 mm/h. For the measurements samples were cut with diamond saw and oriented with a precision better than 0.5° by using an X-ray beam. Polished (110) TeO_2 surfaces were prepared by standard method using SiC for grinding and AB Alpha Alumina (Buehler Linde A 0.3 μm) for polishing.

2.2. Laser Ablation of c-TeO_2

2.2.1. The Nanosecond Laser Processing Setup

A schematic illustration of the experimental setup for TeO_2 ablation by a molecular fluorine (F_2) laser (λ = 157 nm, τ = 20 ns, 1-5 Hz) is shown in Figure 1.

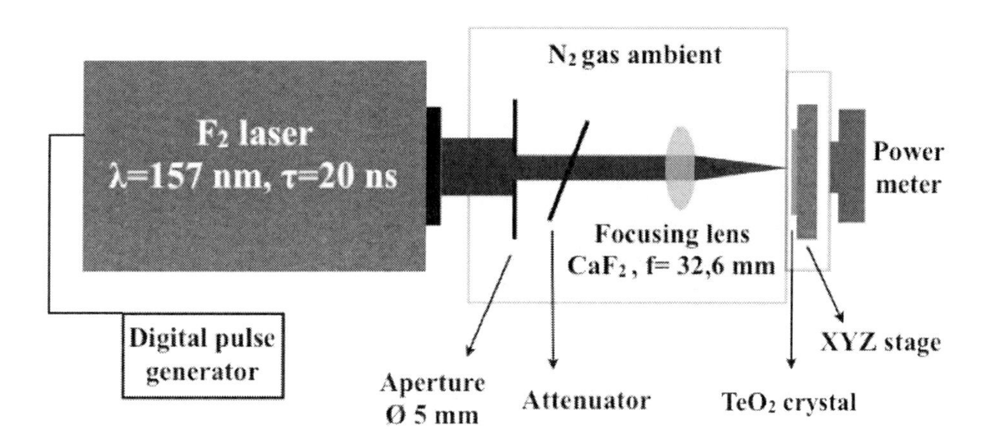

Figure 1. Schematic representation of the experimental set up for TeO_2 ablation in N_2 gas by a ns laser.

For the laser processing, the sample was placed on a computer-controlled x–y–z translation stage in a 3.5×5×5 cm^3 ablation chamber. The chamber was filled with dry nitrogen gas to prevent the absorption of the F_2 laser beam from the oxygen contained in the air atmosphere. The F_2 laser beam was spatially shaped to a circular flat-top like beam by using a hard aperture of 5 mm diameter which is then image-projected to the sample surface by a CaF_2 lens with a focal length of 32.6 mm. The repetition rate of F_2 laser was varied from 1 to 5 Hz depending on the applied number of pulses. The laser-irradiated spots were characterized by means of scanning electron microscopy (SEM: JEOL JSM-6330F). The ablated crater depths were measured by a surface profiler (α-step 500, KLA Tencor Co. Ltd.).

2.2.2. The Femtosecond Laser Processing Setup

The samples were irradiated by a commercial fs-laser processing workstation (an 800 mW regenerative amplified laser, Clark-MXR, CPA 2001) emitting linearly polarized 150 fs laser pulses at a center wavelength of 775 nm (photon energy ~1.6 eV) with a repetition rate of 1 kHz. The schematic representation of the workstation is shown in Figure 2.

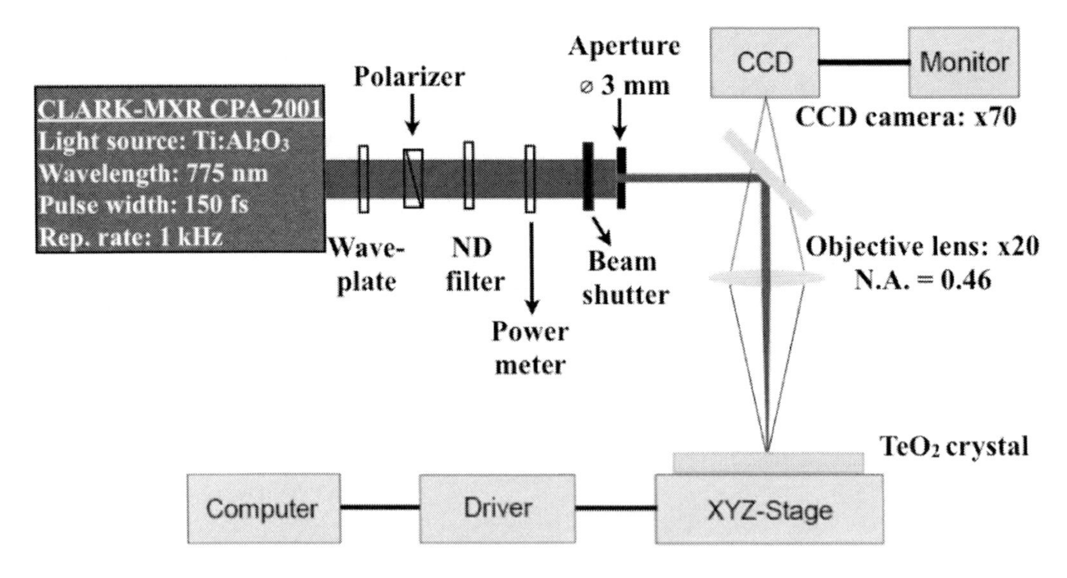

Figure 2. Schematic representation of the experimental set up for TeO_2 ablation by a fs laser.

The pulse energy in the sample plane was measured by a pyroelectric detector. Its value was adjusted by using a half-wave plate and a polarizer combination along with a set of neutral density (ND) filters. To improve the spatial beam shape in the focal plane, the 6 mm diameter of the output laser beam was reduced to 3 mm by passing it through a circular aperture in front of the focusing system. The apertured laser beam was focused by a 20x microscope objective with a numerical aperture (NA) of 0.46 and a working distance of 3.1 mm onto the sample surface to a spot diameter ($1/e^2$) of about $2w_0 < 5.1\pm0.1$ μm (as discussed in section 3.2.1). In case of fs-laser pulse irradiation, we refer to the peak fluence as the maximum value at the centre of the Gaussian beam in front of the sample surface.

The samples were mounted on a motorized x–y–z translation stage with a positioning resolution of better than 0.5 μm. They were placed perpendicular to the direction of the laser beam incidence. The irradiation process was monitored (in-situ) by a charge-coupled device (CCD) camera. An electro-mechanical shutter was used to select the desired number of pulses from a pulse train with a repetition rate of 1 kHz. Since the shortest shutter opening time was limited to 0.01 s, the minimum number of laser pulses that can be applied to the same spot in this way is 10. However, we generated single-pulse irradiations on the sample without any laser spot overlapping by using a translational scanning speed of 20000 μm/s during repetitive irradiation.

The laser-irradiated spots were characterized by means of SEM and optical microscopy (OM) in bright field imaging mode (Olympus BX51). The ablated crater depths were measured by atomic force microscopy (AFM) in tapping mode (Veeco Instruments, NanoScope V).

3. Laser Processing Results: Surface Analyses

3.1. Nanosecond Laser Processing of c-TeO_2

As the optical band gap of TeO_2 is significantly lower than the photon energy of the applied F_2 laser (<7.9 eV) a linear single photon absorption process is expected to take place. Figure 3 shows a typical ablated surface area which has been irradiated by ten (N = 10) subsequent ns-laser pulses, indicating an ablation spot diameter of ~320 μm. This spot diameter along with the measured laser pulse energy allows an estimation of the average fluence at the sample surface, which is ~145 mJ cm^{-2} for the spot shown in Figure 3.

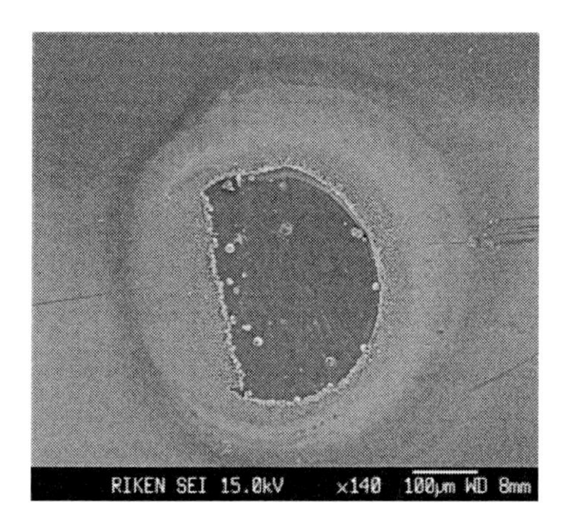

Figure 3. SEM image of an ablated crater in case of ten (N = 10) subsequent VUV ns-laser pulses at an average fluence of 145 mJ cm^{-2}.

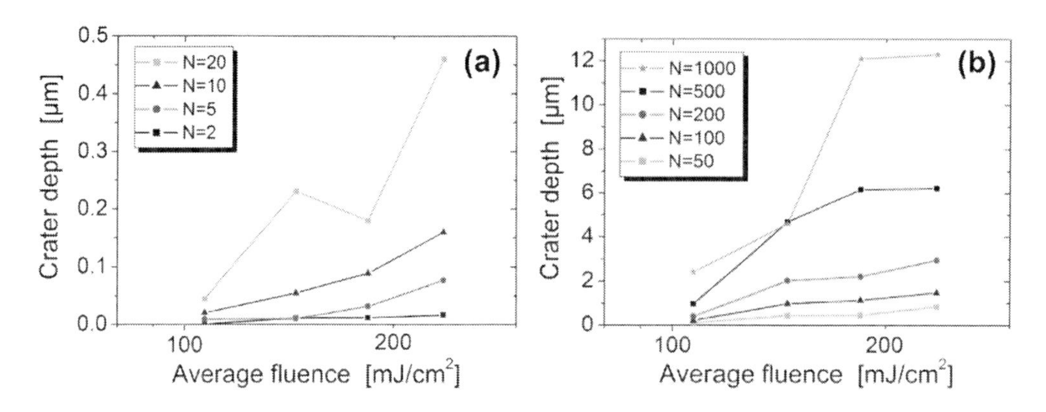

Figure 4. Crater depths in of c-TeO_2 surfaces after VUV ns-laser pulse irradiation as a function of the average laser fluence for N = 2-20 (a) and N = 50-1000 (b). Note the logarithmic average fluence scale.

Since the pulse energy of the F_2 laser often lacks of stability, we refrained from applying single shot irradiations in our experiments. N = 2-1000 laser pulses at 4 different average fluence values (107 - 236 mJ cm^{-2}) were used to reveal the correlation between the crater depth and the average fluence. The ablated crater depth as a function of the average laser fluence is presented in Figure 4(a) and 4(b) in a semilogarithmic representation. For each number of pulses per spot (N), the curves demonstrate a roughly logarithmic dependence of the crater depth on the applied laser fluence (as it would be expected for linear absorption of the radiation in the sample). In case of double pulse irradiation (N = 2) very shallow ablation craters and, therefore, a very high precision of ablation can be realized. For example at an average fluence level of 236 mJ cm^{-2} a shallow layer of maximum 17 nm is removed from the surface [see Figure 4(a)]. However, when applying higher pulse numbers (N=5-1000) significant more material can be removed and crater depths of more than 10 μm are demonstrated [Figure 4(b)].

The comparison of all curves shown in Figure 4 suggests that the ablation threshold of TeO_2 single-crystals upon VUV ns-laser irradiation is close to 100 mJ cm^{-2} and does not depend very strongly on the number of laser pulses applied to the same spot at the surface which means that little incubation effect is present.

3.2. Femtosecond Laser Processing of c-TeO_2

The TeO_2 single-crystals were irradiated by a tightly focused fs laser beam (Figure 2) using pulse number N = 1, 10, 20, 50 and 1000 subsequent laser pulses with different pulse energies [9]. The morphology, the diameters and depths of the fs-laser pulse generated ablation craters in c-TeO_2 surfaces have been investigated by SEM, AFM and OM. The combination of these techniques provides an almost complete picture of the surface state after fs pulse laser radiation.

Figure 5 shows SEM images of the morphology of the ablation craters formed at a pulse energy (at the surface) of E_{pulse} = 0.24 μJ (2.33 J cm^{-2}).

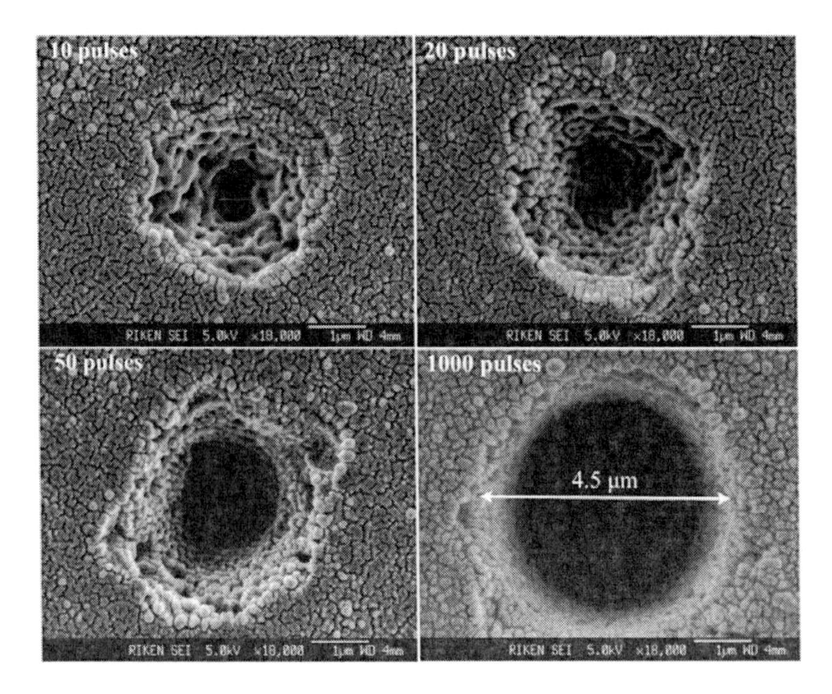

Figure 5. SEM pictures of c-TeO_2 (110) surfaces after irradiation by N = 10, 20, 50, 1000 fs laser pulses in air (pulse energy 0.24 μJ, pulse duration 150 fs, wavelength 775 nm). [reprinted from [9] with permission of IOP Publishing Ltd.]

The ablation craters have a diameter of about 4 μm in the case of 10–50 pulses. When applying a significantly larger number of pulses (N = 1000), somewhat wider ablation craters (4.5 μm) can be observed, which indicates the contribution of incubation effects. Namely, the incubation effect induced by multiple pulse irradiation can reduce the ablation threshold resulting in ablation even at the edge regions of Gaussian beam where ablation cannot take place at smaller number of pulses.

For every ablated spot, micro- and nanostructures resulting from a molten and solidified surface layer can be seen in the crater walls. In addition, at very large pulse numbers (N = 1000), redeposited material in the form of nanoparticles can be observed in the surrounding of the ablation crater. Complementary information on the crater depth can be obtained by AFM investigations. Figure 6 shows a collage of an intercomparison of SEM morphology (upper row), AFM topography images (middle row) and the corresponding optical micrographs (lower row) in the case of different pulse numbers (N = 1, 10 and 50) in c-TeO_2 at a peak fluence value of 2.33 J cm^{-2} (E_{pulse} = 0.24 μJ). The corresponding crater depths obtained by the AFM measurements are 195 nm (N = 1), 1.86 μm (N = 10) and 2.10 μm (N = 50), respectively (see the cross sectional crater profiles indicated in the middle row).

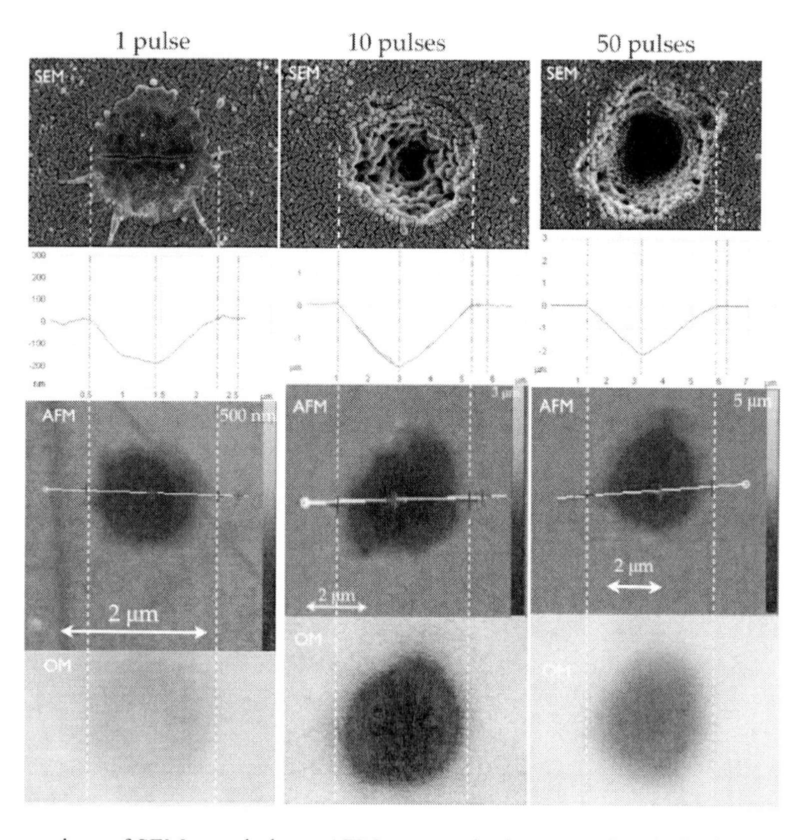

Figure 6. Comparison of SEM morphology, AFM topography images and optical micrographs (OM) of c-TeO_2 (110) surfaces after irradiation by N = 1, 10, and 50 fs laser pulses in air (pulse energy 0.24 μJ, pulse duration 150 fs, wavelength 775 nm). [reprinted from [9] with permission of IOP Publishing Ltd.]

In the SEM picture, in the case of a single-pulse irradiation one can observe a horizontal line in the central area of cavity (parallel to the <001> direction), which is attributed to a microcrack formation along the intersection of {100} cleavage planes. (TeO_2 is mechanically highly anisotropic.) It should be noted that the grainy structure of the non-irradiated TeO_2 surface and the redeposited material (as seen in the SEM images) around the craters are not visible in the AFM images. This is due to the fact that the AFM images have a limited color resolution when displaying deep crater structures. Consequently, very shallow surface corrugations in the range of a few nanometers cannot be seen in the AFM images.

3.2.1. Determination of the Ablation Threshold Fluence and the Precision of the Ablation

Laser pulse illumination to dielectrics can create various material processing including ablation [10-17]. In general, the laser fluence (ϕ_0) has to exceed a certain threshold (ϕ_{th}) value to cause an irreversible change (modification, damage) at the surface. This damage threshold fluence essentially depends on the material and the number of laser pulses N applied to the same spot. For spatially Gaussian laser beams, it can be determined from the diameter of the laser-generated surface modifications processed with different (peak) laser fluences [18, 19] as it will be briefly summarized in the following.

For laser pulses with a Gaussian spatial beam profile, the maximum laser fluence ϕ_0 at the sample surface and the diameter D of the damaged area are related by

$$D^2 = 2w_0^2 \ln\left(\frac{\phi_0}{\phi_{th}}\right) \quad , \qquad (1)$$

where w_0 is the Gaussian $1/e^2$ beam radius. The maximum laser fluence ϕ_0, which is defined as the laser fluence at the center of a Gaussian beam, can be obtained from the Gaussian beam radius w_0 and the measured pulse energy (E_{pulse}) by

$$\phi_0 = \frac{2E_{pulse}}{\pi w_0^2} \quad . \qquad (2)$$

Therefore, it is possible to determine the Gaussian beam radius w_0 by measuring the diameters of the damaged areas D versus the applied pulse energies E_{pulse}. Due to the linearity between the energy and the maximum laser fluence, one can determine w_0 by a linear-least-squares-fit in the representation of D^2 as a function of $\ln(E_{pulse})$. The slope of a linear fit according to (1) and (2) then yields the Gaussian beam radius w_0.

In order to get the most accurate determination of the Gaussian beam radius w_0 and to avoid any problems potentially arising from nonlinear interaction and incubation effects in TeO_2, we have used single-pulse ablation of single-crystalline silicon wafers here, resulting in a value of $w_0 = 2.56$ μm (data not shown here). Using this value along with the experimentally measured laser pulse energy, we calculated all peak fluences according to Eq. (2).

Figure 7 shows the squared damage diameter versus the peak fluence for single pulse (N = 1, full squares) and multiple pulse (N = 10, open circles) irradiations in a semi-logarithmic representation. The back extrapolation of $D^2 \rightarrow 0$ leads to damage threshold values of $\phi_{th}(N = 1) = 2.08$ and ϕ_{th} (N = 10) = 1.07 J/cm^2, respectively – indicating that the damage threshold significantly decreases with an increasing number of laser pulses (i.e., incubation takes place).

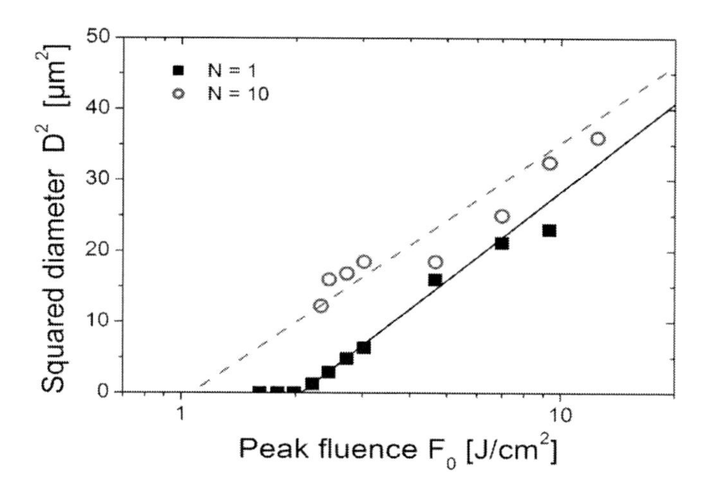

Figure 7. Squared damage diameters as a function of the laser peak fluence for single (N = 1) and multiple (N = 10) fs laser pulse irradiation of c-TeO_2 (110) in air (pulse duration 150 fs, wavelength 775 nm). Note the logarithmic peak fluence scale. [reprinted from [9] with permission of IOP Publishing Ltd.]

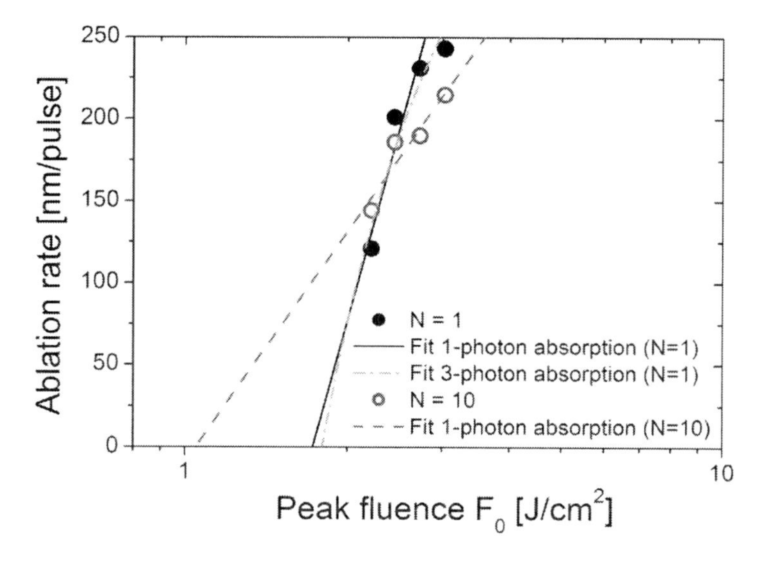

Figure 8. Average ablation depth per pulse as a function of the laser peak fluence for single (N = 1, full circles) and multiple (N = 10, open circles) fs laser pulse irradiation of c-TeO_2 (110) in air (pulse duration 150 fs, wavelength 775 nm). Note the logarithmic peak fluence scale. (The additional lines represent least-squares-fits to different ablation models, for details refer to the text). [reprinted from [9] with permission of IOP Publishing Ltd.]

Figure 8 demonstrates the high precision of the fs laser pulse ablation of tellurium dioxide as characterized by AFM measurements. In order to compare the results obtained for single-pulse (N = 1) and multiple-pulse (N = 10) ablation, the ablation rates d (i.e., crater depth/pulse number) are plotted semi-logarithmically as a function of the peak fluence. The data were least-squares-fitted (N = 1: solid black line, N = 10: dashed red line) to a widely used model of ablation $d \sim \ln(\phi_0/\phi_{abl})$ based on pure linear absorption processes, in other words, single photon absorption [20]. The back-extrapolation of the data to $D^2 \rightarrow 0$ gives ablation threshold values of $\phi_{abl}(N = 1) = 1.70$ and $\phi_{abl}(N = 10) = 1.06$ J/cm^2, in reasonable agreement with the damage threshold data obtained from the D^2-method (compare Figure 7).

For the single-pulse ablation data (N = 1), a small mismatch between the experimental data points and the fit to the linear-ablation model can be observed at fluences higher than ~2.3 J/cm^2. Therefore, the single-pulse data have been additionally analyzed using an ablation model based on pure 3-photon absorption (since the band gap of TeO_2 is ~4.05 eV, at least 3 photons at 775 nm wavelength are needed to bridge the gap). This model [21] predicts a scaling behavior of the ablation rate according to $d = B - A/(\phi_0)^2$. The two least squares fit parameters A and B can be used to determine the threshold fluence via $\phi_{th} = \sqrt{A/B} \sim 1.80$ J/cm^2 (see Figure 8). As seen in the dash-dotted green line, the least squares-fit matches better to the experimental data here indicating that the 3-photon absorption dominates the fs-laser beam interaction with c-TeO_2 at 775 nm wavelength. The deviation of the experimental data from the linear absorption model at high fluences can be explained by the reduced energy deposition depth upon the simultaneous 3-photon absorption (compare the two fits).

For the case of multi-pulse ablation (N = 10), the least-squares fit based on the linear absorption model seems to be appropriate. This would be consistent with the phenomenon of incubation (as observed previously in Figure 7) and most likely arises from the formation and accumulation of defects / color centers [16] (absorbing optical radiation at 775 nm wavelengths) during the first ten laser pulses, which can then change the dominant order of optical absorption from 3-photon absorption (N = 1) to single-photon absorption (N = 10). In fact, the fit by the linear absorption provided a better agreement rather the one by a three-photon absorption.

3.2.2. Incubation Behavior

In order to characterize the incubation behavior of c-TeO_2 more quantitatively, the observed threshold values of damage (as obtained from the D^2-method) and the ablation threshold values (as obtained by the AFM study of the crater depths) are shown in Figure 9. Apparently, both thresholds agree very well indicating that the damage of the material is dominated by ablation. Moreover, the most significant changes of the thresholds occur already during the first ten fs laser pulses.

We have fitted the data obtained by optical microscopy to the equation based on the empirical incubation model proposed by Ashkenasi et al. [22]

$$\phi_{th}(N) = \phi_{th}(\infty) + [\phi_{th}(1) - \phi_{th}(\infty)]e^{-k(N-1)} \quad (3)$$

The least-squares fit gives fitting parameters of the multi-pulse threshold $\phi_{th}(N = \infty) = 0.78$ J/cm^2, a single-pulse damage threshold of $\phi_{th}(1) = 2.08$ J/cm^2, and an exponent $k = 0.185$, where k is an empirical parameter characterizing the strength of the incubation.

The single-pulse damage threshold obtained from the least-squares-fit agrees well with the experimentally determined value of 2.08 J/cm^2. Moreover, the data clearly show that incubation effects in c-TeO_2 essentially take place during the first ten laser pulses. For more than 20 pulse numbers per spot, the damage threshold fluence is essentially constant.

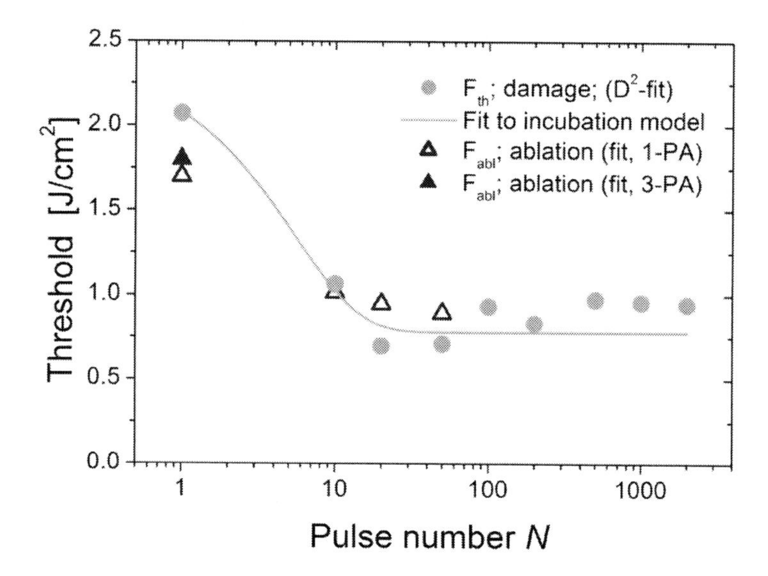

Figure 9. Ablation threshold values as a function of the applied pulse numbers for N=1-1000 fs laser pulse irradiation of c-TeO_2 (110) in air (pulse duration 150 fs, wavelength 775 nm). Note the logarithmic pulse number scale. [reprinted from [9] with permission of IOP Publishing Ltd.]

4. Time-of-Flight Mass Spectroscopy during Laser Ablation of TeO_2 Crystals: Analysis of the Ablation Products

In the previous sections, we presented that fs laser ablation has a great potential for the surface microprocessing of c-TeO_2. However, in the dynamics and mechanism of fs-laser ablation of TeO_2, there are still many open questions. For a more detailed understanding of the mechanism, an *in-situ* study is indispensable [23].

Using TOFMS one can analyze the intensity ratios of different ionic species generated in the ablated plasma precisely, reproducibly and accurately [24]. In fact, TOFMS has been widely applied for laser ablation studies of various materials [25-31] so far. Here we take benefit of the excellent (single-pulse) sensitivity of our reflectron TOFMS setup in order to study the ionic species produced by the laser ablation with fs- and ns-laser pulses of similar laser wavelengths and focusing conditions throughout a wide range of mass-to-charge ratios. This process provides a universal detection method of ions over a wide mass range. In this study, we use this system to investigate *in-situ* the fs-laser ablation of TeO_2 in comparison with ns laser ablation.

Since Paratellurite crystals have a band gap energy of 4.05 eV, we assume 2-photon absorption processes (as the single photon energies are 3.1 eV and 3.5 eV for the two relevant wavelengths of 398 nm and 355 nm used in this study). In fact, impurities and crystal defects can cause local changes in the band structure with additional allowed energy levels in the band gap. For instance, oxygen vacancies may result in electron-donor states below the valence band [32].

4.1. Experimental Aspects of the Time-of-Flight Analysis

Figure 10 shows the schematic of the experimental setup used for the TOFMS experiments allowing the use of different laser sources. Femtosecond laser pulses were supplied from a Ti:Sapphire laser system (Hurricane; Spectra-Physics) operating at 500 Hz repetition rate. The wavelength of the laser radiation was 796 nm with 120 fs pulse duration. The laser pulses were frequency-doubled using a 1 mm thick second harmonic generation crystal (BiBO crystal) resulting in a 398 nm laser pulse of ~200 fs duration. A spectral band-pass filter (03FCG033; Melles Griot, transparent to the 398 nm wavelength) was placed to suppress the non-converted fraction of the NIR fs-laser pulses. The filter was used only in the fs-laser irradiation experiments.

The typical fs pulse energy of 115 μJ after passing through the band-pass filter was adjusted in the 0.1–7.3 μJ range by passing through attenuating neutral density (ND) filters placed in front of a view port of the TOF vacuum chamber. A fast-response electromechanical shutter (Uniblitz LS6; Vincent Associates) was placed in the beam path to perform single-pulse measurements. The laser pulses were then focused by an achromatic lens of 70 mm focal length onto the sample surface at an incident angle of 45°. The laser spot diameter ($1/e^2$) was approx. 20 μm at the sample surface in the short direction of the elliptical spot. The 2 mm thick c-TeO_2 sample was mounted on an XY-translation stage and was moved by 150 μm after each irradiation (data acquisition) event. The laser pulse ablation was carried out under high vacuum conditions of 10^{-7} Torr.

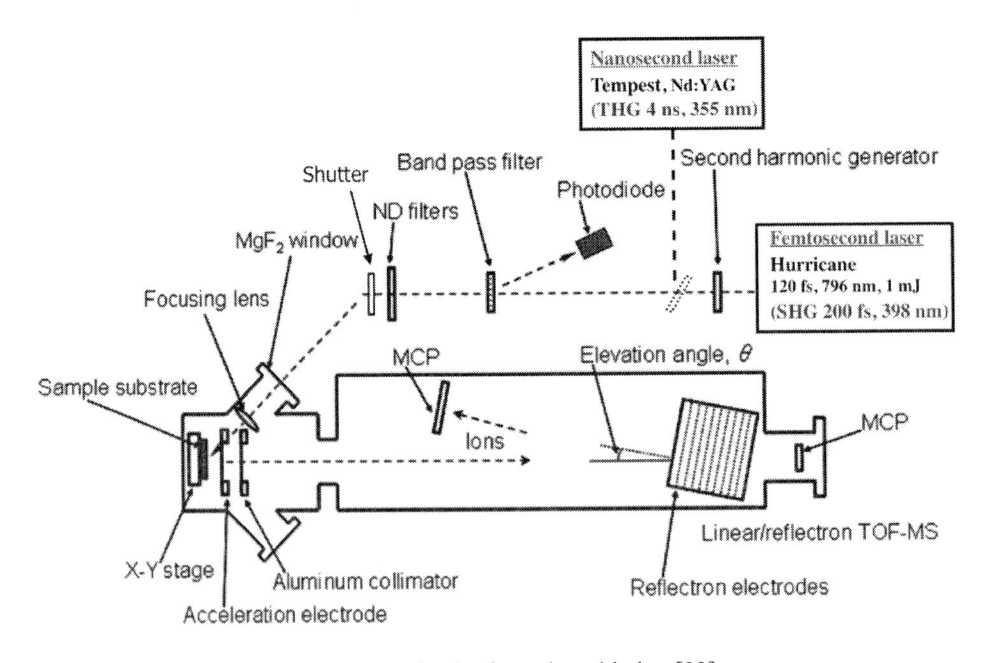

Figure 10. Schematic of the TOFMS set up for the fs- and ns ablation [23].

The TOFMS was operated in a reflectron mode. In this mode, positive ions are accelerated by an external electric field in a direction normal to the sample substrate. The maximum voltage applied to the sample mount was +5 kV. A cathode mesh was placed at a distance of 6 mm from the sample and kept at 0 V. Ions traveling from the center of the flight path were blocked by an aluminum collimator with a hole (6 mm diameter). The accelerated ions were reflected from a potential gradient formed by reflection electrodes (elevation angle $\theta = 3.5°$). The termination voltage was +6 kV. Those ions were temporally separated by their mass-to-charge ratios during their flights and were detected using a three-stage MCP. The total flight path length was 1.7 m. Output signals from the MCP were recorded with a digital oscilloscope (WaveRunner 6050; LeCroy Corp.) through a fast preamplifier (model 9305; Ortec). The TOF measurements were synchronized to the laser pulse irradiation by means of a photodiode signal triggered from a part of the optical pulse reflected at the band-pass filter. The data of each individually recorded spectrum were smoothened by adjacent averaging over 20 data points to reduce the noise from the MCP detector.

In some complementary experiments, in order to compare with the results obtained by the fs laser, a nanosecond laser (New Wave Research "Tempest", Third Harmonic Generation (THG) of a Nd:YAG laser: 355 nm wavelength, 4 ns pulse duration) beam was passing through almost the same beam path as the fs laser pulses just by placing a steering mirror into the setup and by replacing the band-pass filter. Care was taken to realize a similar laser spot diameter on the sample surface as in the fs-case. In this case, the wavelength of ns laser is close to that of fs laser, while the peak intensity is significantly different by the order of three, which should produce different specific species in the ablation plume.

Table. 1. Time-of-flight calculations of ablated species based on the equation of

$$TOF = k\sqrt{m/q}$$

Species	*q* [electron charges]	*m* [atomar units]	TOF [μs]
O^+	1	16.0	9.3
Te^+	1	126.7	26.3
TeO^+	1	143.6	27.9
TeO_2^+	1	159.6	29.4
Au^+	1	197.0	32.7
Te_2^+	1	255.2	37.2
$Te_2O_2^+$	1	287.2	39.5
Te^{2+}	2	127.6	18.6
Te^{3+}	3	127.6	15.2
TeO_2^{3+}	3	159.6	17.0

Since electrical charging of the surface critically affects the proper acquisition of TOFMS spectra, prior to the laser irradiation experiments the c-TeO_2 samples were sputter-coated by a thin conductive layer of gold (~20 nm thickness). Therefore, a pre-irradiation sequence of 5-10 laser pulses per spot was required to locally remove the gold film and prepare a clean TeO_2 surface. This film removal was monitored *in-situ* during the TOFMS experiment by tracking the Au^+ peak until it vanished almost completely. However, in some spectra it still can be noticed at 32 μs time-of-flight.

The experimentally measured times of flight have been used to identify the different positively charged ionic species which are generated during laser ablation and which are passing through the TOFMS without recombination. Based on the energy balance of ions being accelerated in an external electric field, the time-of-flight (TOF) can be calculated according to $\mathrm{TOF} = \mathrm{k}\sqrt{\mathrm{m}/\mathrm{q}}$, with m and q being the mass and the charge of the ions, respectively. The proportionality constant k includes experimental factors related to the instrument settings and characteristics. Once k is known from the calibration, the latter equation can be applied to extract the mass-to-charge ratio directly from the TOF. The maximum mass-resolution of the TOFMS under the given conditions was estimated to be $m/\Delta m$~300. The results of our calculations are presented in Table 1.

4.2. Time-of-Flight Analysis Upon Femtosecond Laser Ablation of c-TeO_2

Figure 11 shows typical TOFMS spectrum, recorded upon ablation of TeO_2 by the single fs-laser pulse at laser pulse energy of 3.0 μJ (fluence 1.0 Jcm^{-2}, peak intensity 6.4×10^{12} Wcm^{-2}). For a better visibility of the peaks, the spectrum is divided into three different temporal ranges (a: 0-20 μs, b: 20-30 μs, c: 30-45 μs). The peaks between 2 and 10 μs [Figure 11(a)] are attributed to light elements such as H^+, C^+ and O^+ arriving the earliest at the detector. At 13.1 μs, 15.1 μs and 17 μs, the O_2^+, Te^{3+} and TeO_2^{3+} peaks are observed, respectively. The peaks ranging from 20 to 30 μs [Figure 11(b)] are identified as Te^+, TeO^+ and TeO_2^+ ions at 25.6 μs, 27.2 μs and 29 μs, respectively.

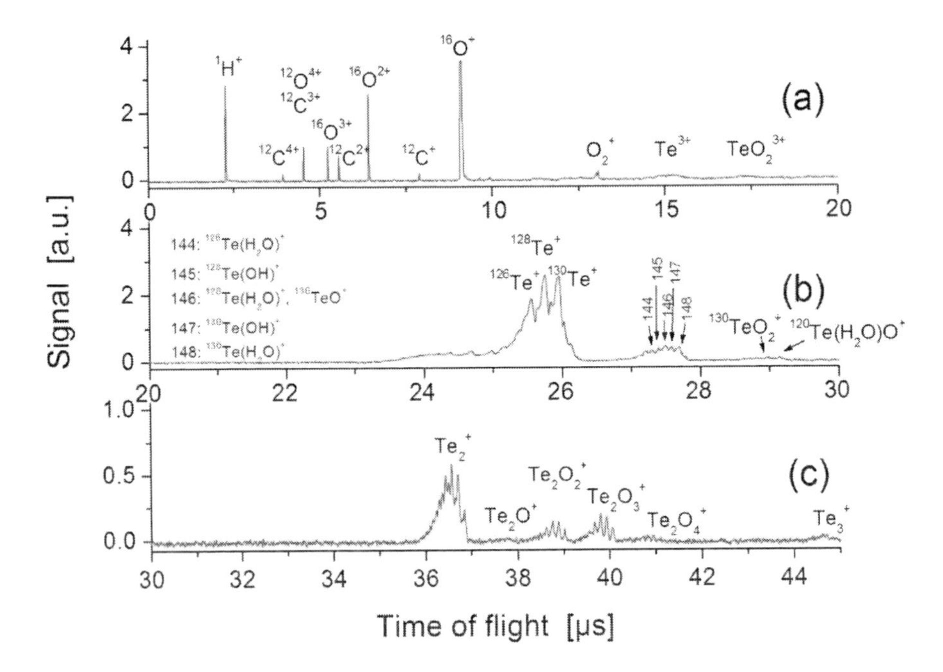

Figure 11. Typical TOFMS spectra of the TeO_2 using a single-pulse fs irradiation in the range of TOF: (a) 1-20 μs, (b) 20–30 μs and (c) 30-45 μs. The pulse energy of the fs-laser was 3.0 μJ/pulse (fluence 1.0 Jcm^{-2}, peak intensity 6.4×10^{12} Wcm^{-2}) [23].

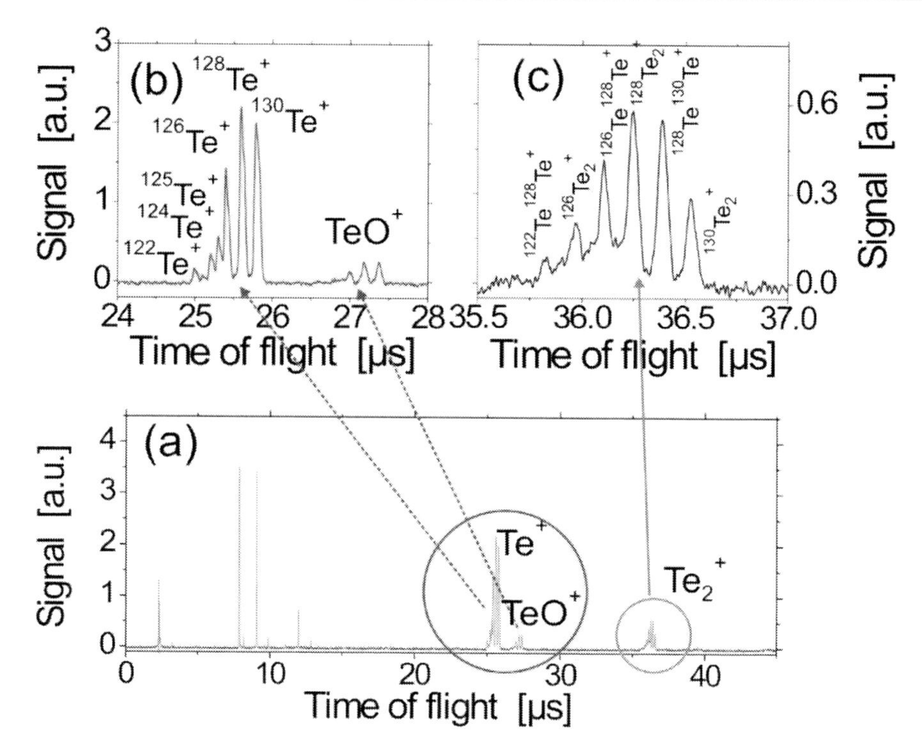

Figure 12. Close up of the Te monomer and dimer peaks in the TOFMS spectrum indicating the presence of several isotopes in the ablation plume (fs pulse energy 2.1 μJ, fluence 0.7 Jcm^{-2}, peak intensity 4.5×10^{12} Wcm^{-2}) [23].

Figure 11(c) shows the 30-45 μs domain in which the heaviest species appear such as the singly charged Te dimer (Te_2^+), Te_2O^+, $Te_2O_2^+$, $Te_2O_3^+$, $Te_2O_4^+$ and the singly charged Te trimer (Te_3^+) between 36 and 45 μs.

Figure 12 presents a close up of the Te monomer and dimer peaks when using 2.1 μJ/pulse energy (fluence 0.7 Jcm^{-2}, peak intensity 4.5×10^{12} Wcm^{-2}). The high-resolution view of Te^+ peak ion reveals that 6 different isotopes (^{122}Te, ^{124}Te, ^{125}Te, ^{126}Te, ^{128}Te, ^{130}Te) are clearly present in the spectrum [Figure 12(b)]. Tellurium has eight natural isotopes of which six are recognizable in our spectrum. The other two peaks are not visible due to the low relative quantity ratio. The close up of the Te_2^+ peak shows the combinations of the isotopes. Taking into account the probability of the different Te isotopes (taken from ref. [33]), the peaks are assigned as shown in Figure 12(c).

Increasing the fs laser pulse energy can give rise to the formation of new peaks and species in the spectrum as demonstrated in Figure 13. The laser pulse energy was systematically increased until the first ion signals appeared. Below 0.7 μJ/pulse (fluence 0.2 Jcm^{-2}, peak intensity 1.5×10^{12} Wcm^{-2}) no relevant signal was detected. The first TeO_2 ablation related ion signals appear at a threshold pulse energy of 0.7 μJ/pulse. The signal can be observed at ~25.6 μs and it is associated with the detection of Te^+ [Figure 13(a)]. The peak is broadened due to the presence of the different Te isotopes. From 1.2 μJ laser pulse energy (fluence 0.4 Jcm^{-2}, peak intensity 2.6×10^{12} Wcm^{-2}), the TeO^+ peak together with the Te dimer (Te_2^+) peak start to evolve at ~27.1 μs and ~36.5 μs [Figures 13(b)–(g)]. The O^+ peak at 9.4 μs is present throughout the whole energy range indicating that singly ionized oxygen is always produced in the course of fs laser ablation.

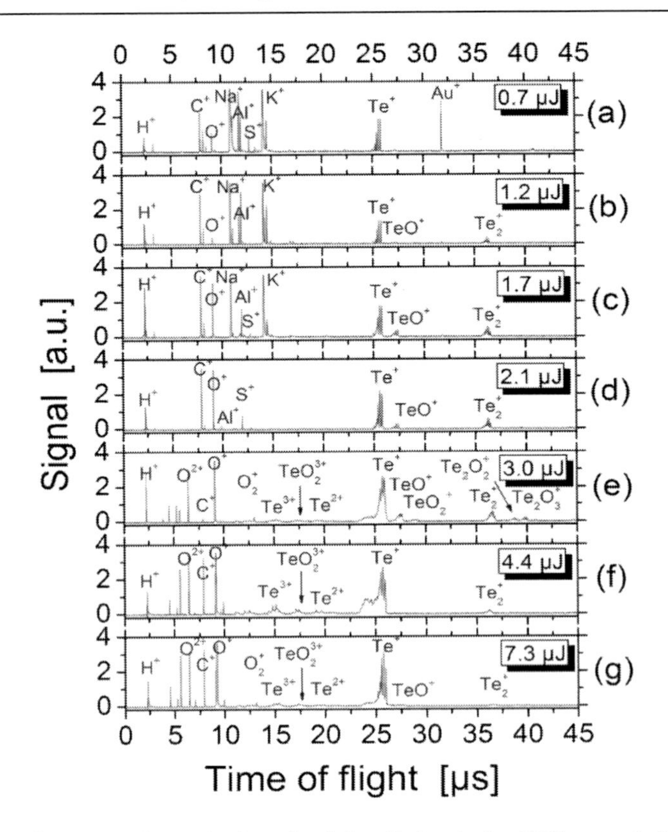

Figure 13. TOFMS of c-TeO_2 using a single-pulse fs irradiation in the TOF range 1-45 μs. The laser pulse energy was changed from 0.7 to 7.3 μJ/pulse [23].

At 3 μJ laser pulse energy (fluence 1.0 Jcm^{-2}, peak intensity 6.4×10^{12} Wcm^{-2}), the $Te_2O_2^+$ and the $Te_2O_3^+$ peaks appear at ~38.9 μs and 39.8 μs, respectively [Figure 13(e)]. Te trimer (Te_3^+) at 44.7 μs becomes visible [a higher magnification of this peak can be seen in Figure 11(c)]. Up to 3 μJ/pulse, the amplitude of the Te dimer peak slightly increases, however at higher energies (above 3 μJ) it starts to decrease indicating that high laser pulse energies do not favor the Te dimer formation. The TeO^+ peak signal also decreases with higher pulse energy.

Above 3 μJ pulse energy double and triple charged species (Te^{2+}, TeO_2^{3+}) are also observed, which can be explained by the high peak intensity. In the case of fs laser irradiation light elements are also observed in the spectra (in the 1-15 μs range) since fs pulse can disintegrate materials into atoms more efficiently than ns laser pulses due to much higher peak intensity [34]. When low laser pulse energies were used (0.7-1.7 μJ), several different peaks of Na^+, Al^+, Si^+, S^+ and K^+ ions can be observed originating from contaminations on the surface. The Gold ions ($^{197}Au^+$) can be recognized at 32 μs as shown in Figure 13(a) deriving from the gold thin films deposited on TeO_2 to avoid the electronic charging.

Te^+ is produced by the dissociation of the Te dimer (Te_2^+) during its flight (metastable dissociation). The shifted time depends on the position where metastable dissociation takes place. The shoulder observed at around 24.5 μs in case of high pulse energies is associated with the distortion of the Te^+ peaks. This happens often when high laser pulse energies are used. Along with this phenomenon, the mass resolution becomes worse as it can be observed in Figures 13(e) and (f). The distortion is caused by the shielding of the acceleration voltage due to the high-density plasma induced by laser ablation. It does not appear for light ions such as H^+ and C^+ since they travel away quickly before the high-density part of the plasma cloud develops.

4.3. Time-of-Flight Analysis Upon Nanosecond Laser Ablation of c-TeO_2

In case of ns laser irradiation, the same method was used as in case of the fs irradiation to determine the threshold laser pulse energy. The results are shown in Figure 14. The first ion signals were observed at laser pulse energy of 43 μJ (fluence 14 Jcm^{-2}, peak intensity 3.4×10^9 Wcm^{-2}) [Figure 14(a)]. It is interesting to note that there is almost no formation of TeO^+ and multiple charged species are observed in the course of ns ablation even at large pulse energy – in contrast to the fs ablation. The three-order difference in the peak intensity between the fs- and ns laser pulses seems crucial in the formation of these species.

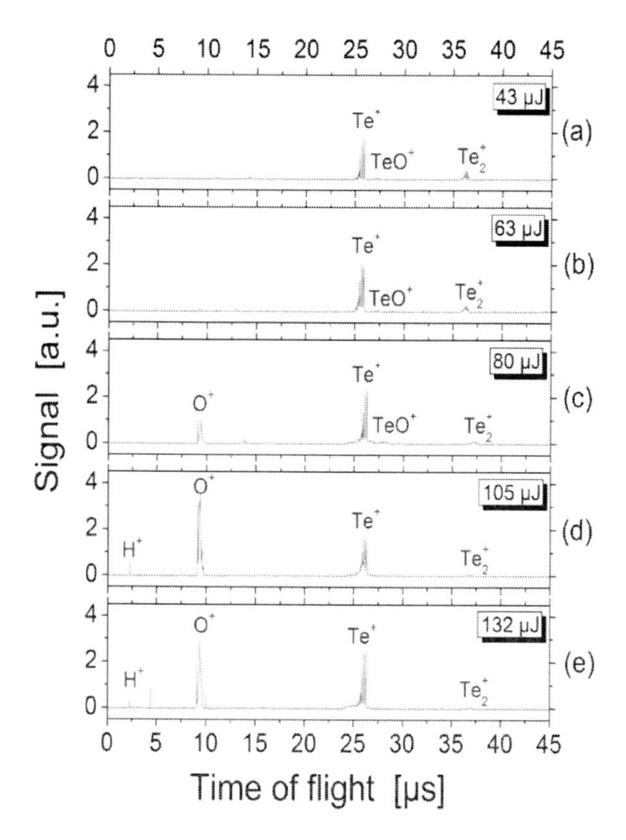

Figure 14. TOFMS of c-TeO_2 using a single-pulse ns irradiation in the TOF range of 1-45 μs. The laser pulse energy was changed from 43 to 132 μJ/pulse [23].

Oxygen ions $^{16}O^+$ (9.3 μs) are produced for laser pulse energies exceeding 80 μJ (fluence 26.1 Jcm^{-2}, peak intensity 6.4×10^9 Wcm^{-2}) [see Figs. 14(c)-(e)]. The main species observed in the ns ablation are Te^+ and O^+. The signal of the singly charged Te dimer (Te_2^+) (37.2 μs) gradually decreases with increasing pulse energy. It disappears almost completely at 132 μJ (fluence 43 Jcm^{-2}, peak intensity 10.5×10^9 Wcm^{-2}) [Figure 14(e)] indicating that high laser energy restrains the Te dimer formation. It is assumed here that high laser fluence disables the formation of Te dimer ions and only singly charged O^+ and Te^+ are produced as it was also observed in case of the fs laser pulse irradiation.

The shoulder at around 24.5 μs is visible also in case of the ns ablation when the pulse energy is higher than 105 μJ (fluence 34.2 Jcm^{-2}, peak intensity 8.4×10^9 Wcm^{-2}), indicating that its origin does not depend on the laser pulse duration and the peak intensity, however it happens in both cases when using high laser pulse energies.

SUMMARY

In sections 1-3, the surface microprocessing of single-crystalline TeO_2 induced by ns (pulse duration 20 ns, wavelength 157 nm) and fs laser pulses (pulse duration 150 fs, wavelength 775 nm) was investigated. Several observation techniques, i.e., SEM, AFM and optical microscopy were employed to reveal the topographical alterations as well as the crater dimensions (diameters, depths) in order to determine the single- and multiple-pulse damage threshold fluences for the fs pulse irradiation.

In case of the ns VUV laser irradiation, a remarkable precision of ablation with ablation rates less than 9 nm/pulse was observed. The ablation threshold is around 0.1 Jcm^{-2} and does not depend very strongly on the number of laser pulses applied to the same spot. Linear absorption dominates the energy deposition in c-TeO_2 upon VUV ns laser of here.

In case of the fs NIR laser irradiation in the transparency region of the Paratellurite, significant incubation effect was observed to occur already during the first ten laser pulses reducing the threshold from the single pulse value of ~2 Jcm^{-2} by approximately 60% down to ~0.8 Jcm^{-2} in the multi-pulse case, which also changed the absorption nature of the ablation from the nonlinear multiphoton absorption towards the linear single photon absorption.

In section 4, single-pulse fs (pulse duration ~200 fs, wavelength 398 nm) and ns (pulse duration 4 ns, wavelength 355 nm) laser ablation of single-crystalline TeO_2 was studied using a reflectron TOFMS. Due to the three-order of magnitude different peak intensities of the ns- and fs laser pulses, significant differences were observed regarding the laser-induced species in the plasma plume. Singly charged positive Te ions in the form of six isotopes were detected predominantly in the ablation plume in case of both ns- and fs laser irradiations. In case of the ns laser ablation the TeO^+ formation was negligible compared to the fs case and there was no Te trimer (Te_3^+) formation observed. The Te ion peak intensities in the TOFMS strongly depended on the applied laser pulse energy. Low laser pulse energies favor the formation of Te dimer species, however with a continuously decreasing signal when the laser pulse energy was increased. Tellurium trimer species were also observed when fs laser (3 μJ/pulse) was used. O^+ was usually detected as a byproduct of the ns- and fs laser ablation. This TOFMS study also confirmed that fs laser pulses could disintegrate materials into atoms more efficiently than ns laser pulses due to the much higher peak intensity.

ACKNOWLEDGMENTS

S. Beke is thankful for the financial support of the Japan Society for the Promotion of Science. The authors are grateful for the AFM support, TeO_2 crystals preparation and discussions for Kowashi Watanabe, Ágnes Péter and László Nánai, respectively. Authors are thankful to Tohru Kobayashi for his support in TOFMS measurements.

REFERENCES

[1] Uchida, N. and Ochmachi, Y., J. Appl. Phys. 40, 4692 (1969).
[2] Ochmachi, Y. and Uchida., N., Rev. Electr. Commun. Lab. 20, 529 (1972).
[3] Stegeman, R., et. al. Opt. Lett. 28, 1126 (2003).
[4] Greenwood, N.N. and Earnshaw, A., 1984 Chemistry of the Elements (Oxford: Pergamon) p. 911, ISBN 0-08-022057-6.
[5] McWhinnie, W.R., 1994 Tellurium—inorganic chemistry Encyclopedia of Inorganic Chemistry ed. R Bruce King (New York: Wiley).
[6] Yasutake, K., et. al. J. Mater. Sci. 32, 6595 (1997).
[7] Sundarajan, K., Sankaran, K. and Kavitha, V., J. Mol. Struct. 876, 240 (2008).
[8] Schmidt, F., and Voszka, R., Cryst. Res. Technol. 16, K127 (1981).
[9] Beke, S., Sugioka, K., Midorikawa, K., Peter, A., Nanai, L. and Bonse, J., J. Phys. D: Appl. Phys. 43, 025401 (2010).
[10] Beke, S., Giorgio, S., Korosi, L., Nanai, L. and Marine, W., Thin Solid Films 516, 4659 (2008).
[11] Nánai, L., Vajtai, R. and George, T.F., Thin Solid Films, 298(1-2), 160-164 (1997).
[12] Nánai, L. and Hevesi, I., Spectrochimica Acta A. 48A(1), 19-24 (1992).
[13] Bekesi, J, Klein-Wiele, J.H. and Simon, P., Appl. Phys. A 76, 355 (2003).
[14] Sugioka K., Hanada, Y. and Midorikawa, K. Laser Photon. Rev. 4, 386, (2010).
[15] Stuart, B.C., Feit, M.D., Herman, S., Rubenchik, A.M., Shore, B.W. and Perry, M.D. Phys. Rev. B 53, 1749 (1996).
[16] Mao, S.S., Quere, F., Guizard, S., Mao. X., Russo, R.E., Petite, G. and Martin, P., Appl. Phys. A 79, 1695 (2004).
[17] Siegel, J., Puerto, D., Gawelda, W., Bachelier, G., Solis, J., Ehrentraut, L. and Bonse, J., Appl. Phys. Lett. 91, 082902 (2007).
[18] Liu, J.M., Opt. Lett. 7, 196 (1982).
[19] Bonse, J., Wrobel, J.M., Krüger, J. and Kautek, W., Appl. Phys. A 72, 89 (2001).
[20] Sauerbrey, R. and Pettit, G.H., Appl. Phys. Lett. 55, 421 (1989).
[21] Preuss, S., Späth, M., Zhang, Y. and Stuke, M., Appl. Phys. Lett. 62, 3049 (1993).
[22] Ashkenasi, D., Stoian, R. and Rosenfeld, A., Appl. Surf. Sci.154–155, 40 (2000).
[23] Beke, S., Kobayashi, T., Sugioka, K., Midorikawa, K., and Bonse, J., Int. J. Mass Spectrom. 299, 5-8 (2011)..
[24] Poitrasson, F., Mao, X., Mao, S.S., Freydier, R., and Russo, R.E., Anal. Chem. 75, 6184 (2003).
[25] Matsuo, Y., Kobayashi, T., Kurata-Nishimura, M., Kato, T., Motobayashi, T., Kawai, J., Hayashizaki, Y., J. Phys.: Conf. 59, 555 (2007).

[26] Stoian, R., Varel, H., Rosenfeld, A., Ashkenasi, D., Kelly, R., and Campbell, E.E.B., Appl. Surf. Sci. 165, 44 (2000).
[27] Varel, H., Wähmer, M., Rosenfeld, A., Ashkenasi, D., and Campbell, E.E.B., Appl. Surf. Sci. 127–129, 128, (1998).
[28] Kobayashi, T., Kato, T., Matsuo, Y., Kurata-Nishimura, M., Hayashizaki, Y., and Kawai, J., Appl. Phys. A 92, 777-780 (2008).
[29] Kobayashi, T., Matsuo, Y., Kurata-Nishimura M., Hayashizaki, Y., and Kawai, J., Appl. Surf. Sci. 255, 9652 (2009).
[30] Kobayashi, T., Kato, T., Matsuo, Y., Kurata-Nishimura, M., Hayashizaki, Y., and Kawai, J., J. Chem. Phys. 127, 061101 (2007).
[31] Kobayashi, T., Kato, T., Matsuo, Y., Kurata-Nishimura, M., Kawai, J., and Hayashizaki, Y., J. Chem. Phys. 126, 111101 (2007).
[32] Anderson Janotti, G. Chris, Van de Walle, Appl. Phys. Lett. 87, 122102 (2005).
[33] Chronological Scientific Tables 2006, Edited by National Astronomical Observatory, published by Maruzen Co., Ltd., 2005.
[34] Kurata-Nishimura, M., Tokanai, F., Matsuo, Y., Kobayashi, T., Kawai, J., Kumagai, H., Midorikawa, K., Tanihata, I., and Hayashizaki, Y., Appl. Surf. Sci. 197-198, 715 (2002).

In: Laser-Induced Plasmas
Editor: Ethan J. Hemsworth, pp. 97-124
ISBN 978-1-61324-851-5

Chapter 5

RESONANT HARMONIC GENERATION OF SHORT PULSE LASER IN PLASMA

***Niti Kant*[*1] *and A. K. Sharma*[2]**
[1]Department of Physics, Lovely Professional University,
Phgwara-144402, Punjab, India
[2]Centre for Energy Studies, IIT Delhi, Hauz Khas,
New Delhi-16, India

ABSTRACT

The process of second harmonic generation of an intense short pulse laser in a plasma is resonantly enhanced by the application of a magnetic wiggler. The wiggler of suitable wave number k_0 provides necessary momentum to second harmonic photons to make harmonic generation a resonant process. Harmonic generation can also be made resonant in the presence of density ripple $(0, \vec{k}_0)$. Density ripple also provides necessary momentum to second harmonic photons to make harmonic generation a resonant process. In both the processes resonant second harmonic is produced. However, the group velocity of the second harmonic wave is greater than that of the fundamental wave, hence, the generated pulse slips out of the fundamental laser pulse and its amplitude saturates.

The process of third harmonic generation of an intense short pulse laser in plasma is resonantly enhanced by the application of a magnetic wiggler. The laser exerts a ponderomotive force at second harmonic driving density oscillations. The second harmonic oscillations coupled with electron velocity at the laser frequency, produces a non-linear current, driving the third harmonic. Third harmonic pulse generates in the fundamental pulse domain. However, the group velocity of the third harmonic wave is greater than the fundamental wave. Hence, the third harmonic pulse saturates strongly and moves forward from the fundamental pulse at shorter distance than the second harmonic pulse.

---

[*] E-mail: nitikant@yahoo.com

1. Introduction

Harmonic generation of an intense laser radiation in plasma is an important nonlinear process [1-5]. It is a process whereby laser light of frequency ω can be used to generate new frequencies. The newly generated frequencies are integer multiples of the fundamental frequency. The highly nonlinear natures of the interaction of the laser pulse with plasmas imply that harmonic light should be a significant feature of such interaction. When the plasma has a pre-existing density gradient or the one induced by the laser, one obtains even harmonics [1]. The harmonics are a powerful tool for diagnostics of laser produced plasmas and also an efficient source of ultraviolet and soft X-ray radiation. Esarey *et al.* [2], and Gibbon [3], have given elegant reviews of laser harmonic generation in plasmas.

Matsumoto [6] has presented both static and dynamical analysis of quasi-phase matched second harmonic generation by backward propagating interaction, where the second harmonic wave is generated in reflection. Wilks *et al.* [7] have used simulations to explore the interaction of ultra-intense laser light pulses with overdense plasmas. It has been shown that by reflecting an ultra-intense laser pulse from sharp vacuum plasma interface the efficient generation of odd harmonics of the incident laser frequency can be achieved. Zondy *et al.* [8] have investigated second harmonic generation in a periodic structure made from N pairs of optically contacted, birefringence phase-matched, walk-off-compensating bulk plates theoretically.

The two models (heuristic and new explicit second order split-step BPM) were found to be in excellent agreement near the optimal value of the phase-mismatch parameter, whereas a qualitative agreement was found for a larger phase-mismatch. A closed form expression for the conversion efficiency was derived in the plane wave limit that highlights an unexpected filtering effect brought about by walk-off compensation. Rotermund *et al.* [9] have demonstrated the potential of noncollinear parametric interaction in a traveling wave Mgo:$LiNbO_3$ optical parametric amplifier operating in the mid infrared near $3\mu m$. The calculations indicate that for pump wavelengths below 816 nm (the wavelength near which collinear group matching is achieved at a fixed single wavelength of $1064\,\mu m$), the group velocity mismatch can be reduced and the parametric gain can be increased by noncollinear propagation of the three beams.

2. Second Harmonic Generation

The phenomenon that an input wave in a nonlinear material can generate a wave with twice the optical frequency. Crystal materials lacking inversion symmetry can exhibit a so-called $\chi^{(2)}$ nonlinearity. This can give rise to the phenomenon of frequency doubling, where an input (pump) wave generates another wave with twice the optical frequency in the medium. This process is also called second-harmonic generation. In most cases, the pump wave is delivered in the form of a laser beam, and the frequency-doubled (second-harmonic) wave is generated in the form of a beam propagating in a similar direction.

In plasma second harmonic is generated from the interaction between intense light at pump frequency with the plasma. Three type of nonlinearity may arise namely, ponderomotive force (or $\vec{v}\times\vec{B}$ Lorentz force), relativistic electron mass, and the product of oscillatory electron density with drift velocity. In the process of second harmonic generation, two photons having energy $\hbar\omega_2$ and momentum $\hbar\vec{k}_1$, combine to generate a photon of second harmonic radiation of energy $\hbar\omega_2$ and momentum $\hbar\vec{k}_2$, where $\omega_1,\vec{k}_1$ and $\omega_2,\vec{k}_2$ satisfy the linear dispersion relation for electromagnetic waves. Energy and momentum conservation in a second harmonic process demand $\omega_2 = 2\omega_1$, $\vec{k}_2 = 2\vec{k}_1$. However, due to dispersive nature of the plasma medium, refractive index of the plasma increases with the frequency of the wave. Hence, $\vec{k}_2 > 2\vec{k}_1$ and resonance condition cannot be satisfied.

Parashar and Pandey [4,5] have proposed two interesting schemes to achieve phase matching condition, one, a plasma has a density ripple that provides additional momentum required for second harmonic photon. Second, a transverse wiggler magnetic field is introduced in the medium to provide the required additional momentum ($\hbar\vec{k}_2 - 2\hbar\vec{k}_1$) for the harmonic photon to make the process resonant. Wiggler magnetic field can be generated by placing bar magnets of alternate polarity over the plasma. A wiggler magnetic field aids the generation of the second harmonic in two ways. First it produces a transverse second harmonic current. Second, it provides additional momentum $\hbar\vec{k}_0$ to the second harmonic photon, making harmonic generation a resonant process. Another possibility of resonant second harmonic generation arises in inhomogeneous plasma where laser mode converts itself into a plasma wave near the critical layer [10]. A nonlinear coupling between the plasma wave and the laser can produce second harmonic resonantly. Kim *et al.* [11] have studied two photon resonant second harmonic generation from potassium vapor, using tunable picosecond pulses. Berger [12] has theoretically analyzed doubly resonant second harmonic generation in a monolithic cavity. The general expression for SHG in a monolithic cavity has been calculated. There is yet another limiting mechanism for second harmonic generation. The group velocity of the harmonic pulse is bigger than that of fundamental pulse, consequently the harmonic pulse slips away from the main pulse.

The second harmonic is efficiently generated near the critical layer via two schemes, first; p-polarized light propagating at a small angle to the density gradient undergoes linear mode conversion into a Langmuir wave near the critical layer. The density oscillations arising from the Langmuir wave beat with the oscillatory velocity due to the laser to produce a second harmonic current giving rise to second harmonic radiation. Second; the laser wave undergoes decay instability into a Langmuir wave and an ion acoustic wave near the critical layer. The Langmuir wave beats with the laser to produce a second harmonic down shifted in frequency. Second harmonic have been observed in the presence of density gradients [13-14]. This is due to the laser-induced quiver motion of the electrons across a density gradient at the laser frequency [15].

Nitikant *et al.* [16] have studied the resonant second harmonic generation of a short pulse laser in plasma channel and observed the generation of second harmonic in plasma in the presence of a magnetic wiggler. They have analyzed the effect of self- focusing on resonant second harmonic generation under wiggler magnetic field. Sajal *et al.* [17] have calculated the efficiency of phase-matched second and third harmonic generation in plasmas with density ripple. Singh *et al.* [18] show a density ripple in a plasma could be properly employed for resonant second harmonic generation and the efficiency of the process is sensitive to the angle between the density ripple and the incident laser and beam energy. The ripple density provides the additional momentum required by the second-harmonic for phase matching. Liu *et al.* [19] have related third harmonic generation with periodic wave vector in case when an intense machining laser beam, impinged on a gas jet target, causes space periodic ionization of the gas and heats the electrons. The non uniform plasma pressure leads to atomic density redistribution. When, after a suitable time delay, a second more intense laser pulse is launched along the periodicity wave vector $\vec{q}$, a plasma density ripple n_q is instantly created, leading to resonant third harmonic generation when $q = 4\omega p^2 /(3\omega c\gamma_0)$, where ω_p is the plasma frequency, ω is the laser frequency, and γ_0 is the electron Lorentz factor. Nitikant and Sharma [20] have seen the effect of pulse slippage on a resonant second harmonic generation of a short pulse laser in plasma by application of magnetic wiggler and found the group velocity of second harmonic wave is greater than that of fundamental wave, and hence generated pulse slips out of the main laser pulse and its amplitude saturates. Hence they found that the wiggler magnetic field plays a dynamic role in producing the traverse harmonic current and a kinematical one is ensuring phase-matching. Askari and Noroozi [21] have considered the effect of a wiggler magnetic field and the ponderomotive force collectively on the second generation in the laser plasma interactions. They considered this second harmonic generation and it's phase-matching in underdense plasma in the presence of a wiggler magnetic field. It is shown that the effect of inertial ponderomotive force in second harmonic generation is considerable at definite range of frequency and negligible at rest. Wiggler magnetic field plays both a dynamic role in producing the traverse harmonic current and a kinematical role in ensuring phase-matching. The inertial ponderomotive force $\rho(\vec{u}.\vec{\nabla})\vec{u}$ is a source of harmonic generation. The inertial ponderomotive force can also affect the efficiency of second harmonic generation; and its effect on second harmonic generation is also considered.

Resonant second harmonic generation of an intense short pulse laser in a plasma embedded with a magnetic wiggler is analyzed. The issue of mismatch between the group velocities of the pump and harmonic wave is addressed here. The second harmonic pulse has higher group velocity than the laser pump and hence, it may slip out of the domain of the pump.

3. Third Harmonic Generation

Third harmonic generation is a very useful technique that can convert output of infrared lasers to shorter wavelengths in the visible and near ultraviolet. The usual model of third harmonic generation takes the fundamental beam to be Gaussian and has third harmonic generation as sole linearity. Because of the homogeneous nature of the medium, only odd harmonics can be generated in the air with third harmonic generation being dominant. The third harmonic of the fundamental laser wavelengths at 800 nm lies in the ultraviolet region, which makes intense laser filaments appealing for remote sensing applications. Rickes *et al.* [22] have demonstrated strong enhancement of third harmonic generation in a nonlinear medium, prepared in maximum coherence by Stark chirped adiabatic passage. Many mechanisms can generate laser harmonics in plasmas; the main mechanism is in the presence of density gradient in the plasma [23].

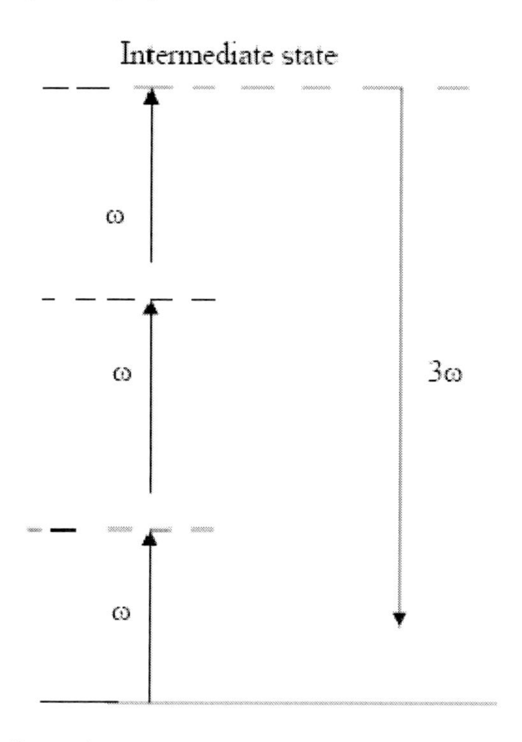

Figure 1. Third Harmonic Generation.

Third harmonic generation is possible in any media including centrosymmetric or isotropic material (i.e. crystals which exhibit inversion symmetry) even in the electric dipole approximation. It is based on the non-linear polarization process. The first optical third harmonic generation was observed in 1962 in a calcite crystal possessing centrosymmetry. The essential feature of third harmonic generation is the parametric interaction between four photons in third order non-linear media. The third harmonic generation can be explained in terms of quantum theory of radiation based on the concept of intermediate state and representation of virtual energy-levels.

When a large amplitude wave propagates through the nonlinear medium it produces electromagnetic wave at a harmonic frequencies, say third harmonic. However, the wave vector of third harmonic $k_3 = (3\omega / c)\eta\,(3\omega)$ is not equal to $3k$. In the third harmonic generation process three photons of frequency ω , with energy $\hbar\omega$ and momentum $\hbar\vec{k}$ combine to generate a photon of frequency 3ω with energy $3\hbar\omega$ and momentum $\hbar\vec{k}_3$. Since $\vec{k}_3 \neq 3\vec{k}$, the momentum of the third harmonic photon is different than the sum of the momenta of combining photons. This difference in momentum is provided by the medium. In a homogeneous medium this share of momentum is very weak hence the process of harmonic generation is non-resonant process.

As it was discussed earlier that there are two schemes for k-matching for second harmonic generation [4,5]. These two schemes can also be applied to third harmonic generation. Rajput *et al.* [24] have examined the effect of pulse slippage on a resonant third harmonic generation of a short pulse laser in plasma by application of external wiggler and found that group velocity of third harmonic wave is greater than second harmonic wave's group velocity which is further greater than group velocity of that of fundamental wave. They have seen that in comparison with second harmonic pulse, third harmonic pulse saturates strongly and moves forward from fundamental pulse at a shorter distance than second harmonic pulse. Enhancement in the third harmonic efficiency strongly depends on strength of wiggler magnetic field. First, plasma has a density ripple that provides additional momentum required for second harmonic generation, second a wiggler transverse magnetic field introduced in the medium. A wiggler magnetic field provides additional momentum to third harmonic photon, thereby making third harmonic generation a resonant process. Small mismatch can also be beneficial for third harmonic generation [25]. Ferrante *et al.* [26] have obtained harmonic generation in the skin layer of hot dense plasma and found the explicit dependencies of the third harmonic generation efficiency on the plasma and pump field parameters. Therberge *et al.* [27] have performed experiments on long range third harmonic generation in air using Ti-Sapphire chirped pulse amplification laser system and observed co-filamentation of high intensity fundamental and third harmonic pulses over long propagation distance using the Lidar technique. Liu *et al.* [28] have seen harmonic generation in neutral and ionized gases and obtained results of harmonic generation in a simple gas (hydrogen) using 1- ps, 1-μm laser pulses with a range of intensities exceeding from below to far above laser ionization saturation threshold. Akozbek *et al.* [29] have got third harmonic generation and self channeling in air using high power femtosecond laser and showed both theoretically and experimentally, that during laser filamentation in air, an intense ultra short third harmonic is generated forming two colored filament. Richard *et al.* [30] experimentally observed pump pulses and third harmonic generation with Kerr effects and formulate self consistent and complete set of non-linear Schrödinger equations for a pair of coupled beams- fundamental and its third harmonic.

Shibu and Tripathi [31] have studied phase-matched third harmonic generation of a laser beam propagating through plasma channel and showed that the presence of a background density perturbation can account for phase-matching. Yang *et al.* [32] observed strong third harmonic emission with a conversion efficiency higher than 0.1% from plasma channel formed by self-guided femtosecond laser pulses propagating in air. They found an optimized condition under which third harmonic conversion efficiency is maximized. Their experimental results show that radiation of emission in ultraviolet wavelength range makes a major attribution to third harmonic emission, whereas the effects of self-phase modulation are not important when intense laser pulse interacts with gaseous media. The model has exact solution in which third harmonic beam takes a Gaussian profile.

In this chapter, we will see the effect of pulse slippage on resonant second and third harmonic generation of an intense short pulse laser in plasma embedded with a magnetic wiggler and, the same effect on resonant second harmonic generation in the presence of density ripple. Wiggler magnetic field provides the additional angular momentum to the third harmonic photon to make the process resonant. The second/ third harmonic pulse has a higher group velocity than the laser pump and hence, it slips out of the domain of the pump. Wiggler magnetic field also enhances the efficiency of harmonic generation.

4. Pulse Slippage Effect of Second Harmonic Wave in Plasma in the Presence of Wiggler Magnetic Field

Consider the propagation of an intense short laser pulse in a plasma in the presence of a wiggler magnetic field $\vec{B}_w$,

$$\vec{E}_1 = \hat{x}A_1(z,t)\exp[-i(\omega_1 t - k_1 z)] ,$$

$$\vec{B}_1 = \frac{c\vec{k}_1 \times \vec{E}_1}{\omega_1}, \tag{1a}$$

$$\vec{B}_w = \hat{y}B_0 \exp(ik_0 z), \tag{1b}$$

where $A_1(z,t) = F(z - v_{g1}t)$, $v_{g1} = c(1-\omega_p^2/\omega_1^2)^{1/2}$. The fundamental wave and the second harmonic wave satisfy the linear dispersion relation, $k^2 \approx (\omega^2/c^2)(1-\omega_p^2/\omega^2)$, where $\omega_p = (4\pi n_0 e^2/m)^{1/2}$ is the plasma frequency, c is the velocity of light in vacuum, n_0 is the equilibrium electron density and -e and m are the charge and mass of an electron. The wave vector $\vec{k}$ increases more than linearly with frequency ω, hence $\vec{k}_2 > 2\vec{k}_1$. The difference of momentum can be provided to the second harmonic photon by the wiggler when $\vec{k}_2 = 2\vec{k}_1 + \vec{k}_0$,

The equation of motion for electron is

$$m\frac{\partial\vec{v}}{\partial t} = -e\vec{E} - m\nu\vec{v}_1 . \tag{2}$$

The laser imparts an oscillatory velocity to electrons.

$$\vec{v}_1 = \frac{e\vec{E}_1}{mi(\omega_1 + i\nu)} \tag{3}$$

$\vec{v}_1$ and $\vec{B}_w$ beat to exert a ponderomotive force $\vec{F}_1 = -(e/2c)\vec{v}_1 \times \vec{B}_w$, on the electrons at $(\omega_1, \vec{k}_1 + \vec{k}_0)$, giving an oscillatory velocity $\vec{v}_1'$ at $(\omega_1, \vec{k}_1 + \vec{k}_0)$,

$$\vec{v}_1' = -\frac{e^2 E_1 B_w}{2c\omega_1 m^2(\omega_1 + i\nu)}\hat{z} . \tag{4}$$

It also produces density perturbation at $(\omega_1, \vec{k}_1 + \vec{k}_0)$, in compliance with the equation of continuity. We get,

$$n_1 = \frac{(\vec{k}_1 + \vec{k}_0)n_0}{\omega_1} \cdot \vec{v}_1'. \tag{5}$$

$\vec{v}_1$ and $\vec{B}_1$ also beat to exert a ponderomotive force on electrons at $(2\omega_{1,} 2\vec{k}_1)$, $\vec{F}_2 = -(e/2c)(\vec{v}_1 \times \vec{B}_1)$, giving oscillatory velocity at $(2\omega_1, 2\vec{k}_1)$,

$$\vec{v}_2 = -\frac{e^2 E_1^{\,2} k_1}{4m^2 \omega_1^{\,2} (\omega_1 + i\nu)} \hat{z}. \tag{6}$$

$\vec{v}_1'$ and $\vec{v}_2$beat with $\vec{B}_1$ and $\vec{B}_w$ respectively to produce a transverse second harmonic ponderomotive force at $(2\omega_{1,} 2\vec{k}_1 + \vec{k}_0)$, $\vec{F}_2' = -(e/2c)(\vec{v}_1' \times \vec{B}_1) - (e/2c)(\vec{v}_2 \times \vec{B}_w)$ which gives an oscillatory velocity at $(2\omega_1, 2\vec{k}_1 + \vec{k}_0)$,

$$\vec{v}_2'^{\,NL} = -\frac{3e^3 B_w k_1 E_1^{\,2}}{16cim^3 \omega_1^{\,3} (\omega_1 + i\nu)} \hat{x}. \tag{7}$$

The nonlinear current density at the second harmonic turns out to be

$$\vec{J}_2^{\,NL} = -n_0 e \vec{v}_2'^{\,NL} - \frac{1}{2} n_1 e \vec{v}_1 = \frac{n_0 e^4 B_w E_1^{\,2}}{4ic\omega_1^{\,2} m^3 (\omega_1 + i\nu)} \left(\frac{3k_1}{4\omega_1} + \frac{k_1 + k_0}{\omega_1 + i\nu}\right) \hat{x}.$$

In weakly collisional plasma, we may ignore ν.

4.1. Second Harmonic Field

The wave equation governing the propagation of electromagnetic wave can be deduced from Maxwell's equations

$$\nabla . \vec{D} = 4\pi\rho\ , \tag{8a}$$

$$\nabla . \vec{B} = 0\ , \tag{8b}$$

$$\nabla \times \vec{E} = -\frac{1}{c}\frac{\partial \vec{B}}{\partial t}, \tag{8c}$$

$$\nabla \times \vec{H} = \frac{4\pi}{c}\vec{J} + \frac{1}{c}\frac{\partial \vec{D}}{\partial t}. \tag{8d}$$

In a plasma one may take $\vec{D} = \in \vec{E}$, $\vec{B} = \mu\vec{H}$. Taking the curl of Eq. (8c) and using $\vec{B} = \mu\vec{H}$ we obtain

$$\nabla \times \nabla \times \vec{E} + \frac{1}{c}\frac{\partial}{\partial t}\left(\nabla \times \vec{H}\right) = 0$$

Differentiating Eq. (8d) with respect to time one may obtain

$$\nabla^2 \vec{E} - \nabla\left(\nabla . \vec{E}\right) = +\frac{4\pi}{c^2}\frac{\partial \vec{J}}{\partial t} + \frac{\in}{c^2}\frac{\partial^2 \vec{E}}{\partial t^2}$$

For transverse electromagnetic waves $\nabla . \vec{E} = 0$. We obtain the wave equation governing by the propagation of the laser

$$\nabla^2 \vec{E} = \frac{4\pi}{c^2}\frac{\partial \vec{J}}{\partial t} + \frac{1}{c^2}\frac{\partial \vec{E}}{\partial t}, \tag{9}$$

where $\vec{J} = -ne\vec{v}$ is the current density.

The wave equation for the second harmonic field $\vec{E}_2$ is written as

$$\nabla^2 \vec{E}_2 = \frac{4\pi}{c^2}\frac{\partial \vec{J}_2}{\partial t} + \frac{1}{c^2}\frac{\partial^2 \vec{E}_2}{\partial t^2},$$

where $\vec{J}_2 = \vec{J}_2{}^{L} + \vec{J}_2{}^{NL}$, and $\vec{J}_2{}^{L}$ is the linear current density due to the self consistent field $\vec{E}_2$; $\vec{J}_2{}^{L} = -n_0 e\vec{v}_2{}^{L}$, where $\vec{v}_2{}^{L}$ is governed by the equation of motion $\partial\vec{v}_2{}^{L}/\partial t = -e\vec{E}_2/m$, giving $\frac{\partial \vec{J}_2}{\partial t} = \frac{n_0 e^2}{m}\vec{E}_2$,

$$\frac{\partial A_2}{\partial z} + \frac{1}{v_{g2}}\frac{\partial A_2}{\partial t} = \alpha_2[F(z - v_{g1}t)]^2, \tag{10}$$

where $v_{g2} = c(1-\omega_p^2/4\omega_1^2)^{1/2}$ and $\alpha_2 = (i\omega_p^2 e^2 B_0/4c^3 m^2 \omega_1^3 k_2)[(7k_1/4)+k_0]$. Now we specify the temporal profile of the laser pulse to be Gaussian, $F(z-v_{g1}t) = A_0 \exp[-(z-v_{g1}t)^2/\tau^2 v_{g1}^2]$, where τ is the laser pulse length. Introducing a new set of variables $z - v_{g1}t = \xi$ and $z = \eta$, we can write

$$\frac{\partial}{\partial z} = \frac{\partial}{\partial \xi} + \frac{\partial}{\partial \eta} \text{ and } \frac{\partial}{\partial t} = -v_{g1}\frac{\partial}{\partial \xi}$$

and equation (10) can be written as

$$\beta\frac{\partial A_2}{\partial \xi} + \frac{\partial A_2}{\partial \eta} = \alpha_2 A_0^2 e^{-\frac{\xi^2}{\xi_0^2}}, \tag{11}$$

where $\beta = (1 - v_{g1}/v_{g2})$, $\xi_0 = \tau v_{g1}$. The complementary solution of the above equation is $A_2 = -(\alpha_2 A_0^2/\beta) f(\xi - \beta\eta)$ while the particular integral is $A_2 = (\alpha_2 A_0^2/\beta) f(\xi)$ where $f(\xi) = \int_{-\infty}^{\xi} e^{-\xi^2/\xi_0^2} d\xi$. Hence the complete solution of equation (11) can be written as

$$A_2 = -\frac{\alpha_2 A_0^2}{\beta} f(\xi - \beta\eta) + \frac{\alpha_2 A_0^2}{\beta} f(\xi)$$

$$A_2 = -\frac{\alpha_2 A_0^2}{\beta}\left(\int_{-\infty}^{\xi-\beta\eta} e^{-\xi^2/\xi_0^2} d\xi - \int_{-\infty}^{\xi} e^{-\xi^2/\xi_0^2} d\xi\right)$$

$$A_2 = \frac{\alpha_2 \xi_0 A_0^2 \sqrt{\pi}}{2\beta}\left[\mathrm{erf}\left(\frac{\xi}{\xi_0}\right) - \mathrm{erf}\left(\frac{\xi - \beta\eta}{\xi_0}\right)\right]$$

where erf(ψ) is the error function of argument ψ. The ratio of amplitudes of the second harmonic and the fundamental is

$$\left|\frac{A_2}{A_0}\right| = \left|\frac{\alpha_2 A_0 \xi_0 \sqrt{\pi}}{2\beta}[\mathrm{erf}(z' - t') - \mathrm{erf}\{(1-\beta)z' - t'\}]\right|, \tag{12}$$

where, $z' = z/\xi_0$, $t' = v_{g1}t/\xi_0$ are dimensionless quantities.

4.2. Numerical Analysis

In figures 2-5, normalized amplitude of the second harmonic with the normalized propagation distance z' is presented for $\omega_p^2/\omega_1^2 = 0.8$, $\left|\alpha_2 A_0 \xi_0 \sqrt{\pi}/2\beta\right| \approx 0.1$, $\epsilon_L = 1$ and different values of t'.

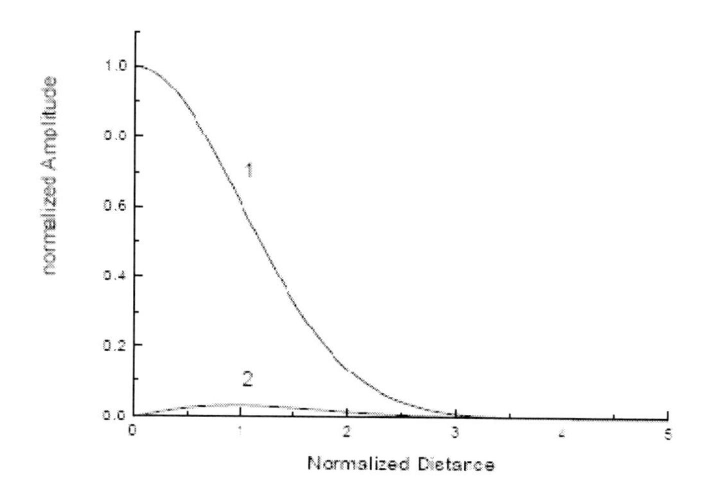

Figure 2. Variation of normalized amplitude of second harmonic pulse $\left|A_1 / A_0\right|$ denoted by 2 and incident laser pulse $\left|A_1 / A_0\right|$ denoted by 1 with normalized propagation distance $z' = z/\xi_0$ for time $t' = 0$ with $(\omega_p^2/\omega_1^2) = 0.8$, $\left|\alpha_2 A_0 \xi_0 \sqrt{\pi}/2\beta\right| \approx 0.1$.

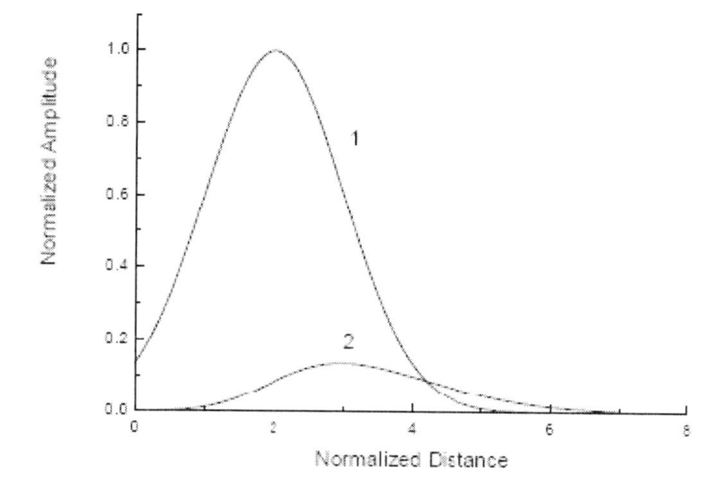

Figure 3. Variation of normalized amplitude of second harmonic pulse $\left|A_2 / A_0\right|$ denoted by 2 and incident laser pulse $\left|A_1 / A_0\right|$ denoted by 1 with normalized propagation distance $z' = z/\xi_0$ for time $t' = 2$ with $(\omega_p^2/\omega_1^2) = 0.8$, $\left|\alpha_2 A_0 \xi_0 \sqrt{\pi}/2\beta\right| \approx 0.1$.

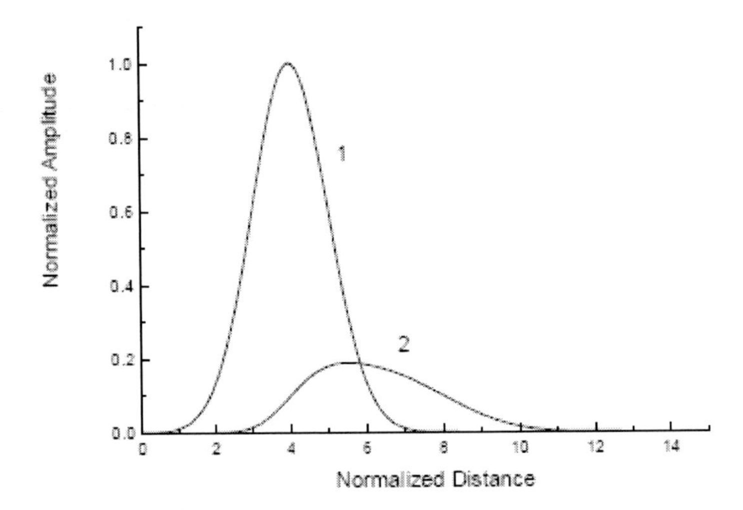

Figure 4. Variation of normalized amplitude of second harmonic pulse $|A_2 / A_0|$ denoted by and incident laser pulse $|A_1 / A_0|$ denoted by 1 with normalized propagation distance $z' = z / \xi_0$ for time $t' = 4$ with $(\omega_p^2 / \omega_1^2) = 0.8$, $|\alpha_2 A_0 \xi_0 \sqrt{\pi} / 2\beta| \approx 0.1$.

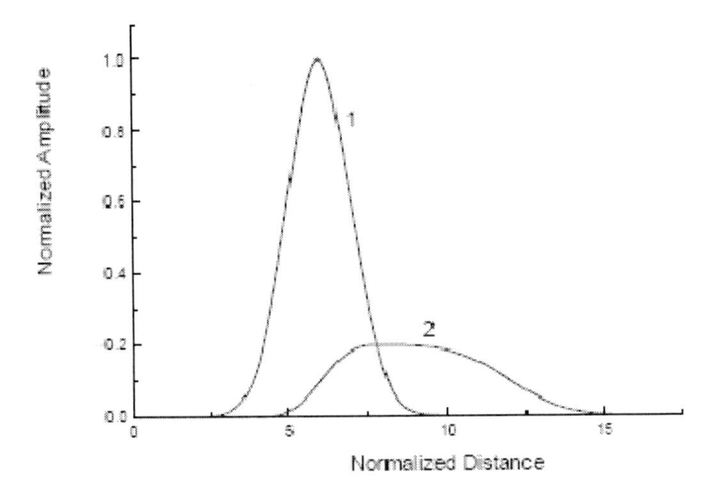

Figure 5. Variation of normalized amplitude of second harmonic pulse $|A_2 / A_0|$ denoted by and incident laser pulse $|A_1 / A_0|$ denoted by 1 with normalized propagation distance $z' = z / \xi_0$ for time $t' = 6$ with $(\omega_p^2 / \omega_1^2) = 0.8$, $|\alpha_2 A_0 \xi_0 \sqrt{\pi} / 2\beta| \approx 0.1$.

At $t' = 0$, the second harmonic pulse of small amplitude is generated in the domain of the fundamental laser pulse. Fundamental laser pul se and second harmonic pulse are denoted by 1 and 2 respectively. At $t' = 2$, the amplitude of the second harmonic increases and second harmonic pulse starts slipping out of the fundamental laser pulse. At $t' = 4$, the second harmonic pulse moves forward with increasing amplitude. At $t' = 6$, the generated pulse slips out of the main laser pulse and its amplitude saturates. Then the second harmonic pulse propagates with saturated amplitude.

5. Pulse Slippage Effect of Second Harmonic Wave in Plasma in the Presence of Density Ripple

Consider a plasma with density ripple and an electromagnetic wave is propagating in the plasma with electric and magnetic fields.

$$\vec{E}_1 = \hat{x}A_1(z,t)\exp[-i(\omega_1 t - k_1 z)], \tag{13}$$

$$\vec{B}_1 = c\frac{\vec{k}_1 \times \vec{E}_1}{\omega_1}, \tag{14}$$

$$k_1 \cong \frac{\omega_1}{c}\left(1 - \frac{\omega_p^2}{\omega_1^2}\right)^{1/2} \tag{15}$$

in an un magnetized plasma where $\omega_p = (4\pi n_0^0 e^2 / m)^{1/2}$ is plasma frequency and n_0^0 is the average electron density. The plasma also contains a density ripple

$$n_0^0 = n_0 \exp(i\vec{k}_0 \cdot \vec{x}) \tag{16}$$

whose wave vector $\vec{k}_0$ makes an angle θ with the z axis. In response to the field of the electromagnetic wave, the electrons acquire an oscillatory velocity; $\vec{v}_1 = \vec{v}_1 \exp - i(\omega_1 t - \vec{k}_1 z)$

$$\vec{v}_1 = \frac{e\vec{E}_1}{mi(\omega_1 + i\nu)} \tag{17}$$

Due to density ripple, current density $\vec{J}_2$ can be written as:

$$\vec{J}_2^L = -\frac{n_0^0 e^2 \vec{E}_2}{mi(2\omega_1 + i\nu)} \tag{18}$$

$$\vec{J}_2^{NL} = -\frac{en_0 v_1^2}{4\omega_1}(\hat{x}k_0 x + \hat{z}\frac{k_1}{2}) \tag{19}$$

using (18) and (19) in the Poisson equation, we obtain the second harmonic density perturbation n_2

$$n_2 = -\frac{1}{e}\frac{\vec{\nabla}.\vec{J}_2}{2i\omega_1} \tag{20}$$

where $\vec{J}_2 = \vec{J}_2^L + \vec{J}_2^{N.L}$.

The wave equation for second harmonic generation can be written as

$$\nabla^2 \vec{E}_2 = \frac{4\pi}{c^2}\frac{\partial \vec{J}_2}{\partial t} + \frac{1}{c^2}\frac{\partial^2 \vec{E}_2}{\partial t^2} \tag{21}$$

and $\vec{E}_2 = A_2(z,t)\exp[-i(2\omega_1 t - k_2 z)]$

On solving for (21) assuming $\vec{J}_2 = \vec{J}_2^L + \vec{J}_2^{NL}$, we get,

$$\frac{\partial A_2}{\partial z} + \frac{1}{v_{g2}}\frac{\partial A_2}{\partial t} = \alpha_2\left[F(z - v_{g1}t)\right]^2 \tag{22}$$

where $v_{g2} = c(1 - \omega_p^2 / 4\omega_1^2)^{1/2}$ and $\alpha_2 = -e\omega_p^2 / 8mc^2 k_2 \omega_1^2 (2k_{0x} + k_1)$

Now we specify the temporal profile of laser pulse to be Gaussian, F ($z - v_{g1}t) = A_0 \exp\left[-(z - v_{g1}t)^2 / z_2 v_{g1}^2\right]$, where [illegible] is the laser pulse length.

Introducing a new set of variables $z - v_{g1}t = \xi$ and $z = n$ we can write

$$\frac{\partial}{\partial z} = \frac{\partial}{\partial \xi} + \frac{\partial}{\partial n} \text{ and } \frac{\partial}{\partial t} = -v_{g1}\frac{\partial}{\partial \xi}$$

and equation (22) can be written as

$$\beta\frac{\partial A_2}{\partial \xi} + \frac{\partial A_2}{\partial n} = \alpha_2 A_0^2 \exp(-\frac{\xi^2}{\xi_0^2}) \tag{23}$$

where $\beta = (1 - v_{g1} / v_{g2})$, $\xi_0 = \tau\, v_{g1}$. The complimentary solution of the above equation is $A_2 = -(\alpha_2 A_0^2 / \beta) f(\xi - \beta n)$ while the particular integral is

$$A_2 = (\alpha_2 A_0^2 / \beta) f(\xi - \beta n)$$

where $f(\xi) = \int_{-\infty}^{\xi} \exp(-\xi^2 / \xi_0^2) d\xi$. Hence the complimentary solution of equation (23) can be written as $A_2 = -\frac{\alpha_2 A_0^2}{\beta} f(\xi - \beta\, n) + \frac{\alpha_2 A_0^2}{\beta} f(\xi)$

$$= \frac{\alpha_2 \xi_0 A_0^2 \sqrt{\pi}}{2\beta} \left[erf\left(\frac{\xi}{\xi_0}\right) - erf\left(\frac{\xi - \beta n}{\xi_0}\right) \right] \tag{24}$$

where erf (ψ) is the error function of argument ψ . The ratio of amplitudes of second harmonic and the fundamental is

$$\left|\frac{A_2}{A_0}\right| = \left| \frac{\alpha_2 A_0 \xi_0 \sqrt{\pi}}{2\beta} \left[erf(z' - t) - erf\{(1-\beta) z' - t\} \right] \right| \tag{25}$$

where $z' = z / \xi_0$, $t' = v_{g1} t / \xi_0$ are dimensionless quantities.

$$\alpha_2 = -\frac{4\pi n_0 e^3}{4m^2 c^2 k_2 \omega_1^2} \left(k_{0x} + \frac{k_1}{2} \right), \text{ here } n_0 \text{ is density ripple, and } n_0 = n_0\ e^{i\vec{k}_o . \vec{x}} \tag{26}$$

$\beta = 1 - v_{g1} / v_{g2}$, here v_{g1} is group velocity of fundamental pulse and v_{g2} is the group velocity of second harmonic pulse. In terms of electron density we can deduce the expression for α_2 in eq. (26) as

$$\alpha_2 = -\frac{4\pi n_0^0 e^3}{4m^2 c^2 k^2 \omega_1^2} \left(\frac{n_0}{n_0^0} \right) \left(k_{0x} + \frac{k_1}{2} \right), \text{ here } n_0^0 \text{ is average electron density.} \tag{27}$$

So Eq. (25) can be written in normalized form as

$$\left|\frac{A_2}{A_0}\right| = \left(\frac{\omega_p^2}{\omega_1^2} \right) \left(\frac{n_0}{n_0^0} \right) \left(\frac{eA}{4mc^2 k_2} \right) \left(k_{0x} + \frac{k_1}{2} \right) \frac{\xi \sqrt{\pi}}{2\beta} \left(erf(z' - t') - erf(1-\beta)(z' - t') \right), \tag{28}$$

where $erf(\psi)$ is the error function of argument ψ . (n_0 / n_0^0) can be taken as a constant n which is given different values in the graphs (Figure 7) while observing effect of density ripple on intensity and amplitude of second harmonic pulse.

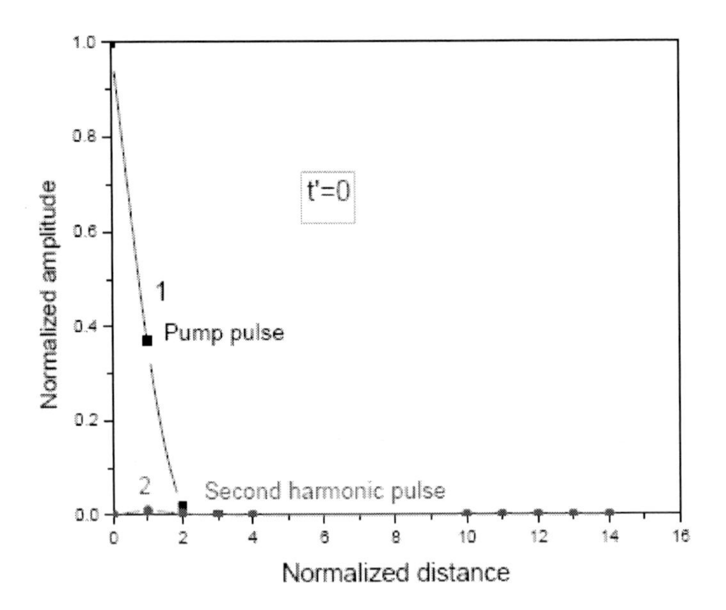

Figure 6. Variation of normalized amplitude of second harmonic pulse $|A_2 / A_0|$ denoted by 2 and incident laser pulse $|A_1 / A_0|$ denoted by 1 with normalized propagation distance z' for time t' =0 with , $\beta = 0.39$ $|\alpha_2 A_0 \xi_0 \sqrt{\pi} /2 \beta| \approx 0.04$.

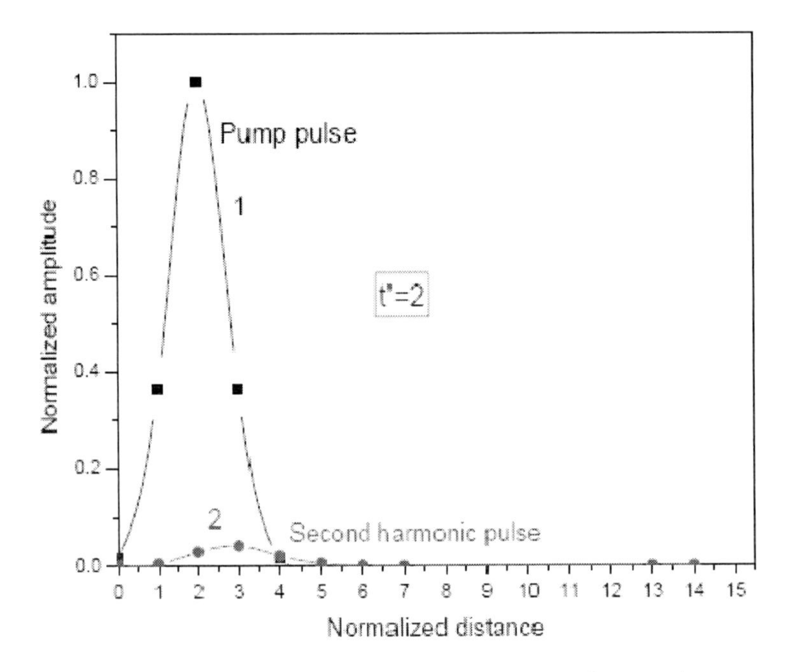

Figure 7. Variation of normalized amplitude of second harmonic pulse $|A_2 / A_0|$ denoted by 2 and incident laser pulse $|A_1 / A_0|$ denoted by 1 with normalized propagation distance z' for time t' =2 with , $\beta = 0.39$ $|\alpha_2 A_0 \xi_0 \sqrt{\pi} /2 \beta| \approx 0.04$.

5.1. Numerical Analysis

In Figures 6-10, normalized amplitude of second harmonic pulse with the normalized propagation distance z' is presented for $\beta = 0.39$ $|\alpha_2 A_0 \xi_0 \sqrt{\pi}/2\beta| \approx 0.04$ at different values of t'. At $t' = 0$, the second harmonic pulse of small amplitude generates in the domain of the fundamental laser pulse as depicted in the Figure 6. The fundamental pulse is denoted by 1 and the second harmonic pulse is denoted by 2. At $t' = 2$, the amplitude of the second harmonic pulse increases and second harmonic pulse starts slipping out of the domain of the fundamental laser pulse.

At $t' = 4$, the second harmonic pulse amplitude increases and moves forward and then its amplitude increases up to $t' = 6$ and then saturates at $t' = 8$ as depicted in Figure 10. We also see the effect of density ripple on second harmonic generation by taking values, $\beta = 0.33$, $\omega_p^2/\omega_1^2 = 0.4$, $|\alpha_2 A_0 \xi_0 \sqrt{\pi}/2\beta| \approx 0.01$ and assuming time constant.

If we increase the value of density ripple, the amplitude and intensity of second harmonic pulse increases. One can clearly see from Figure 12(a) and (b), the significant increase in amplitude of second harmonic pulse for the higher values of density ripple. In Figure 12(c) and (d) at $t' = 4$ and $t' = 6$, the amplitude starts increasing with increase in density ripple but as in Fig 12(e) at $t' = 8$ and beyond it, the amplitude of second harmonic pulse saturates.

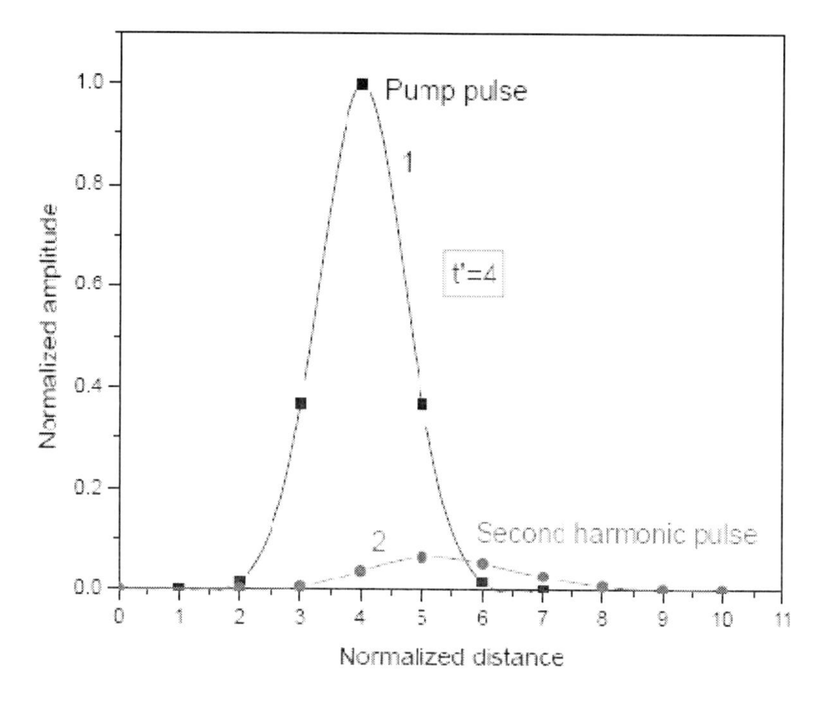

Figure 8. Variation of normalized amplitude of second harmonic pulse $|A_2/A_0|$ denoted by 2 and incident laser pulse $|A1/A^0|$ denoted by 1 with normalized propagation distance z' for time t' =4 with , $\beta = 0.39$ $|\alpha_2 A_0 \xi_0 \sqrt{\pi}/2\beta| \approx 0.04$.

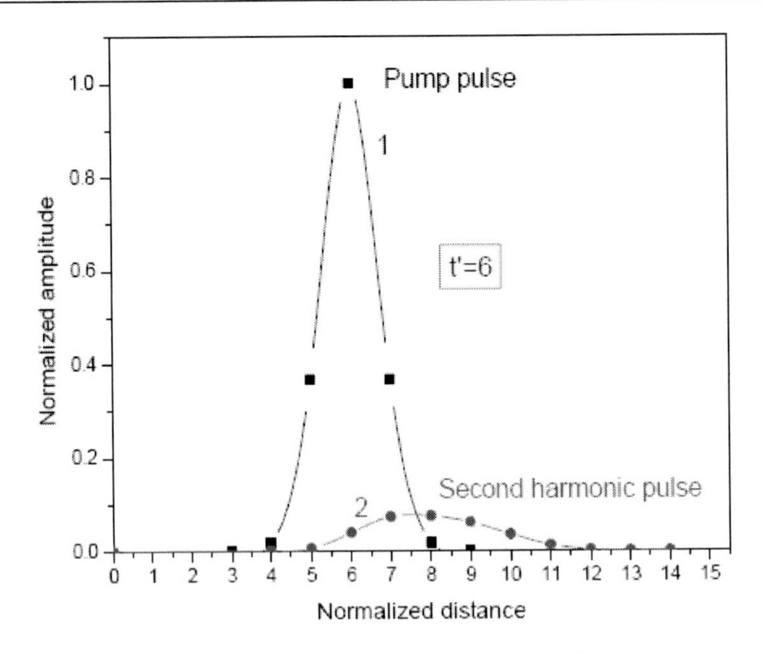

Figure 9. Variation of normalized amplitude of second harmonic pulse $\left|A_2 / A_0\right|$ denoted by 2 and incident laser pulse $|A_1 / A_0|$ denoted by 1 with normalized propagation distance z' for time t' =6 with , $\beta = 0.39$ $|\alpha_2 A_0 \xi_0 \sqrt{\pi} /2 \beta| \approx 0.04$.

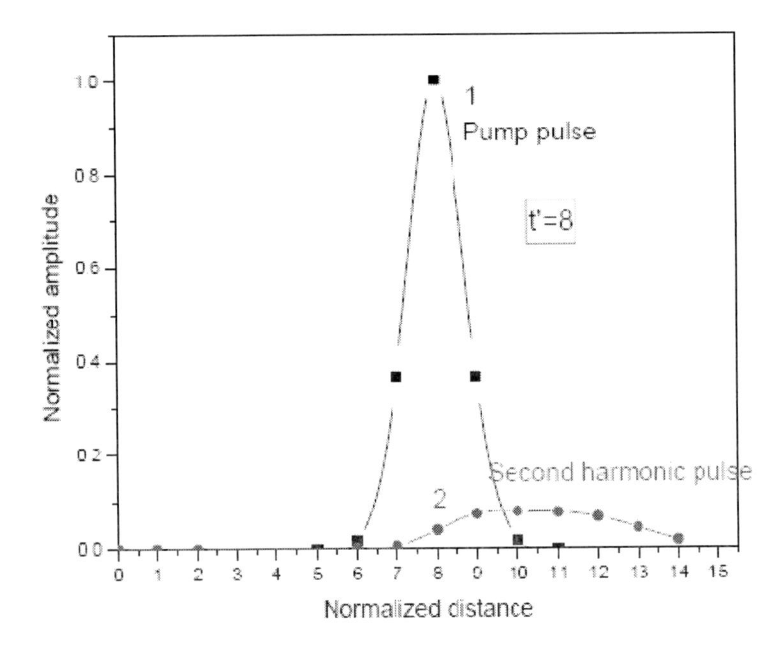

Figure 10. Variation of normalized amplitude of second harmonic pulse $\left|A_2 / A_0\right|$ denoted by 2 and incident laser pulse $|A_1 / A_0|$ denoted by 1 with normalized propagation distance z' for time t' =8 with , $\beta = 0.39$ $|\alpha_2 A_0 \xi_0 \sqrt{\pi} /2 \beta| \approx 0.04$.

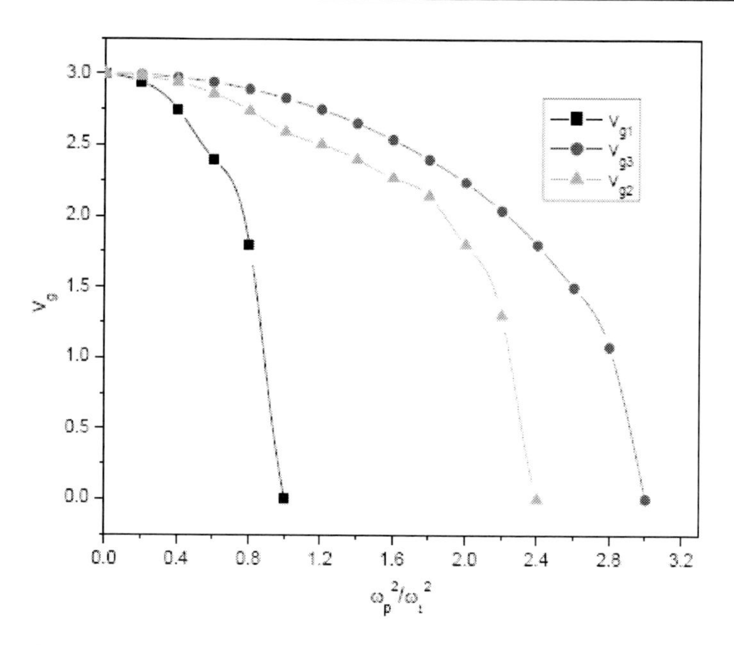

Figure 11. Comparison of group velocities of pump pulse, second and third harmonic pulses with ω_p^2 / ω_1^2.

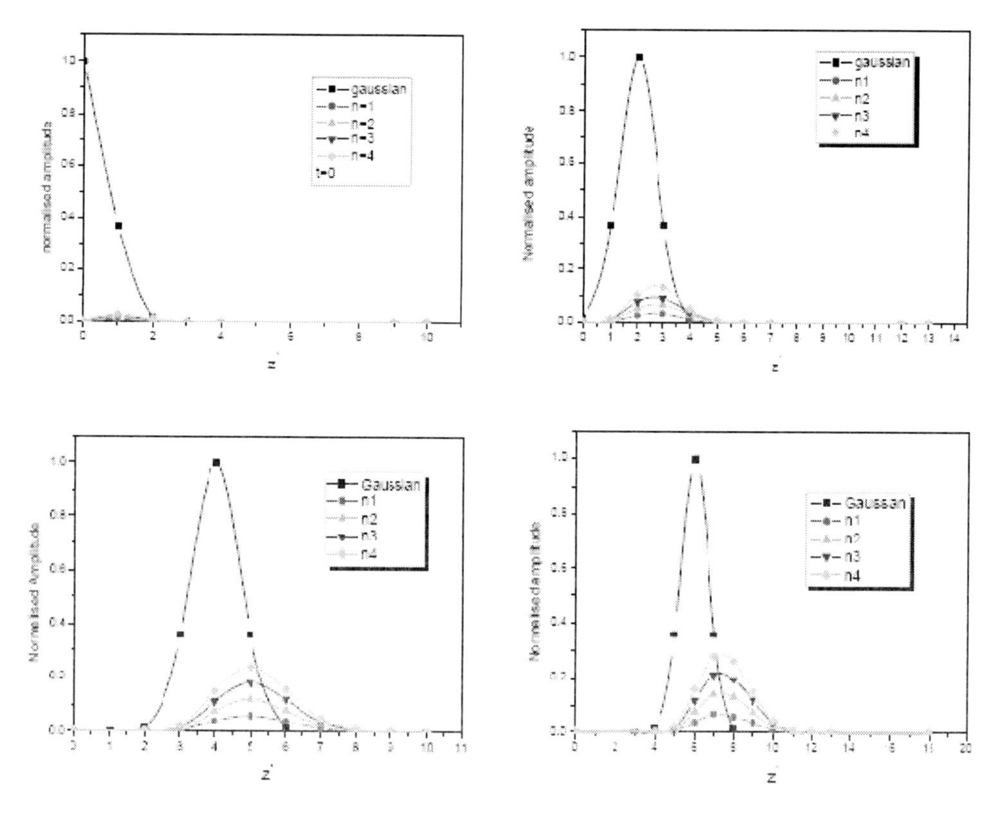

Figure 12. (Continued).

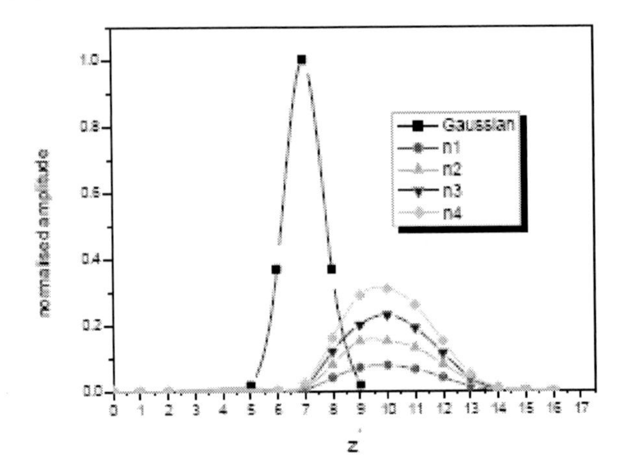

Figure 12. Variation of normalized amplitude of an incident laser pulse $|A_1/A_0|$ and second harmonic pulse $|A_2/A_0|$ with normalized propagation distance $z' = z/\xi_0$ by taking variations of density ripple $n = n_0/n_o^0$ =1, 2, 3 and 4, , $\beta = 0.33$, $\omega_p^2/\omega_1^2 = 0.4$, $\left|\alpha_2 A_0 \xi \sqrt{\pi}/2\beta\right| \approx 0.1$ at time t' = 0, 2, 4, 6 and 8 respectively.

6. Pulse Slippage Effect of Third Harmonic Generation in Plasma in the Presence of Wiggler Magnetic Field

6.1. Non Linear Current Density

Consider the propagation of an intense short laser pulse in plasma in the presence of a wiggler magnetic field $\vec{B}_W$.

$$\vec{E}_1 = \hat{x}A_1(z,t)\exp[-i(\omega_1 t - k_1 z)], \tag{29}$$

$$\vec{B}_1 = \frac{c\vec{k}_1 \times \vec{E}_1}{\omega_1}, \tag{29(a)}$$

$$\vec{B}_W = \hat{y}B_0 \exp(ik_0 z), \tag{29(b)}$$

where $A_1(z,t) = F(z - v_{g1}t)$, $v_{g1} = c\left(1 - \omega_P^2/\omega_1^2\right)^{1/2}$. The laser pump and third harmonic electromagnetic wave obey the linear dispersion relation $k^2 \approx \left(\omega^2/c^2\right)\left(1 - \omega_P^2/\omega^2\right)$. The wave vector $\vec{k}$ increases more than linearly with frequency ω, hence $\vec{k}_3 > 3\vec{k}_1$. The difference of momentum can be provided to the third harmonic photon by the magnetic wiggler when $\vec{k}_3 = 3\vec{k}_1 + \vec{k}_0$. The laser imparts an oscillatory velocity to electrons.

$$\vec{v}_1 = \frac{e\vec{E}_1}{mi(\omega_1 + i\nu)}. \tag{30}$$

$\vec{v}_1$ and $\vec{B}_W$ beat to exert a ponderomotive force $\vec{F}_1 = \left(\frac{-e}{2c}\right)\left(\vec{v}_1 \times \vec{B}_W\right)$, on all electrons at $(\omega_1, \vec{k}_1 + \vec{k}_0)$ giving an oscillatory velocity

$$\vec{v}_1' = -\frac{e^2 E_1 B_0}{2c\omega_1 m^2(\omega_1 + i\nu)}\hat{z} \tag{31}$$

It also produces density perturbation at $(\omega_1, \vec{k}_1 + \vec{k}_0)$, in compliance with the equation of continuity, we get,

$$n_1 = \frac{(\vec{k}_1 + \vec{k}_0)}{\omega_1} v_1' \tag{32}$$

$\vec{v}_1'$ and $\vec{B}_1$ also beat to exert a ponderomotive force on electrons at $(2\omega_1, 2\vec{k}_1)$,

$\vec{F}_2 = -(e/2c)\ (\vec{v}_1' \times \vec{B}_1) - (e/2c)\ (\vec{v}_2 \times \vec{B}_W)$ which gives an oscillatory velocity at $(2\omega_1, 2\vec{k}_1)$.

$$\vec{v}_2 = -\frac{e^2 E_1^2 k_1}{4m^2\omega_1^2(\omega_1 + i\nu)}\hat{z}, \tag{33}$$

$\vec{v}_1'$ And $\vec{v}_2$ beat with $\vec{B}_1$ and $\vec{B}_W$, respectively, to produce a transverse second harmonic ponderomotive force at $(2\omega_1, 2\vec{k}_1 + \vec{k}_0)$,

$\vec{F}_2' = -(e/2c)(\vec{v}_2' \times \vec{B}_1) - (e/2c)\left(\vec{v}_2' \times \vec{B}_W\right)$, which gives an oscillatory velocity at $(2\omega_1, 2\vec{k}_1 + \vec{k}_0)$,

$$\vec{v}_2'^{N.L} = \frac{3e_3 B_0 k_1 E_1^2}{16cim_3\omega_1^3(\omega_1 + i\nu)}\hat{x} \tag{34}$$

$\vec{v}_2$ and $\vec{B}_1$ also beat to exert a ponderomotive force on electrons at (3 ω_1, $3\vec{k}_1$) $\vec{F}_3 = -(e/2c)(\vec{v}_2 \times \vec{B}_1)$, giving oscillatory velocity $\vec{v}_3$ at (3 ω_1, 3 $\vec{k}_1$),

$$\vec{v}_3 = -\frac{e^3 E_1^{\ 3} k_1^{\ 2}}{24 i m^3 \omega_1^4 (\omega_1 + i\nu)} \hat{z}. \tag{35}$$

$\vec{v}_2$ and $\vec{B}_W$ beat to exert a ponderomotive force $\vec{F}_2 = -(e/2c)(\vec{v}_2 \times \vec{B}_W)$, on electrons at (2 ω_1, $2\vec{k}_1 + \vec{k}_0$), giving an oscillatory velocity $\vec{v}_2'$,

$$\vec{v}_2' = -\frac{e^3 E_1^{\ 2} B_0 k_1}{16 i c m^3 \omega_1^{\ 3} (\omega_1 + i\nu)} \hat{z}. \tag{36}$$

$\vec{v}_2'$ and $\vec{v}_3$ beat with $\vec{B}_1$ and $\vec{B}_W$, respectively to produce a transverse third harmonic ponderomotive force at (3 ω_1,$3\vec{k}_1 + \vec{k}_0$), $\vec{F}_3' = (e/2c)(\vec{v}_2' \times \vec{B}_1)$-$(e/2c)(\vec{v}_3 \times \vec{B}_W)$ which gives non-linear oscillatory velocity at $(3\omega_1, 3\vec{k}_1 + \vec{k}_0)$,

$$\vec{v}_3'^{N.L} = \frac{5}{88} \frac{e^4 B_0 k_1^{\ 2} E_1^{\ 3}}{c m^4 \omega_1^{\ 5} (\omega_1 + i\nu)} \hat{x}. \tag{37}$$

The non-linear current density at the third harmonic turns out to be

$$\vec{J}_3^{\ N.L} = -n_0 e \vec{v}_3'^{N.L} - \frac{1}{2} n_1 e \vec{v}_2$$

$$= -\frac{n_0 e^5 B_0 k_1 E_1^{\ 3}}{16 m^4 \omega_1^{\ 4} (\omega_1 + i\nu)} \left[\frac{5k_1}{18\omega_1} + \frac{k_1 + k_0}{\omega_1 + i\nu} \right] \hat{x}. \tag{38}$$

In weakly collisional plasma, we may ignore ν. The wave equation for third harmonic field $\vec{E}_3$ is written as

$$\nabla^2 \vec{E}_3 = \frac{4\pi}{c^2} \frac{\partial \vec{J}_3}{\partial t} + \frac{1}{c^2} \frac{\partial^2 \vec{E}_3}{\partial t^2}, \tag{39}$$

where $\vec{J}_3 = \vec{J}_3^L + \vec{J}_3^{N.L}$, and $\vec{J}_3^L$ is the linear density due to self consistent field $\vec{E}_3$; $\vec{J}_3^L = -n_0 e \vec{v}_3^L$, where $\vec{v}_3^L$ is governed by the equation of motion $\partial \vec{v}_3^L / \partial t = -e\vec{E}_3 / m$, giving,

$$\frac{\partial \vec{J}_3}{\partial t} = \frac{n_0 e^2 \vec{E}_3}{m} \ . \tag{40}$$

$$\frac{\partial A_3}{\partial z} + \frac{1}{v_{g3}} \frac{\partial A_3}{\partial t} = \alpha_3 \left[F(z - v_{g1} t)\right]^3 , \tag{41}$$

where $v_{g3} = c(1 - \omega_p^2 / 9\omega_1^2)^{1/2}$ and $\alpha_3 = (3\omega_p^2 e^3 B_0 k_1 / 32 m^3 \omega_1^3 c k_3)$ $((23k_1/18) + k_0)$. Now we specify the temporal profile of laser pulse to be Gaussian, $F(z - v_{g1}t) = A_0 \exp(-(z - v_{g1}t)^2 / \tau^2 v_{g1}^2)$, where τ is the laser pulse length. Introducing a new set of variables $z - v_{g1}\,t = \xi$, and $z = \eta$, we can write

$$\frac{\partial}{\partial z} = \frac{\partial}{\partial \xi} + \frac{\partial}{\partial \eta} \text{ and } \frac{\partial}{\partial t} = -v_{g1} \frac{\partial}{\partial \xi}$$

and equation (41) can be written as

$$\beta \frac{\partial A_3}{\partial \xi} + \frac{\partial A_3}{\partial \eta} = \alpha_3 A_0^3 \exp\left(\frac{-\xi^2}{\xi_0^2}\right) , \tag{42}$$

where $\beta = (1 - v_{g1} / v_{g3})$, $\xi_0 = \tau v_{g1}$. The complimentary solution of the above equation is $A_3 = -(\alpha_3 A_0^3 / \beta) f(\xi - \beta\eta)$ while the particular integral is $A_3 = (\alpha_3 A_0^3 / \beta) f(\xi)$ where $f(\xi) = \int_{-}^{\xi} \exp(-\xi^2 / \xi_0^2) d\xi$. Hence the complete solution of equation (42) can be written as

$$A_3 = \frac{\alpha_3 \xi_0 A_0^3 \sqrt{\pi}}{2\beta} \left[erf(\frac{\xi}{\xi_0}) - erf(\frac{\xi - \beta\eta}{\xi_0}) \right] , \tag{43}$$

where erf(ψ) is the error function of argument ψ . The ratio of amplitudes of third harmonic and the fundamental is

$$\left| \frac{A_3}{A_0} \right| = \left| \frac{\alpha_3 A_0^2 \xi_0 \sqrt{\pi}}{2\beta} [erf(z' - t') - erf\{(1-\beta)z' - t'\}] \right| , \tag{44}$$

where $z' = z / \xi_0$ and $t' = v_{g1} t / \xi_0$ are dimensionless quantities.

6.2. Numerical Analysis

For a typical case, plasma irradiated by a 1.06 μm CO_2 laser ($\omega = 1.8 \times 10^{14}$ rad/s) of intensity 10^{15} W/cm^2, and plasma density $n_0 = 2 \times 10^{15}$ cm^{-3}, the normalized amplitude of the third harmonic with the normalized propagation distance z' for $\omega_p^2 / \omega_1^2 = 0.8$, $\left|\alpha_3 A_0 \xi_0 \sqrt{\pi} / 2\beta\right| \approx 0.05$, $\in_L = 1$ and at different values of t' is presented in Figure 13.

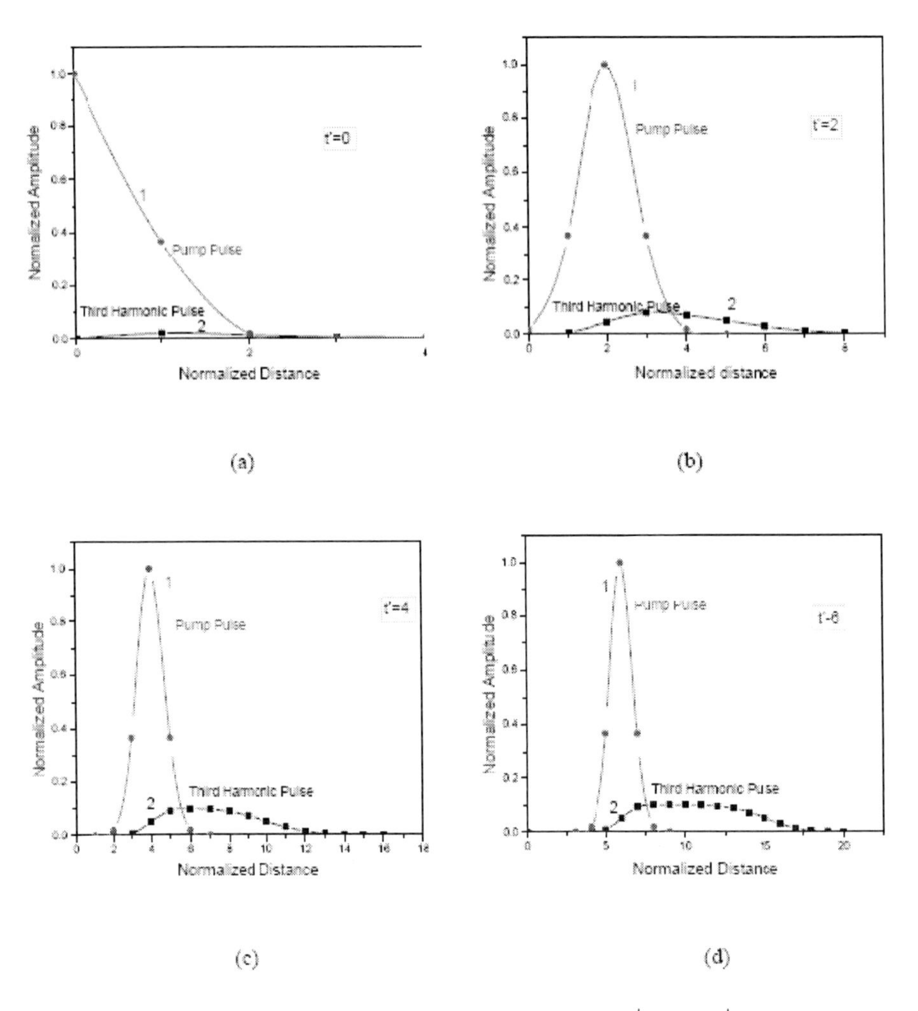

Figure 13. Variation of normalized amplitude of third harmonic pulse $\left|A_3 / A_0\right|$ denoted by 2 and incident laser pulse $|A_1 / A_0|$ denoted by 1 with normalized propagation distance $z' = z / \xi_0$ for (ω_p^2 / ω_1^2) = 0.8, $| \alpha_2 A_0 \xi_0 \sqrt{\pi} /2 \beta | \approx 0.05$ at (a) t' =0, (b) t' =2, (c) t' =4, and (d) t' =6.

At $t' = 0$, the third harmonic pulse of small amplitude generates in the domain of the fundamental laser pulse as depicted in Figure 13(a). The fundamental laser pulse is denoted by 1 and third harmonic pulse is denoted by 2. At $t' = 2$, the amplitude of the third harmonic pulse increases and third harmonic pulse starts slipping out of the domain of the fundamental laser pulse. At $t' = 4$, the third harmonic pulse moves forward and its amplitude saturates at $t' = 6$ as depicted in Figure 13(c and d).

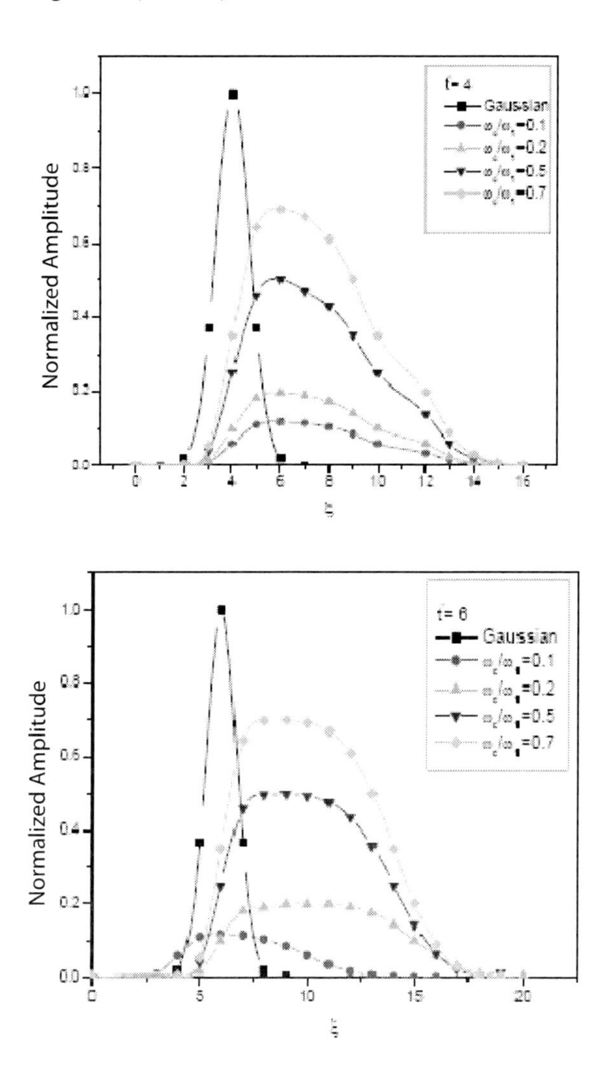

Figure 14. Variation of normalized amplitude of an incident laser pulse $|A_1 / A_0|$ and third harmonic pulse $|A_3 / A_0|$ with normalized propagation distance $z' = z / \xi_0$ by taking variations of external applied wiggler magnetic field for $\omega_c / \omega_1 = 0.1$, $\omega_c / \omega_1 = 0.2$, $\omega_c / \omega_1 = 0.5$, and $\omega_c / \omega_1 = 0.7$ at (a) t' =4, and (b) t' =6.

Wiggler magnetic field also plays an important role in the enhancement of resonant third harmonic generation of an intense pulse laser in plasma. We also see the effect of wiggler magnetic field on third harmonic generation. If we increase the value of external wiggler magnetic field, the amplitude and the intensity of third harmonic pulse increases. One can clearly see from the Figure 14 (a and b) that the significant increase in amplitude of third harmonic pulse for the higher value of wiggler magnetic field. At $t'=4$, for $\omega_c/\omega_1 = 0.7$; where $\omega_c = eB_0/mc$ is the angular frequency of the wiggler magnetic field. The third harmonic amplitude attains the highest value. As time passes i.e. at $t'=6$, the third harmonic amplitude saturates.

CONCLUSION

Harmonic generation has been a subject of great interest since the advent of high intensity, short pulse lasers. It is an important non-linear process as it acts as a valuable diagnostics in short pulse laser plasma experiments and also has applications in frequency up-conversion, signal processing etc. The advantage of using short laser pulses is that the atoms get exposed to high laser intensity before they ionize. This leads to electrons with higher kinetic energy, a larger amount of non-sequential ionization and higher order harmonics.

The group velocity mismatch between the fundamental laser and the second harmonic is significant in high-density plasma. It leads to the escape of the second harmonic pulse from the fundamental laser pulse, limiting the yield of harmonic generation and its amplitude saturate. The effect becomes more pronounced as the plasma density increase. The theory given here is also applicable to second harmonic generation in semiconductors with k_1 replaced by $(\omega_1/c)(\in_L - \omega_p^2/\omega_1^2)^{1/2}$, k_2 replaced by $(2\omega_1/c)(\in_L - \omega_p^2/4\omega_1^2)^{1/2}$ and β replaced by $(1 - v_{g1}/v_{g2})$, where $\in_L$ is the lattice permittivity of semiconductor. Semiconductors and plasmas are dispersive media in which the refractive index increases with the frequency of wave, hence $k_3 > 3k_1$ and resonance condition is not satisfied. Wiggler magnetic field is introduced in the medium to provide additional momentum for the harmonic photon and hence make the harmonic generation a resonant process. It could be effective in generating higher harmonics also. The wiggler acts as a photon of zero energy and momentum $\hbar k_0$ that compensate for the unbalanced momentum between the third harmonic and fundamental photons.

The wiggler wave number for the perfect phase matching decreases with the frequency of the laser. In addition to the wiggler field, if one applies strong guide magnetic field, one could improve the efficiency of harmonic generation a great deal through the cyclotron resonance. For the short laser pulses, ν is taken to be constant. However, for longer pulses of pulse duration exceeding the energy relaxation time, the collision frequency could be significantly changed due to ohmic heating of electrons, affecting the output power.

Resonant second harmonic is produced through the coupling of density oscillations with the oscillatory velocity of the electrons. The energy conversion efficiency scales as the square of plasma density and square of depth of density ripple, and is $\approx 0.2\%$ for normalized laser amplitude $a_0 \approx 1$ in a plasma of 1% critical density with 20% density ripple. Plasma with a density ripple could be a suitable medium for efficient generation of electromagnetic radiation. The density ripple can be produced in many ways. The laser itself can undergo a filamentation instability producing transverse density ripple with $k_0 \ll \omega_1/c$. One may also launch a sound wave in the plasma with frequency much smaller than the laser frequency but with a wave number k_0 comparable to k_1. The efficiency of harmonic generation in excess of 10% can be achieved for an Nd: glass laser at 10^{15} W/cm ^{2} with a density ripple of a few percent. The attractive feature of the scheme is that it is operational even in the underdense region.

The group velocity mismatch between the fundamental laser and the third harmonic is significant in high-density plasma. Effect increases with increase in plasma density. In the presence of laser wave, the electrons acquire an oscillatory velocity at third harmonic due to the pondromotive force exerted by the laser. These electrons then beat with the wiggler field and produce third harmonic current at (3ω, $3\vec{k}_1 + \vec{k}_0$) and produces the third harmonic radiation. Third harmonic group velocity is found to be greater than the fundamental velocity, hence slips out of the fundamental wave. Here we have shown graphically the comparison of the fundamental pulse and third harmonic pulse at different times. One can see that the third harmonic pulse moves forward and saturate for a long distance. Wiggler strength ≥ 10 kG is required to achieve high efficiency of third harmonic generation. In addition to wiggler field, if one applies a strong guide magnetic field, one could improve the efficiency of harmonic generation. In comparison with the second harmonic pulse [20], third harmonic pulse saturates strongly and moves forward from the fundamental pulse at shorter distance than the second harmonic pulse. Enhancement in the efficiency of the third harmonic generation strongly depends on the strength of the wiggler magnetic field [24].

REFERENCES

[1] Castillo-Herrera, C. I.; Johnston, T. W. IEEE Trans. Plasma Sci. 1993, 21, 120.
[2] Esarey, E. et al.; IEEE Trans. Plasma Sci.1993, 21, 95.
[3] Gibbon, P, IEEE J. Quantum Electron.1997, 33, 1915.
[4] Prashar, J. et al.; J. Plasma Phys.1993, 50, 339.
[5] Prashar, J.; Pandey, H. D. IEEE Trans. Plasma Sci. 1992, 20, 996.
[6] Matsomoto, M.; Tanaka, K. IEEE J. Quantum Electron. 1995, 31, 700.
[7] Wilks, S.C.; Kruer, W. L.; Mori, W.B. IEEE Trans. Plasma Sci.1993, 21, 125.
[8] Zondy, J. J. et al.; J. Opt. Soc. Am. B.2003, 20, 1675.
[9] Rotermund, F. et al.; Optics Communication. 1999, 169, 183.
[10] Ramchandran, K.; IEEE Trans. Plasma Sci.1996, 24, 487.
[11] Doseok, K.; Mullin, S. C.; Shen, Y. R. J. Opt. Soc. Am. B. 1997, 14, 2530.
[12] Berger,V. J. Opt. Soc. Am. B. 1997, 14, 1351.
[13] Esarey, E.; Ting, A.; Liu, X.; Sprangle.; Umstader.; IEEE Trans Plasma Sci. 1993, 21, 95.
[14] Malka, V.; Modena, J.; Nazmudin, Z. et al., Phys. Plasmas 1997, 4, 1127.
[15] Jha, P.; Mishra, K.; Raj, G.; Upadhyay. Phys. Plasmas 2007, 14, 53107.
[16] Nitikant.; Sharma, A. K. J. Phys. D: Appl. Phys 2004, 37, 2395-2398.
[17] Dahiya, D.; Sajal, V.; Sharma, A. K. Phys Plasmas 2007, 14, 123104.
[18] Singh, M.; Jain, A. P.; Parashar, J. J. Indian. Inst. of Sci. 2002, 82, 183.
[19] Liu, C.S.; Tripathi, V. K. Phys Plasmas 2008, 15, 023106.

[20] Nitikant, Sharma, A. K. J. Phys. D: Appl. Phys 2004, 37, 998-1001.
[21] Askari, H. R.; Noroozi, M. Turk J Phys. 2009, 33, 299-310.
[22] Rickes, T.; Marangos, J. P.; Halfmann, T. Opt. Commun. 2003, 227, 133.
[23] Alexeev, I.; Ting, A. C.; Gordon, D. F. Opt. Soc. Am. 2005, 30, 12.
[24] Rajput, J.; Kant, N.; Singh, H.; Nanda, V. Opt. commun. 2009, 282, 4614-4617.
[25] Trippenbach,M.; Matuszewski, M.; Infeld, E.; Van, C. L.; Tasgal, R. S.; Band, Y. B Opt. Commun. 2004, 229, 391.
[26] Ferrante, G.; Zarcone, M. Phys. Rev .E 2004, 70, 16403.
[27] Theberge, F.; Luo, Q.; Liu, W.; Hosseini, S. A.; Sharifi, M.; Chin, S. L. App. Phys. Lett. 2005, 87, 081108.
[28] Liu, X.; Umastader, D.; Esarey, E.; Ting, A. IEEE Trans. Plasma Sci. 1993, 21, 90.
[29] Akozbek, N.; Jwasaki, A.; Becker, A.; Scalora, M.; Chin, SS. L.; Bowden, C. M. Phys. Rev. Lett. 2002, 89, 143901.
[30] Tasgal, R. S.; Trippenbach, M.; Matuszewski, M.; Band, Y. B. Phys .Rev. A 2004, 69,13809.
[31] Shibu, S.; Tripathi, V. K. Phys. Lett. A 1998,99, 239.
[32] Yang, H.; Zhang, J.; Zhao, L. Z.; Li, Y. J.; Teng, H.; Li, Y. T.; Wang, Z. H.; Chen, Z. L.; Wei, Y. Z.; Ma, J. X.; Yu, W.; Sheng, Z.M. Phys. Rev. E 2003, 67, 015401.

In: Laser-Induced Plasmas
Editor: Ethan J. Hemsworth, pp. 125-143
ISBN 978-1-61324-851-5

Chapter 6

INFLUENCE OF HETEROGENEOUS NATURE OF LASER ABLATION ON NEAR-SURFACE PLASMA FORMATION AND PROPAGATION

Yu Chivel[*]
Institute of Applied Physical Problems. Minsk, Belarus

ABSTRACT

Experiments investigations and numerical simulation of thermophysical and hydrogasdynamic processes in near surface layer of metalls, irradiated by pulsed laser, allowed to expose low-threshold character of initial destruction of materials, to establish mechanisms responsible for lowering of plasma ignition thresholds.

Keywords: plasma formation, surface destruction, evaporation, defects.

INTRODUCTION

The ablation nature of the near-surface initial plasma formation under the action of pulsed laser radiation in wide region of pulse duration and laser wavelength is established at the present time. By direct experiments the ablation nature of this phenomena has been demonstrated even for action of the microsecond CO_2 - laser [1,2].

Laser ablation under the action of short laser pulses is a well-known and well-studied phenomenon [3]. Several different mechanisms are responsible for the material removal, among them is thermal evaporation that is studied in details [4]. Nano- and micron-sized particles are observed at the early stage of the laser–target interaction and the related mechanisms are not yet fully identified [5]. At present several mechanisms of particles release in nanosecond laser ablation are proposed assuming that most processes are related to

[*] e-mail: yuri-chivel@mail.ru

initiation and decay of the metastable states. For example, existences of under-surface temperature extremes in liquid and solid states were predicted for nanosecond laser irradiation of high-temperature superconductive ceramics [6]. Their existence was explained by volumetric nature of absorption of laser radiation and dynamics of vapour– liquid and liquid–solid phase transitions. In paper [7] critical analysis of basic thermal mechanisms of particle release was carried out. The conclusion was made that explosion boiling due to homogeneous nucleation when superheating occurs and the surface temperature reaches the critical-point region remain a major mechanism of particle release at high fluence. The above mentioned papers apply the hypothesis about isotropy and plane geometry of the near surface layer of ablated material. Alternative mechanisms of laser ablation could be based on initial heterogeneity of the near surface layer of irradiated material.

In papers [8,9] concerned with laser irradiation of metals the phenomenon of low-threshold ablation was observed. Low threshold ablation may occur when, at the moment of the surface destruction, the surface temperature is below the boiling or sublimation points before the moment of the plasma formation. It was found that the destruction of near surface layers is influenced by structural inhomogeneties of solids, generation and structural relaxation of different heterogeneities (pores, voids, gas and other inclusions), concentration of which in metals can be up to 10^{19}–10^{20} cm^{-3} [10]. In general structural defects in material can appear and disappear under intensive laser irradiation [11]. The decisive influence of the heterogeneous structure of solids on ablation thresholds under the exposure of laser pulses of 10 ps to 1 μs was revealed [12-14] . However under the action of millisecond pulses the ejection of nano and micro particles also has been fixed [15] at the beginning of the surface ablation that is associated with volumetric vaporization [16]. In paper [17] a similar low-threshold laser ablation of high-temperature superconductors in the solid state was observed and explained by thermal decomposition of ceramics, gas release and cracks evolution. Similar low-threshold destruction of metals was observed at electrical explosion of conductors [18], at pulsed electrical discharges [19]. Attention is drawn to the fact, that the surface processes under the action of various pulsed energy sources are amazingly similar.

Plasma formation thresholds under such heterogeneous laser ablation are determined not only by initial surface state but the occurrence of surface defects and ejected particles.

1. Heterogeneous Laser Ablation

1.1. Experimental

The comprehensive study of the action of a laser pulses with duration 40 ns and 300 ns with metal surfaces [8,9, 20,21] under control of the surface state during the impact by measuring the dynamic characteristics of the mirror, diffuse and scattered components of the reflected laser radiation , surface temperature and pressure in the laser spot make it possible to establish the explosive – like destruction of the surface with appearance on the surface micro discontinuities with the characteristic size 1-10 microns (Figure 2) and the ejection of particles of condensed dispersed phase (CDP). Such surface structure formation is recorded at the regime with plasma formation, and in preplasma modes of action. It has been found experimentally [8,21] that the appearance of these surface structures is registered as a rule during the duration of the leading edge of impulses (Figure 1) before the maximum temperature and pressure at the irradiation spot and before the plasma formation.

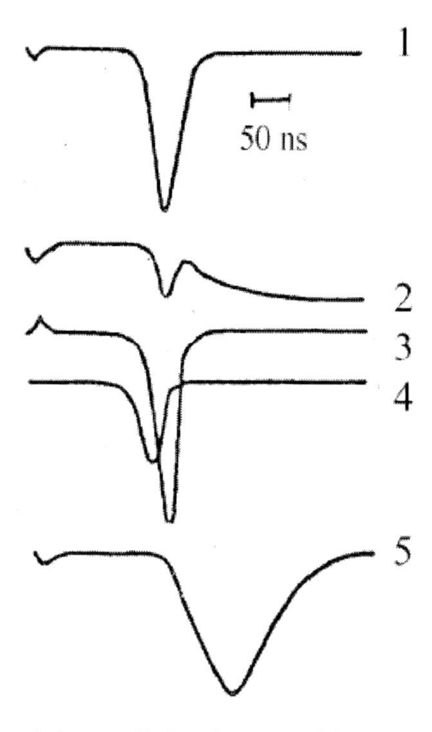

Figure 1 Oscillograms of pulses: 1- laser radiation ;2- scattered laser radiation; 3- diffuse reflected laser radiation; 4 – mirror –like reflected radiation; 5- pressure pulse . Material- Al alloy, q_m = 140 MW/cm^2, t = 40 ns.

The presence of a certain temperature threshold ≥ 2000K was found , above which the explosive destruction of the surface takes place [4], while for some metals it takes place at temperatures below the boiling point of the target material. In order to determine the influence of material structure on the fracture behavior targets of varying degrees of purity and quality of polishing, as well as the targets from powder material were used. It has been found that the family of micro discontinuities (Figure 2) are occurred primarily along the scratches remaining after polishing, but mainly the closed scratches and pores are set off.

1.2. Model of Laser Ablation

In the developed model of heterogeneous ablation , condensed matter is considered as a heterogeneous medium. It is assumed that the subsurface solid's volume contains spherical micro- and nano-sized pores with gas absorbed at the walls. In general, porosity is inherent in all solids. In particular metals contain a pores at most probable size 0.1–1 mm [10]. Let us consider thermal and hydrodynamic processes in the subsurface layer of metal under heating by pulsed ($t = 4 \cdot 10^{-8}$ to $30 \cdot 10^{-8}$ s) laser radiation. An intensive gas desorption takes place when the pores surface is heated up to 2000 K [22].

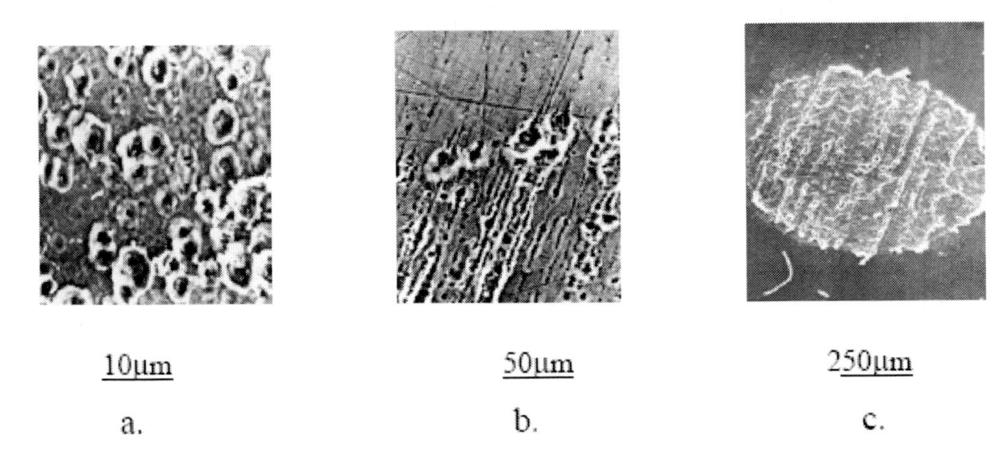

Figure 2. Morphology of the laser irradiated zones: (a) surface of pure Al, power density $q_m = 6 \cdot 10^7$ W/cm^2, pulse duration $t = 4 \cdot 10^{-8}$ s and (b–d) duralumin surface: (b) power density $qm = 1.5 \cdot 10^7$ W/cm^2, (c) $q_m = 4.5 \cdot 10^7$ W/cm^2, (d) $q_m = 25 \cdot 10^7$ W/cm^2, pulse duration is $t = 4 \cdot 10^{-8}$ s.

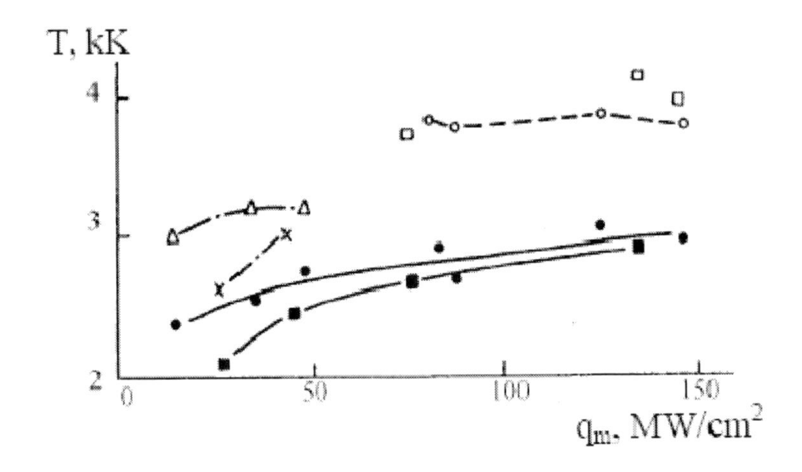

Figure 3. Surface temperature in laser spot at moment of surface destruction (• - D16T, ■ -Zn, ◊ - Bi), at moment plasma ignition (○ - D16T, □ - Zn) versus laser power density.

The gas filled in pores, heats up to pores wall temperature through thermal conductivity in a short time 10^{-9} to 10^{-8} s. The pressure about 10 –100 MPa, developed by gas, could result in surface destruction. The destruction can occur in the solid state when the gas pressure in the pores exceeds the damage threshold of the material, or in the liquid state as the results of gas bubbles growth. For several materials it takes place in the temperature range above 2000 K , but below the corresponding boiling point. The target heating was considered as one-dimensional. For numerical solution of the differential equation of heat transfer with corresponding initial and boundary conditions, the finite element method realized by LS-DYNA program was applied [23]. LS-DYNA is a general purpose finite element code for analyzing the deformation dynamic response of materials under various actions. To decrease the computing time, the target thickness was limited by 10 mm, the segments of triangular mash were in the range 0.1–1 mm. Elastic stresses and plastic strain fields were calculated for the given temperatures (Figure 4a).

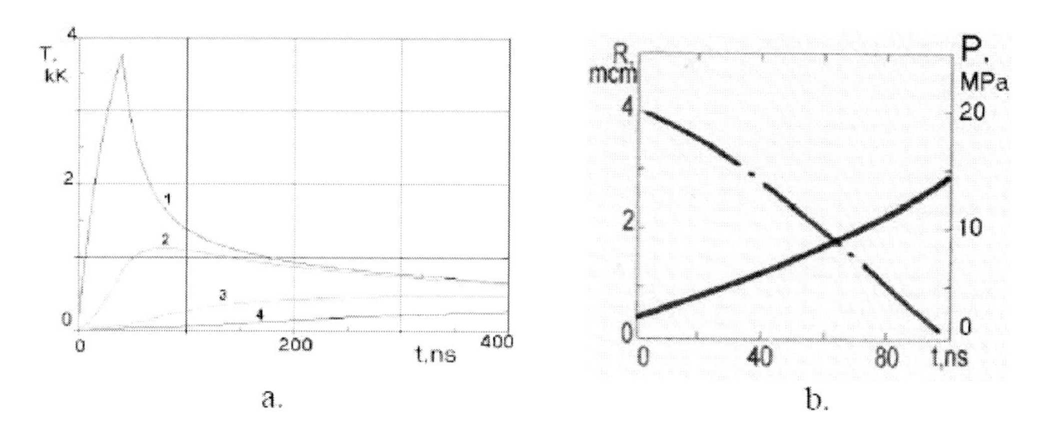

Figure 4. Dynamics of the sample's heating under pulsed laser irradiation (a). Depth interval between points 1,2,3,4 is 1μm, $q_m = 10^8$ W/cm^2 and pulse duration $t = 10^{-8}$ s: .; dynamics of pore gas pressure (dotted line) and the bubble radius (full line) in the melt layer (b). Material - Cu, gas - H_2, $R_0=5\cdot10^{-7}$m , T_0=2000K, $A=10^{20}$ m^{-2}, $B=10^{11}$ K/s.

For this purpose the LSDYNA material model MAT_ELASTIC_PLASTIC_THERMAL [23] was used. The model of the gas bubble growth in the melted layer involves the Rayleigh equation that links gas pressure variation with the bubble radius and the energy equation for gas inside the bubble.

$$R\cdot\frac{d^2R}{dt^2}+\frac{3}{2}\left[\frac{dR}{dt}\right]^2+\frac{4\upsilon}{3R}\cdot\frac{dR}{dt}=\frac{1}{\rho}\cdot\left[P-\frac{2\sigma}{R}\right];$$

$$\frac{dP}{dt}+\frac{3\cdot\gamma\cdot P}{R}\cdot\frac{dR}{dt}=\frac{3(\gamma-1)}{R}\cdot\left[\lambda\cdot\frac{\partial T}{\partial r}\right]_{r=R}; \tag{1}$$

where R – bubble diameter , P – gas pressure , υ- viscosity of the melted metal , ρ- melt density, λ - gas heat conductivity, γ - gas specific heat ratio, T – temperature of gas in bubble. q = λ (dT/dr) - heat flux to bubble which depend on the velocity of surface heating – B, m is the gas molecular mass, R_0 is the initial bubble radius and A is the surface density of the absorbed gas.

$$\rho_1=\left(\frac{3A\cdot m}{R_0}\right)\cdot\left(\frac{R_0}{R}\right)^3 \tag{2}$$

Equation system (1)- (2) was numerically solved under the initial conditions:

$$R(0) = R_0; \left(\frac{dR}{dt}\right)_{t=0} = 0; P(0) = \frac{3(\gamma - 1)Cv \cdot A \cdot m \cdot T_0}{R_0};$$

$$\left(\frac{dP}{dt}\right)_{t=0} = \frac{3(\gamma - 1)Cv \cdot A \cdot m \cdot B}{R_0} \qquad (3)$$

1.3. Growing of Bubbles in Liquid Phase

Assuming the presence of initial submicron pores in the solid metal that will be transformed into bubbles at melting. In addition solid–liquid phase transition causes formation of voids (''nanobubbles'') in liquids [24,25] with the minimum size close to an interatomic distance of about several angstroms. At normal conditions their concentration is up to 10^{18} cm^{-3}. Figure 4 represents results of numerical simulation for submicron pores $R_0 = 5 \cdot 10^{-7}$ m, T_0 = 2000 K, A = 10^{20} m^{-2} and B = 10^{11} K/s. The starting point 2000 K is the temperature above which intensive gas desorption takes place for several metals like Al, Cu, Bi, Zn, W.

A tolerable rise of the bubble diameter to a value exceeding the melt layer thickness is observed at the leading edge of laser pulse (Figure 4 b). The processes slightly depend on the gas nature because the gas parameters enter into the equation in the combination $C_v \cdot m$. The bubbles growth results in ejection of micron-sized droplets, formation of surface irregularities in the form of micro-craters. These phenomena were observed for certain metals as Cu and Al at the early stage of laser irradiation before reaching the evaporation threshold [7]. Photo of Cu surface after pulsed laser irradiation with intensity $8 \cdot 10^7$ W/ cm^2 and pulse duration t = 40 ns is shown in Figure 5 a.

The possibility to reach the critical point under pulsed laser heating is limited by several factors: finite rate of homogeneous nucleation, heterogeneous nucleation, low-threshold ablation and plasma formation. Furthermore, the critical temperatures of most metals under study [26] exceed the plasma formation thresholds.

Frequency of the homogeneous nucleation with regard to its transient character can be written as follows [27]:

$$J=B\,n\,\exp(-\,\Delta G/kT - \tau/t) \qquad (4)$$

where n- atom density, ΔG – work of critical vapour bubble generation, Bfunction, depend weakly on T,P; τ - time constant of transition process.

B is a function weakly depending on T, P and t is the time constant for the transition process. For a large group of metals t = 10^{-8} to 10^{-9} s. At T = 0.9Tcr and t = 10^{-9} s expression (5) gives J = $5 \cdot 10^{25}$ cm^{-3} s^{-1} and in this case the metal liquid–vapour transition time exceeds 10^{-7} s [27]. Thus, it may be concluded that homogeneous nucleation within the laser pulse meets difficulties in the case of nanosecond pulse duration. The results of R. Russo group [28] as well as [29,30] confirm the inability of homogeneous nucleation within a nanosecond laser pulse.

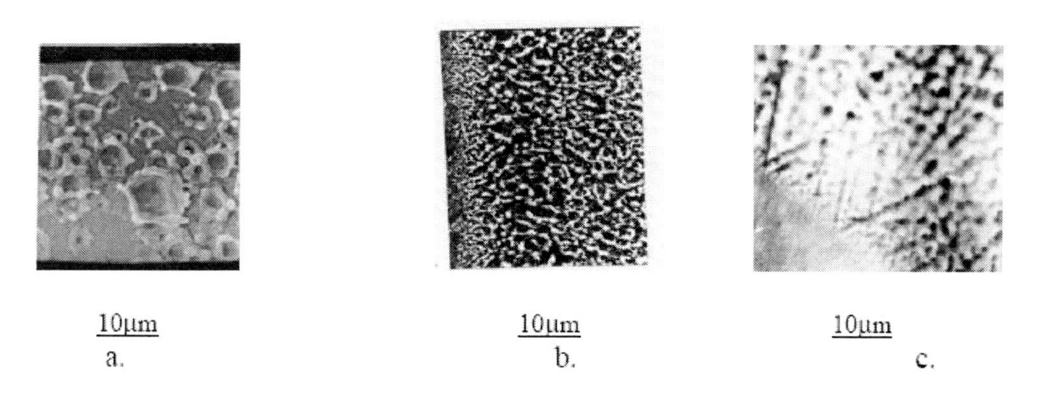

Figure 5. The structure in laser affected spot on Cu- surface (a), Zn –surface (b) Si – surface (c) ; a - $q_m = 8 \cdot 10^7$ W/cm^2, b - $q_m = 4 \cdot 10^7$ W/cm^2 ($\tau = 4 \cdot 10^{-8}$s); c - $q_m = 8 \cdot 10^7$ W/cm^2 ($\tau = 10^{-11}$ s).

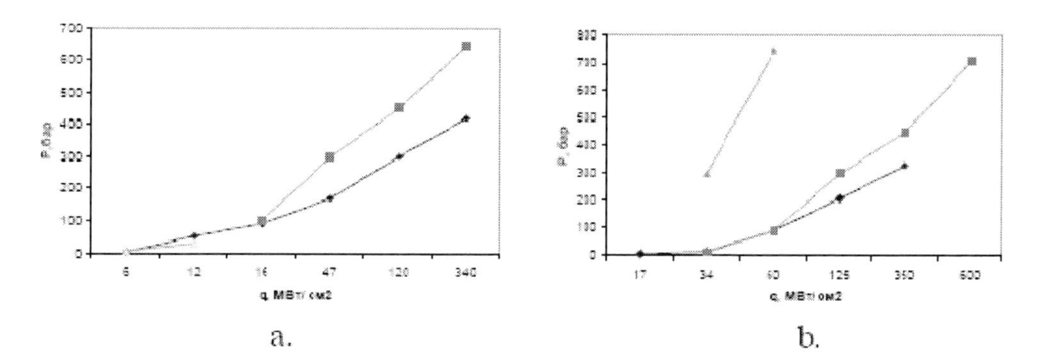

Figure 6. Surface pressure under the action of laser pulse (40 ns, Nd- laser) : ♦ - P= 0,1MPa , ■- P=0,0001MPa (a- Bi , b - Zn), ▲ – saturation vapour pressure.

They watched the ejection of micron-size particles through a 300-400 ns after the laser pulse and to explain it as a manifestation of the homogeneous explosive boiling. But this phenomena may be caused by the target off-loading during the expansion of the plasma.

In the paper [31] as well as in our studies (Figure 6) found a significant decrease in the vapor recoil pressure compared with equilibrium values for surface temperature measurement. But this deviation is also interpreted in the framework of the phenomenon of homogenous explosive boiling.

Accordingly to experiments reported in [13] the critical points were reached during nanosecond laser irradiation (t = 40 ns, 10^7–10^8 W/cm^2) of the metals with critical temperature 2000–3000 K as, for example, Zn. In this case the surface morphology is quite different (Figure 5 b).

So that the homogeneous nucleation during the duration of laser pulse is unlikely on exposure to nanosecond pulse, much less to picosecond pulse, whereas experimental results [12] point of the fact that the character of destruction is analogous under the action of such pulses (Figure 5c).

As already noted concentration of nanosize defects (vacancies, pores) in solids is very high. In such pores the nanoscale effects are to be manifested. These effects arises from the change in interaction potential of electromagnetic fluctuational fields of pore surface in passing from micro to nanopores. As shown in paper [32] above the melt temperature of metals the vaporization rises steeply.

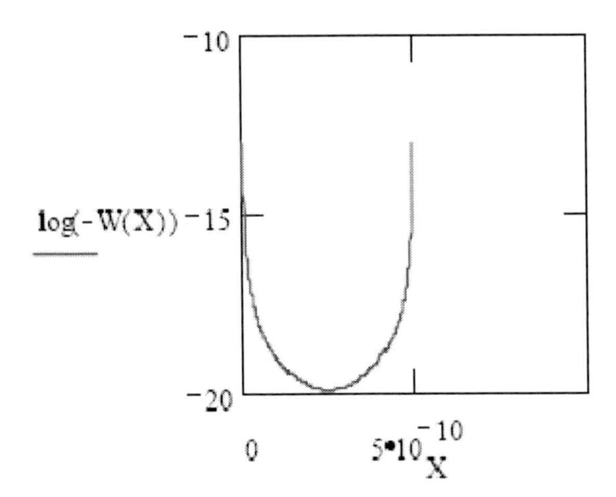

Figure 7. The distribution of dispersion energy in the nanopore.

At T~ 2000K and above the chemical bonds fall off and it is not unreasonable to take into account only Van-de-Waals forces.

Let us consider a pore in slot form h nm wide. It is assumed that molecule of surface and vapour interact with each other through the Van-de-Waals forces [33].

The dispersion energy of molecul in the pore takes the form (Figure 7):

$$W(x) = - A[1/x3 + 1/(h - x)3] \quad (5)$$

where A ~ 10^{-49}Jm^3 for metals, namely, for Al at x_0=1A, A= $5\cdot10^{-49}$Jm^3.

A 25-10% reduction in heat of vaporization occurs when passing from half-space to nanopore h=0,5-1nm wide. According to law of corresponding states [34], as applied to heat of vaporization Q: Q/Tcr=f(T/Tcr).

Reduction in critical temperature comes to 2000K and critical temperature of Al takes the value 6000K rather than 8000K.

By this means the nano-scale heterogeneity of materials can have a pronounced effect on their critical parameters and thresholds of volume destruction and ablation.

1.4. Pore Micro-Explosion in Solid State

For metals with high melting point T_{melt} > 2000 K explosion of micro-pores and solid particles release is possible under pulsed laser irradiation. At gas desorption in pores over 2 $\cdot10^3$ K and heating to such a temperature, pressure in the pores reaches up to P = 10^7 Pa and even higher. For near-surface pores with diameter ~ 1 mm the stress in the thin metal foil (it is assumed that distance between pore and the surface is much less than the pore size) may be written as [35] Ω_{rr} = P/2; Ωgg = PR/2, where R is the pore radius. When P = 10^7–10^8 Pa, the stress attains the values of 10^7–10^8 N/m^2 that are close to the damage thresholds taking into account the temperature dependence Ω = (1 - T/T_{melt}). Modeling was carried out for W sample of 15 mm x 15 mm x 7 mm exposed to a rectangular laser pulse with power density 10^8 W/cm^2 during 10^{-8} s. 3D distributions of plastic strain and effective stresses were analyzed for pores with 0.5– 0.05 mm radius situated at 0.1– 0.5 mm distance from the sample surface (Figure 8). The fact that the yield stresses decay linearly to zero level as the sample temperature increases from 1800 K to 3700 K was taken into account. Pressure of the heated gas in the pores was calculated by formula (2). For the pore radius equals to 0.5 mm the maximum gas pressure at 3000 K is ~26 MPa and P = 260 MPa for the pore radius 0.05 mm.

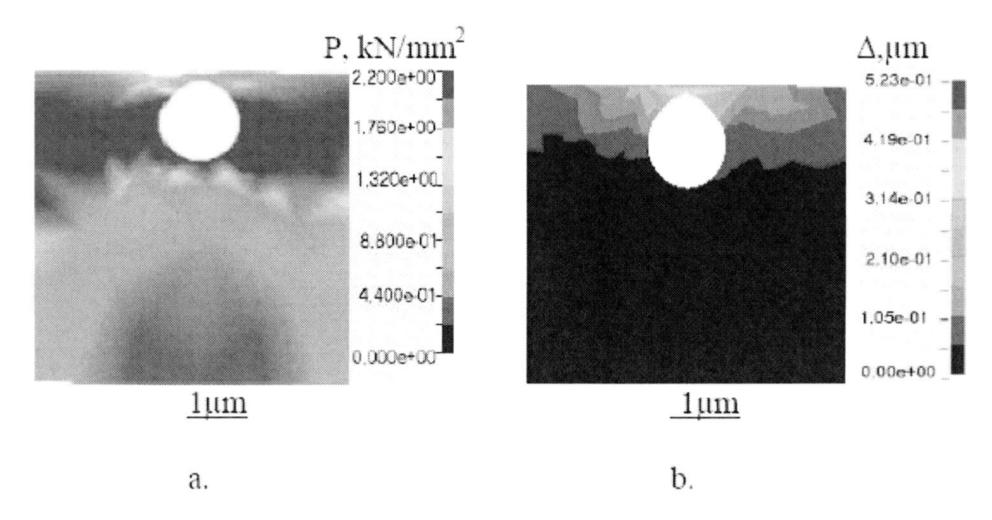

Figure 8. Effective stresses on the pore (1,3) and sample (2) surface points.

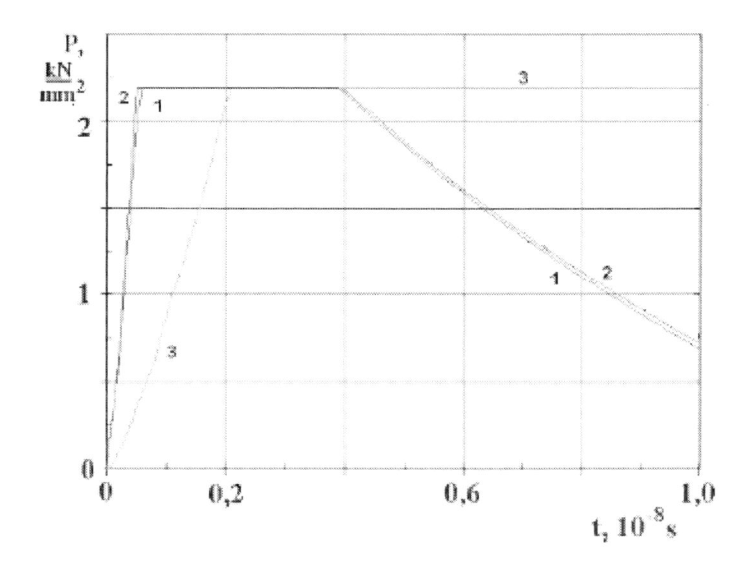

Figure 9. Plastic strain and effective stresses dynamic 3D: a) contours of effective stress distribution in section at 8· 10^{-9} s; b) contours of effective plastic strain distribution in section at 10^{-8} s. Maximum strain is D = 0.2 μm.

Pulsed laser heating results in formation of field of compressive stresses (Figure 8a). Gas desorption from walls of the pore and consequent gas heating cause fast rise of the gas pressure a swell as plastic flow of metal. As the temperature peak is located at the sample surface, fast decay of the compression stresses in the surface layer through the plastic strain (point 2 and 1 are situated at the irradiated surface and at the highest point of pore's surface along the pore's axis; results in deformation of the pore (Figure 8b) and explosive-like destruction of the thin surface layer with condensed particles release (Figure 9). The delay of pore micro-explosion relatively laser pulse beginning is $\sim 10^{-8}$ s for the absorbed radiation intensity $\sim 10^{8}$ W/cm^2.

The model developed gives results in conformity with experimental ones and allows an understanding of some phenomenon among them the near surface plasma formation.

2. Plasma Formation Under Action of Laser Pulses

2.1. Experimental

The experiments were performed at atmospheric pressure and under vacuum using a laser system based on neodymium (Figure 10a) glass, generating a single pulse duration of 40 ns and 300 ns. A number of experiments performed on electronionized CO_2 laser with a pulse ~ 1.0 μs (Figure 10b) and rhodamine laser with a pulse duration of 10 μs (Figure 10c). Also studied the effect of pressure surrounding the target gas in the process of ablation and plasma formation under the influence of quasi-cw laser pulses with a wavelength of 1.06 microns and a duration of 1.5 ms (Figure 10d).

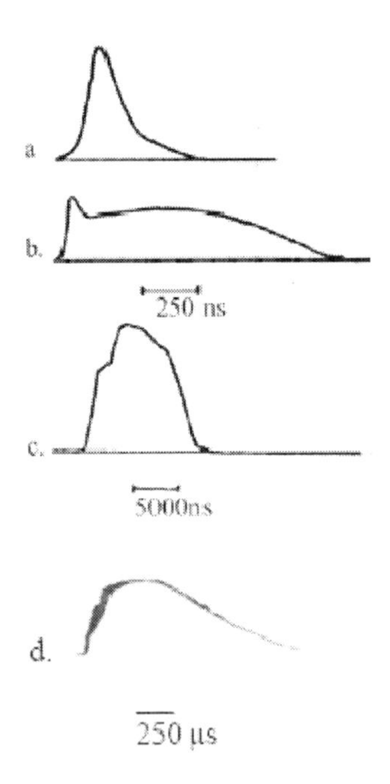

Figure 10. Laser pulses : Nd-laser pulse (a), CO_2 laser pulse (b) , Rd –laser pulse (c), quasi-continuous Nd – laser (d).

The targets of mechanically polished samples of aluminum A99, duralumin D16T, zinc, bismuth and indium. Control of the surface during the impact and determination of the threshold characteristics of plasma was performed by measuring the dynamic characteristics of the mirror, diffuse, scattered component of the reflected laser radiation, brightness surface temperature and 10^{-8} s , surface pressure in the irradiation spot with a time resolution 10^{-8} s. Pressure sensors were used in the regime of current generator . Calibration of pressure sensors was carried out using a laser shock tube [36]. Surface condition before and after treatment was studied by optical and scanning electron microscopy, X-ray analysis was performed of the surface layer of materials. Dynamics of plasma surface studied by high-speed detection of radiation and vapor plasma with high spatial resolution and high-speed spectral measurements with a temporal resolution of 10^{-5}s.

2.2. Results

The morphology and distribution of defects on the surface of mechanically polished samples showed that most of the defects with the characteristic size 0,5 – 1 μm and a concentration of $10^6 \div 10^7 cm^{-2}$ are the scratches left by the abrasive grains and particles of abrasive material (Al_2O_3) are embedded in the surface. The defect density is high enough and it can be assumed that the height of the irregularities of the relief well ~1 μm for the period 1-10 microns. The comprehensive experiments have revealed features of the initial fracture surface. It is shown that local damage in the irradiation spot in evaporation regimes largely determined by the uneven radiation distribution of defects rather than what might be expected, having in view of their high concentration. Established that at the moment of the plasma appearance brightness temperature of the surface reaches 3000 K - 4000 K (Figure 3). Increase in absorption in the vapor begins at this temperature level. Vapor transition from a state of almost complete transparency to the strongly absorbing plasma occurs during 10^{-8}s - 10^{-7}s, depending on the duration of pulse leading edge. A development of plasma in the vapor is substantially influenced by the lateral rarefaction wave. In experiments transition from quasi-one to three-dimensional expansion of the vapor by varying the spot size of irradiation in the range of 0.2 - 4 mm, and duration of laser pulses in the range 40 - 350 ns. In the transition from quasi-to three-dimensional expansion of the vapor increases the delay time and threshold of plasma [37] generation (Figure 11).

The surface layer destruction brings into existence the surface irregularities in the form of microcraters, micropoints, microparticles with the size ~1μm [14]. They facilitates the plasma formation. Under the action of pulsed LR microparticles or insulated defect on the metal surface have impact on plasma formation in the event that theirs heating and evaporation follow rapidly than surface ones. The calculations of particles heating, evaporation and plasma ignition in vapour, having regard to the extremely high laser radiation absorption by micropartieles with the size $\lambda > d > \lambda/6$, have demonstrated that evaporation and plasma ignition take place in a time 10^{-8} s and 10^{-7} s, respectively, during the LR pulse leading edge. This data are in a good agreement with experimental results. The local character of evaporation is evident from the results of pressure measurements in the evaporation regimes and shadowgraphy under the action of Nd-laser (τ=40ns). A measured amplitudes of pulsed pressure much below the magnitude accompanying the equilibrium state of phase transition.

Confluence of separate vapour plume, originated at surface inhomogeneities, occurs at time $\sim 10^{-8}$ s at a sound speed in the vapour $\sim 10^{5}$ cm/s. At a later time geometry of vapour flow, affects on plasma formation thresholds will be determined by the spot size (Figure 11).

It is natural to assume that the increase in pressure of gas surrounding the target will affect the process of heterogeneous ablation target and threshold parameters of plasma.

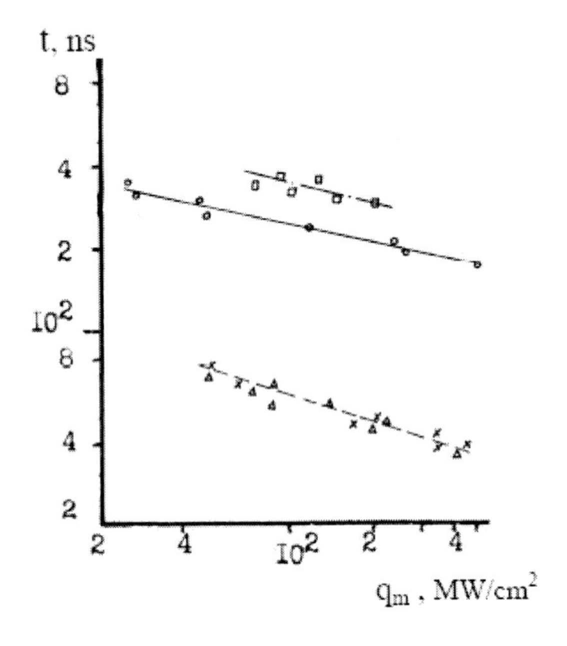

Figure 11. Initial plasma formation time versus laser power density. Target – Al alloy. (λ = 1,06μm): 1. τ = 300 ns: ○ – laser spot – 4 mm; □ – laser spot- 0,25 mm. 2. τ = 40 ns: Δ - laser spot – 4 mm; x – laser spot - 0,25 mm.

2.3. The Influence of External Pressure on the Ablation and Plasma Formation

Considerable attention has been given to high speed photographic and spectroscopic investigations of the dynamic of plume formation versus ambient gas pressures . Neodymium laser capable of generating quasi-continuous pulses with energies of up to 900 J, the total duration of 1.5 ms, and the leading-edge duration of ~300 μs was used. The target is placed in high pressure chamber within ~50mm from the entrance window. The dynamic of near-surface plasma evolution was investigated by the high speed registration of vapour and plasma radiation. Integral in-time, but space resolved spectrum in the 390-700nm region was registered. The time resolved measurements of surface and plasma brightness temperature were conducted as well as high-speed spectra with exposure time$\sim 10^{-5}$s. Al- alloy and Bi targets have been used.

The obtained experimental results point to significant influence of gas pressure on laser ablation processes [38] . Structuring of plasma plume with a target of Al-alloy observed in air of normal density associated with the discrete income of vapor and particle from the surface of the target at the pressures above 1 MPa is completely absent and the plasma plume is "pressed" to the surface (Figure 12). This provides an intensification of the process of laser-plasma action on the target Studies of dynamics of the plume formation shows at the same power densities of laser radiation (LR) at enhanced pressures the initial emergence of vapor and the initial plasma formation in the plume occurs later than that at atmospheric pressure (Figure 13 b). The mass removal is lowered by a factor ~10^2 (Figure 13 a).

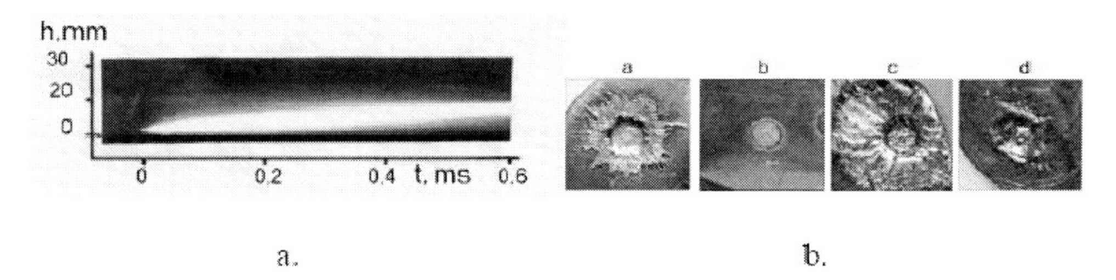

Figure 12. (a) Plasma plume dynamics (q=4MW/cm^2, a- 15 bar, Al-alloy); (b) photographs of irradiation spots after laser exposure: Al- alloy (a, b), Bi (c,d); p = 1 (a,c) и 15 bar (b, d); q = 1.9 (a, b) и 3.2 (c, d) MW/cm^2.

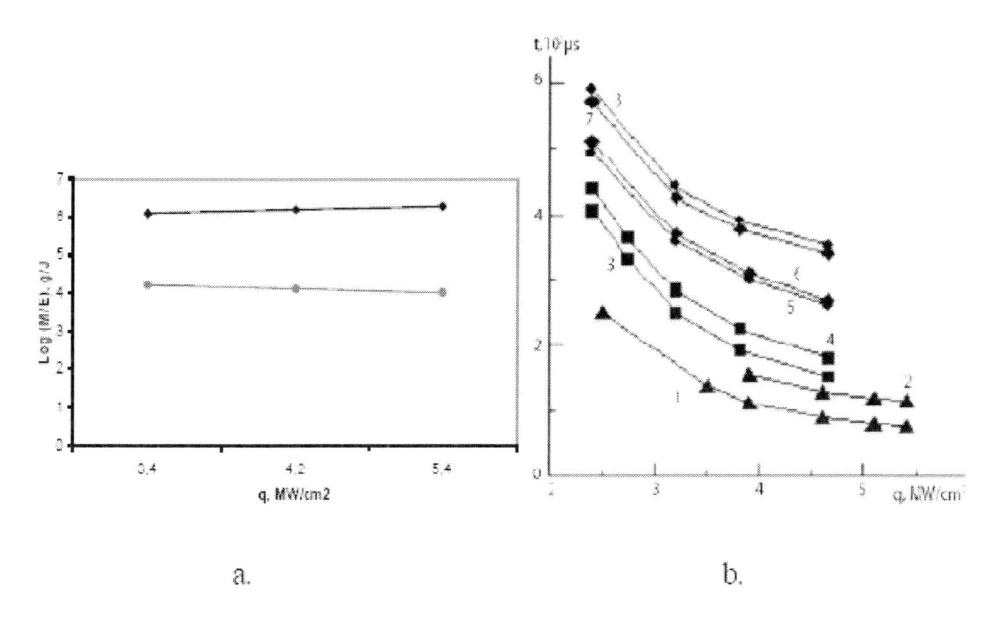

Figure 13. (a) Mass removal at P= 1bar (♦) , P= 10 bar (●); (b) Laser intensity dependence of vapour appearance (1,3,5), erosive plasma ignition (2,4,6), N_2 plasma ignition (8): p = 1 (1, 2), 6 (3, 4), 10 (5, 6) и 15 bar (7, 8). Target – Al- alloy.

Using Clausius-Clapeyron equation pressure dependence of boiling temperature for Al was estimated (Figure 14 b) [39]. Also results of numerical solution of one-dimensional problem of heating and evaporation of Al target at atmospheric pressure [40] under our experimental conditions have been analyzed, see Figure 14 a.

Using the derived velocities of surface heating in the range 1,7- 2,8 10^7 K/s the time delayes for boiling of Al target in going from 1 bar to 10 bar are obtained. These delay times are in the range 15- 40μs. These values are 4-5 times less than those obtained in the experiment.

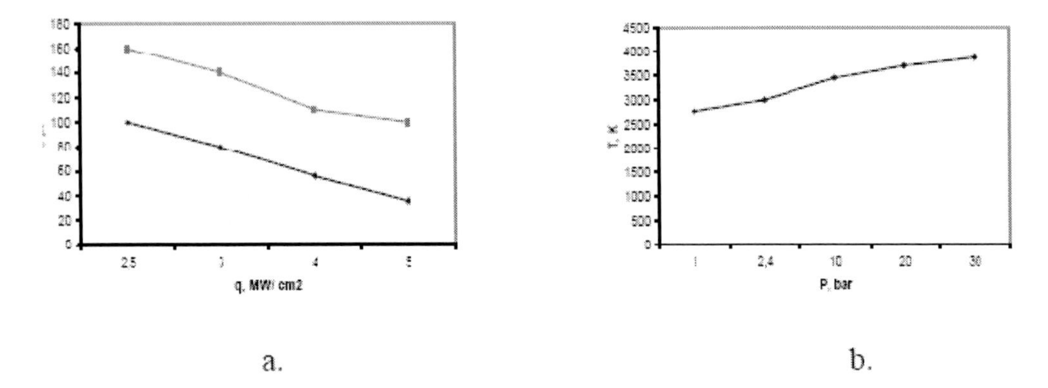

Figure 14. (a) Laser intensity dependence of Al vapour (♦) and plasma (■) appearance at P=1 bar (theory [40] ; (b) Pressure dependence of boiling temperature.

2.3.1. Spectral Diagnostics

Spectroscopic studies show that at ambient gas pressures up to 0.6 MPa the near-surface plasma formation is mainly of erosive nature (Figure 15 a,b). Recorded in such conditions were atomic and ion lines of the target material elements, as well as intense molecular bands. With growth of the ambient gas pressure up to 1.5-1.7 MPa the character of the spectrum drastically changes. At q~4-5 MW/cm^2 on ~ 1-2 mm distance away from surface, plasma front transfers into ambient gas and the laser absorption wave is formed (Figure 15 a,b).

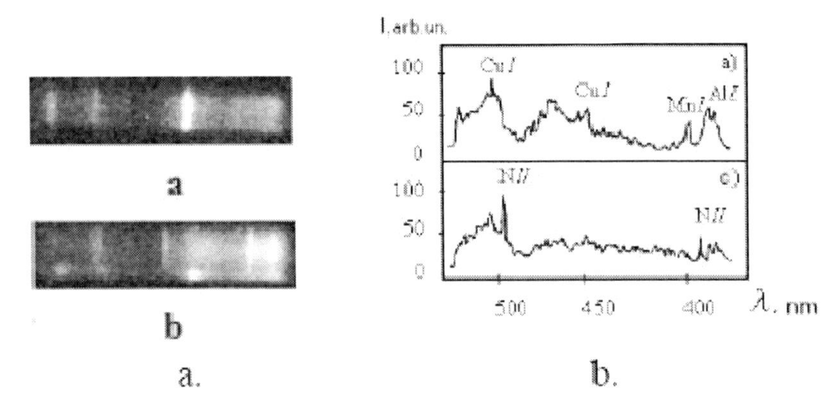

Figure 15. (a) High speed spectrogram of plasma plume (Δt = 180μs, a- P= 1bar - erosive plasma; b- P= 15 bar, N2 –plasma and erosive plasma) ; (b) Densitogram of plasma emission spectra under various pressure (qm=4 MW/cm2: a - P=0.1 MPa, c - P=1.5 MPa).

The spectral nitrogen ion line are registered at all the height of plasma plume excluding ~2-4 mm near-surface area, which erosive plasma occupies. Measurements of surface plasma temperature gives a value T ~ 8- 10·10^3 K. But qualitative assessment of gas plasma temperature on ion lines excitation functions gives value of plasma temperature ~14-16· 10^3

K. Calculated absorption coefficient of erosive and nitrogen plasma under P = 15 bar equal to 2,7 $см^{-1}$, resulting in high level of surface shielding.

As follows from the above to explain the observed delay of surface ablation and plasma formation by the influence of the pressure of the ambient gas on the boiling threshold of the target material is not possible. Numerical calculation according to our model of heterogeneous laser ablation [5] have been conducted. Consider thermal and hydrodynamic processes in near-surface layer of melted Al on exposure to pulsed ($\tau = 10^{-4}$ s) heating by LR ensemble of pores 1-3 μm in diameter, gas filling of which occur due to the desorption of gas layers, covering the porous wall. The pores wih diameter 1-3 μm are most probable for Al and Al-alloy. When surface layer will be melted pores will be transformed into bubbles at melting. Figure 4c represents results of numerical simulation for micron pores $R_0 = 3\cdot 10^{-6}$ m, $T_0 = 2000$ K, $A = 10^{20}$ m^{-2} and $B = 2\cdot 10^7$K/s.

A tolerable rise of the bubble diameter to a value exceeding the melt layer thickness is observed at the leading edge of laser pulse (Figure 16).

From experimental as the pressure of ambient N_2 increases up to 1,0 – 1,5 MPa the particles release is missed. The results of numerical calculation shows that pressure in pores has a value 1,0-2,0 MPa. The ambient gas pressure like that inhibit the heterogeneous volume ablation and suppresses the particle release at initial stage of laser action. The absence of particles in the vapour inhibits the processes of vaporization and plasma formation. The transition of plasma front in ambient high pressure gas creates a shielding of the target that reduces ablation at a later time [41].

Based on the results of measurements of plasma ignition delay time the thresholds parameters were calculated: power density LR at the plasma ignition movement q* and specific consumption of energy on plasma formation E*/S (Figure 17). It was established that under the action of pulsed LR with duration ≤1 μs, the plasma formation takes the non - stationary character, that is, plasma formation occures at LR density much higher then ignition thresholds. A prerequiste to the plasma formation in this conditions is the excess of E*/S above some value, probably, sufficient defect evaporation. The dependence of delay time of plasma formation versus q_m, in this case, describes well by the relationship $\sim t^* \approx \beta\sqrt{\tau}/q_m$. For the longer pulses plasma formation takes quasi - stationary character and plasma form in vapour plume with steady state structure. In this case the excess of q above some threshold value is required (Figure 17).

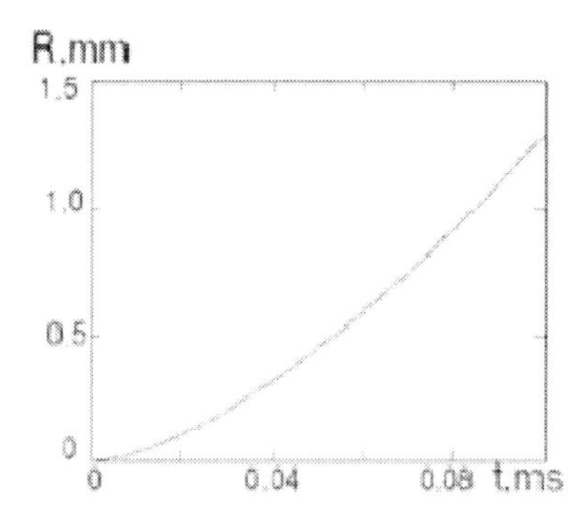

Figure 16. Dynamics of gas bubble in melt layer (material: Al, gas H_2, $R_0=3\cdot10^{-6}$ m , $A=10^{20}$ m^{-2}, dT/dt $=2\cdot10^7$ K/c).

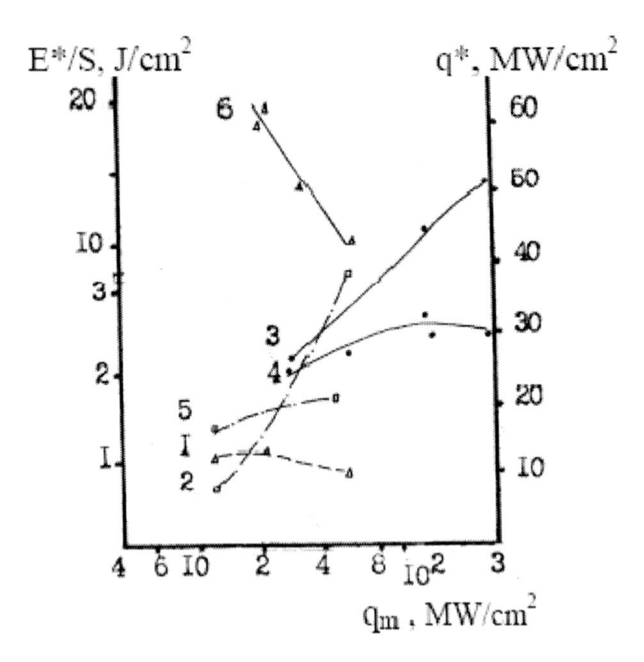

Figure 17. Dependence of plasma parameters on the maximum laser power density: 1. q* - 1- Д16T (λ=0,59 μm -rodamine), 2- D16T (λ=1,06 μm; $\tau_и$=300ns) 3- In (λ=10,6 μm; τ=1 μs); 2. E*/S: 4,5,6 - Д16T (λ=1,06 μm), In (λ=10,6μm), Д16T (λ=0,59 μm).

3. Plasma Front Instability

Local character of evaporation and plasma formation induce instability of interface between erosive plasma and gas, compressed by shock wave, resulting in laser plasma turbulization.

The observed flow structure (Figure 18) is characteristic for the Rayleigh - Taylor (RT) instability [42].

The estimation of the RT instability in conditions of pulse laser action (λ = 1,06 μm τ = 300 ns) on an aluminum target in air at atmospheric pressure has been conducted . With the experimentally obtained velocity of the interface to 3km / s and acceleration up to a = 10^{10}-10^{11} m/s^2 most intensively will rise fashion [42] with a wavelength $\lambda_m = 4\pi\ (\alpha\ v^2 / a)^{1/3}$ equal to 6 -10 μm.

When calculating data on the distribution density of the interface for calculating the kinematic viscosity v and the number of Attwood α were taken from [43]. Linear growth rate of small perturbations has the form: $\tau = (a \cdot 2\pi / \lambda)^{1/2}$ and under our conditions was 10^8 - 10^9 1 / s.

Thus, the observed instability of RT can develop when exposed to pulses of up to 10^{-9} s with the resulting turbulence in the flow of the laser plasma. As shown in [44] for the conditions of this experiment, even at t> 10 ns starting to dominate long-wave disturbances and the union of the bubbles of light fluid, which probably leads to the observed structures with the characteristic size of ~ 100 μm (figure 18).

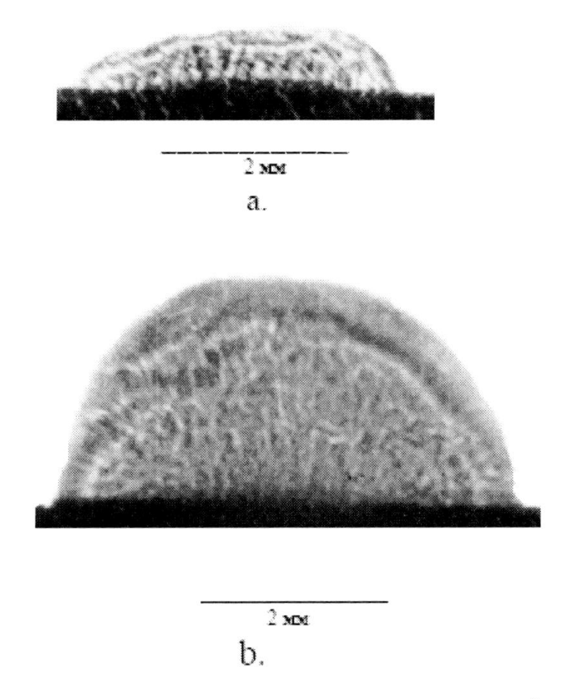

Figure 18. Shadowgraphs. λ=1,06 μm, τ=300 ns . target – Al, q = 8 MW/cm^2 (a), q = 150 MWт/cm^2 (б). Delay time of registration – 600 ns.

Conclusion

The summarize, experiments and numerical simulation of thermophysical and hydrogasdynamic processes in near surface layer of metalls, irradiated by pulsed laser, allowed to expose low-threshold character of initial destruction of materials, to establish mechanisms responsible for lowering of plasma ignition thresholds. The particles and thermoinsulated defects, appearing under destruction processes, contribute to plasma ignition thresholds lowering. Such, low-threshold, character of laser plasma formation near the solid surface take place in wide range of laser wavelengths and LR pulse duration. It is shown that the destruction of the surface with the release of the dispersed phase as a result of a heterogeneous ablation for the majority of the materials studied is observed at surface temperatures far removed from the critical point of the target material and is due to the inherent heterogeneity of the bulk solid state and condensed matter in general under the action of a longer laser pulses (~10μs and above) with slowly increasing edge the heterogeneous steam jet form before plasma ignition. The influence of surface state on plasma formation is much lower in this case, because the particles and plume shield the surface.

References

[1] V.Brunov, A.Gorbunov, V.Kono, *J. Appl.Spectroscopy*.44 (1986), 845.
[2] L.Minko, Yu. Chivel, A. Chumakov. *Quantum Electronics*. 15 (1988) 1619.

[3] Laser Ablation and its Applications. C. Phipps (Ed.), Springer Series in Optical Science, 2007.
[4] A.V. Gusarov, I. Smurov, *J. Appl. Phys*. 97 (2005) 014307.
[5] D.B. Geohegan, A.A. Puretzky, G. Duscher, S.P. Pennycook, *Appl. Phys. Lett*. 73 (1998) 438–440.
[6] V.I. Mazhukin, I. Smurov, G. Flamant, *J. Appl. Phys*. 78 (1995) 1259.
[7] A. Miotello, R. Kelly, *Appl. Phys. Lett*. 67 (1995) 3535.
[8] L. Min'ko, Yu. Chivel, *Journal de physique*. 4 (1994) 175.
[9] Yu. Chivel, L. Min'ko, Proc. *SPIE* 3688 (2000) 206.
[10] P. Cheremskoy, V. Sizov, V. Betekhtin, Pores in Solids, Energoatomizdat, Moscow, 1990, p. 310.
[11] V. Emelyanov, *Laser Phys*. 2 (1992) 389.
[12] Chivel Yu., Chyrvony V., Sazanovich V. Picosecond laser ablation of the metals and Si // *Proc. RAS . Physics*. 2001. v.65. pp.555-557.
[13] Yu.Chivel, M.Petrushina, I.Smurov. *Applied Surface Science*. 2007.V.254.PP.816-820.
[14] L. Minko , Yu. Chivel, V.Chivel, A.Yukechev. Izvestiya RAS, *Phys. Ser*., 61 (1997)1431-1436.
[15] Goncharov V. Engin. *Physical Journal*. 1992. 62. p. 665- 683.
[16] N. Rykalin, A.Uglov, I.Zuev, A. Kokora, Laser and Electron Beam Material Processing, *Mir* (1988) 580p.
[17] E. Sobol, M. Kitai, S. Golberg, A. Zherihkhin, *Appl. Surf. Sci*. 90 (1998) 235.
[18] N.N. Aderdzanov, I.I. Divnov, N.I. Zotov: *The extreme state of matter* (1991) 159.
[19] S. Anders, B. Juttner: *IEEE Trans. Plasma Sci*, 19 (1991) 705.
[20] L.Ya. Min'ko, Yu.A. Chivel: Proc. *SPIE*, 2713 (1996) 361.
[21] L.Minko, Yu.Chivel. *J. Optical Technology*. 63, №2 (1996) 154.
[22] J. Groshovsky, High Vacuum Technology, WNT, Warszawa, 1972, p. 300.
[23] LS-DYNA, Keyword User Manual, Version 970, Livermore Software Technology Corporation, 1993.
[24] A. Hrapak, *JETPh Lett*. 47 (1988) 372.
[25] E. Ohmura, I. Fukumoto, I. Miyamoto, *Proc. LIM*, 2005, p. 105.
[26] V.E.Fortov, A.N. Dremin, A.A. Leontiev. *Teplofiz. Vys. Temp*.13.1072-1074. (1975).
[27] M. Martinyuk, Phase Transitions Under Pulsed Heating, *Nauka*, Moscow, 1999, p. 330.
[28] Q. Lu, S. Mao , X. Mao , R. Russo, *Appl. Phys. Lett.* , 80 (2002) 3072-3074.
[29] K.Song, X. Xu, *Appl. Surf. Sci.* , 127 (1998) 111- 116.
[30] R. Stoian, H. Varel, A. Rosenfeld , et al. , *Appl. Surf. Sci*., 165 (2000) 44- 49.
[31] X.Xu. *Appl.Surf. Sci.* , 197-198 (2002) 61-66.
[32] E.Lazneva. Laser Desorption. L. *LSU Publisher*. (1990).
[33] B.V. Derjaguin, N.V. Churaev, V.M.Muller. Surface Forces.M.*Nauka*. (1985).
[34] L.D. Landau, E.M. Lifshits. Statistical Physics. M. *Nauka*. 2000, L.D. Landau, E.M. Lifchits, Theory of Elasticity, Fizmatlit, Moscow,2000, 150.

[35] Yu. Chivel.Adeviceforcalibration of sensors pulse pressure. Patent № 6749BY.2004.

[36] V. Efremov, L. Minko, Yu. Chivel, A. Chumakov, Proc. III Intern. Conf." Laser and Plasma Surf. Interaction ", (1988)172.

[37] V. Nasonov, Yu.Chivel, *Journ. Appl. Spectroscopy.*, 76 (2009) 898 - 906.

[38] C.Yaws .Yaws Handbook of Vapour Pressure. *Gulf Pub. Co. Huston.* 2007.

[39] L.Minko, G. Romanov, V.Nasonov, Yu.Stankevich, A.Chumakov. *Engin. Physical Journal*, 66 (1994) 449-452.

[40] Yu. Chivel, V. Nasonov. *Physics Procedia.* 5(2010) 255-260.

[41] Chandracekhar S. Hydrodynamic and Hydromagnetic Stability. Oxford Press. 1961.640 p.

[42] V.I.Mazhukin, V.V. Nossov, I.Smurov. *Thin Solid Films.* 453-454 (2004) 353-361.

[43] N.Inogamov, *Journ. Exp. Theoretical Phys. Letter.* 4 (1978) 743-747.

In: Laser-Induced Plasmas
Editor: Ethan J. Hemsworth, pp. 145-203
ISBN 978-1-61324-851-5

Chapter 7

PROSPECTIVES OF LASER-INDUCED BREAKDOWN SPECTROMETRY: MORE SENSITIVE, PRECISE AND FLEXIBLE ANALYSIS

***Timur A. Labutin**[*1]**, Vasily N. Lednev**[≠2]** and Andrey M. Popov**[‡1]*

[1]Department of Chemistry, M.V.Lomonosov Moscow State University, Moscow, Russia
[2]Wave Research Center at Prokhorov General Physics Institute,
Russian Academy of Sciences, Moscow, Russia

ABSTRACT

Laser-induced breakdown spectrometry (LIBS) is one of the most promising and powerful methods for direct spectral analysis of different materials. The method is based on the use of the emission spectrum of laser-induced plasma on the surface or in the bulk of the analyzed sample. Processing of spectral data serves to quickly and effectively obtain information on the elemental composition of the target. The main advantages of this technique are the relatively simple instrumentation, the fast qualitative and quantitative determination of the light and heavy elements, the possibility of local and remote analysis etc. This chapter presents an overview of the actual state of art, main achievements and problems of LIBS over the past years. Progress of the laser systems, spectral devices and detectors for performance of LIBS are shortly discussed with respect to the future LIBS applications. Up to date the weak point of LIBS is insufficient sensitivity of the LIBS determination of microcomponents, particularly, in environmental objects. Methods for LIBS sensitivity enhancement, such as double-pulse ablation, a combination of LIBS with laser-induced fluorescence, the use of additional sources of excitation (spark) and confinement of plasma by magnetic field or shock wave and other specific approaches (microchip lasers, microwave excitation etc), are compared with respect to figures-of-merit. Minimal achievable detection limits of the most elements obtained until now are critically considered. The Achilles' heel of all analytical application of laser plasma is bad reproducibility due to laser energy instability and

---

* e-mail: timurla@laser.chem.msu.ru
≠ e-mail: lednev@kapella.gpi.ru
‡ e-mail: popov@laser.chem.msu.ru

absence of local thermodynamic equilibrium. Modern approaches for improvement of LIBS reproducibility are illustrated and thoroughly discussed.

7.1. Introduction

Laser ablation and laser-induced plasma are widely used in analytical practice For analytical purposes removal of a portion of material by a short intense laser pulse is usually called laser sampling [1]. Laser sampling is intrinsically used in laser-induced breakdown spectrometry (LIBS). Moreover, the laser-induced plasma is used as an atomizer and excitation source in this case. LIBS is one of the most promising and powerful methods for direct spectral analysis of objects of diverse origin. The method is based on the use of the emission spectra of laser plasma obtained on the surface or in the bulk of analyzed sample. The information on the elemental composition of the sample is obtained quickly and reliably by processing of spectral data. This is why, American spectroscopist Winefordner with colleagues believe that LIBS is a future "superstar" of analytical atomic spectrometry [2] Physical basis of the method and the most interesting area of analytical LIBS were described in detail in several monographs [3-5]. On the one hand, the main advantages of LIBS are related with the inherent potential of laser sampling, and on the other hand, with the simple construction of apparatus, including the laser, focusing system, spectrograph with detector. As is known, any material regardless of state of matter can be ablated by laser. Therefore, the range of the analyzed objects is extremely wide: from solids such as alloys and ores to gaseous mixtures and liquid inclusions. A small quantity of material (about 1–100 ng) ablated per pulse allows the quasi-non-destructive analysis, the depth profiling with a resolution of ~10 nm and local analysis with 30 nm [6]. Moreover LIBS is almost indispensable for the determination of light elements such as beryllium and lithium. As far as laser ablation of a material does not a time-consuming process, LIBS provides rapid analysis. For example, the quality control system of steel mountings, developed by Noll with colleagues [7], provides an extremely high productivity — about 120 millions continuous analytical measurements per year (approximately 4 s per one item). LIBS provides an opportunity of *in situ* analysis of the environmental samples and monitoring of industrial systems with commercial available miniature detectors and spectrometers. Since laser beam can propagate in any transparent media, LIBS can be used as a lidar technique for the remote analysis of dangerous and radioactive substances or explosives. Analytical measurements can be carried out even without an operator that opens the prospect of LIBS for planetary studies. This allows NASA to prepare rover for a Mars mission, launching in 2011, with onboard LIBS system for the *in situ* elemental analysis of Martian geological samples [8].

Although the number of publications devoted to LIBS increases from year to year and over 400 ones is published per year, including a series of comprehensive reviews and monographs, in our opinion, the main weak points of the method: a relatively low sensitivity and poor reproducibility has not been considered yet. We have tried to fill this gap in this chapter, devoted to the ways of improving the reproducibility (Labutin T.A.) and increasing the sensitivity (Lednev V.N and Popov A.M). At the beginning, a brief overview of the apparatus features in LIBS is also given (Lednev V.N.).

7.2. INSTRUMENTATION FOR LIBS

The unique characteristic of LIBS is the ability to use only photons for analysis. Analyzed target should be "reacheable only by photons" that make this method very perspective for different applications like analysis of samples in hazard conditions or stand-off analysis.

The typical LIBS setup is presented in Figure 1. Main components of any LIBS system are laser (provides photons for plasma generation) and detection system (register plasma spectra). To vaporize sample, generate and excite plasma a powerful pulsed laser is used. Optical focusing system is used to transfer laser beam from laser and to focus it on sample surface (mirror and lens on Fig. 1). Sample is usually placed in special holder that allows easy manipulation with sample. After plasma formation its irradiation is transported by collection system (lens on Fig. 1) to detection system. Detection system performs light dispersion in spectrum and its quantification. Spectrograph is usually used for light dispersion and obtained formed spectrum is digitized by detector (CCD or ICCD). All timing is performed by synchronization electronics (pulse generator) and data processing is carried out by computer.

7.2.1. Lasers

Laser fundamentals are explained in details in excellent book by Svelto [9] and here we will be discussed shortly. One of the most popular laser for LIBS is a solid state Nd:YAG laser. This laser provides a compact, stable and robust laser source. For this type of laser active media is Nd^{3+} ions doped in yttrium aluminum garnet. Active element usually in form of rod is placed inside laser cavity with two parallel plane mirrors. Flashlamp is traditionally used for pumping in most cases due to good functionality and low price. Recently a new source of pumping, diode pumping, has been developed for solid state lasers. Diode pumped lasers provide greater pulse energy and improved reproducibility but more expansive. Powerful pulses are needed to produce plasma thus laser is usually operated in giant pulse regime with Q-switch. In most cases a Q-switch with electro-optical shutter is used. For this type of Q-switch pulse length is determined by speed of shutter open and is usually in region of 5 to 20 ns. This shutter provides high stability of lasing and allows to generate pulses with energies up to 1 J. Other shutters are used rarely in LIBS: acousto-optic and passive Q-switches. First choice was developed for first laser systems and was unable to produce high power pulses that are needed for LIBS. Passive Q-switch is used in LIBS systems episodically due to lower pulse energy (higher lasing threshold) and high instability of lasing jitter. Pulsed solid state Nd:YAG laser is generating at wavelength 1064 nm. Additional wavelength can be easily obtained with fundamental wavelength convertion to higher harmonics (532, 355 and 266 nm) with appropriate nonlinear crystal.

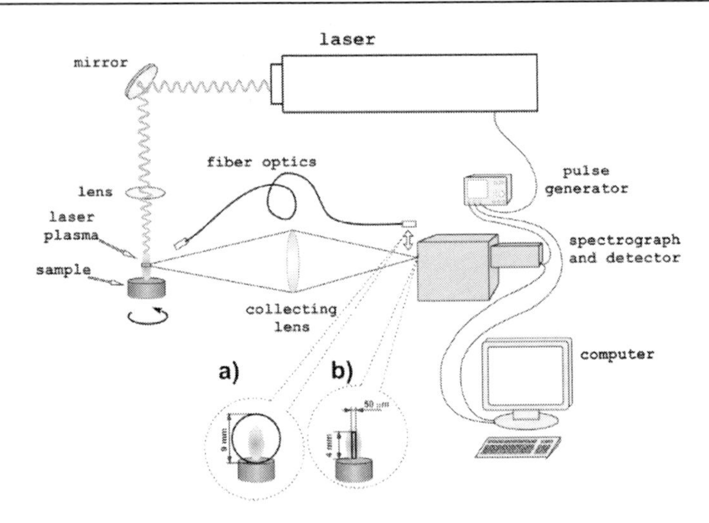

Figure 1. Scheme of typical setup for laser induced breakdown spectroscopy. Two widely used collection schemes are side view (a) and fiber optics (b).

Table 2.1. Specifications of lasers used in LIBS

Laser type	Wavelength, nm	Pulse duration, ns	Pulse energy, mJ/pulse	Repetition rate, Hz
Nd:YAG	1064, 532, 355, 266	5 - 20	1 - 1000	up to 100
Excimer	308, 248, 194	20	100 - 500	up to 200
CO2	10.6 μm	200 – 200000		up to 200
Microchip	1064	<1	<0.1	up to 200k
Fiber laser	1500	200	up to 5	up to 500k
Ti:Sapphire	650 < 1000	50 fs	10	up to 10000k

Several types of laser systems are used in LIBS. They include lasers with UV wavelength and with far IR wavelength. In table 2.1 specifications for laser systems are compared. Comparison of different laser sources have been experimentally performed by [10]. Several laser systems with ultrashort pulse duration have been used in LIBS [11]. It was demonstrated that picosecond (10^{-12} s) [12] and femtosecond (10^{-15} s) [13] laser pulses revealed some perspective results including absence of fractionation. But these laser systems are expansive, sensitive to environment conditions and usually utilized in laboratory. It was proposed to use microchip laser as a portable laser for plasma generation. For such lasers pulses with less 1 ns can obtained with repetition rate above 1 kHz and energy below 100 μJ. Perspectives of such compact lasers for LIBS was studied in [14, 15]. Compact lasers systems (diode pumped solid state lasers) with high repetition rate and low pulse energy was effectively used for analysis of solid samples [16, 17]. A new laser source, fiber laser, was recently introduced for LIBS [18, 19]. These laser systems provide powerful pulses (up to 0.8 mJ/pulse) at high rate (up to 200 kHz) that can be used to improve analytical capabilities of LIBS. Fiber laser systems are good candidates for compact robust laser source in portable LIBS systems.

Properties of Laser Irradiance

Laser irradiation is differed from traditional light sources with its properties that include coherence, directionality and coherence. These properties established a great number of applications for lasers. Properties of laser irradiation important for LIBS includes wavelength, pulsed energy and focused power density. Directionality of laser beams is defined by angle of divergence and is better than few milliradian for solid state laser and several hundreds time smaller for gas laser systems. Divergence is important factor due to capability to focus laser beam in small spot and obtain high power density. This ability of laser pulse is used to transfer energy on large distances for stand-off analysis. Pulse duration is important charachteristics since it determines power density that is important factor describing laser sampling. It is possible to produce laser pulses with short duration in range from tens of nanosecond to several tens femtoseconds. Laser systems usually provided pulses with small energy but due to pulse short duration a power of laser pulse can be extremely high. For example, for solid state laser with 20 ns pulse and 200 mJ/pulse a power will be equal 10 GW. In combination with focusing on sample surface power density can reached high values and any type of sample can be atomized. Laser wavelength is a charachteristics of primary importance for nonconducting materials (glass, cersmics) but for metal samples is not significant. For example, laser sampling of steels is wavelength independent but for ceramics a UV laser should be used to produce suitable plasma for analysis [20]. UV pulses have higher absorption coefficient for ceramics and more material is vaporized and excited in plasma. Laser wavelength can be easily converted to another by generation of higher harmonics. Efficiency of conversion is determined by nonlinear crystal type and power of laser pulse.

To produce plasma a definite power density is needed. For qualitative comparison a term irradiance or energy power density is generally used. The value irradiance on sample surface is determined by laser pulse energy devided on pulse duration and spot size. The unit of energy power density is W/cm^2. Laser spot can be easily changed by adjusting lens to sample distance but smallest spot is depended on laser beam quality. Laser beam quality can be determined by different ways but generally beam parameter product is used. Beam parameter product is usally derermined with beam quality factor M^2. According definition by ISO [21] that is a measure of difference of studied beam profile compared to diffraction limited Gaussian beam . Gaussian beam provide smallest spot and its beam parameter product is equal 1. All other beams are compared to Gaussian thus $M^2 > 1$ and smallest spot dimensions are greater than for Gaussian beam.

The laser beam quality is determined by laser cavity scheme and preferred lasing mode. In most cases a cavity scheme with stable resonator formed by plane mirrors is used. This cavity produce a stationary configuration of electromagnetic waves which are called modes. Two types of modes are known for laser cavity, longitudinal and transverse modes. First type is determined by length of cavity and for solid state laser a great number of longitudinal modes with tiny difference in wavelength will be produced. Second type describes cross section of beam profile. Transverse mode with lowest numbers (TEM00) can be described with Gaussian profile. Higher modes (TEM01, TEM10. etc.) have complex profiles. Usually, multimode lasing is allowed in order to provide high power pulse. Beam profile is a sum of different modes and is called multimode beam. This lasing mode is characterized by instability of lasing, energy fluctuates between different modes thus beam with high energy in pulse but with low reproducibility of energy and beam profile is obtained. To improve laser beam quality another lasing mode with single mode can be allowed to generate. Diffraction losses of high modes greater than for low modes and to produce single mode lasing a diaphragm with small size is placed inside cavity. Stable lasing with Gaussian beam profile and improved reproducibility is achieved in such case. But pulse energy is decreased for more than 10 times compared to multimode lasing. To provide high quality and high energy laser beam usually single mode lasing is used for resonator to obtain high quality beam and amplifiers are used to get higher energy.

Lednev et. al. [22] studied lasing mode influence on analytical capabilities of LIBS. They observed better precision with comparable sensitivity for Gaussian beam (single mode lasing) compared to multimode beam despite ten times lower pulse energy. Chalear et al. [23] indicated that stability of analytical signal can be increased if only central part of inhomogeneous multimode laser beam from excimer laser is used.

Double Pulse Operation

It was observed that two sequential pulses with microsecond delay can substantially increase signal and improved sensitivity for LIBS [24]. This technique is called double pulse method and is widely used. Two pulses can be obtained with single laser system or with two different lasers. First variant is easy to adjust but is difficult to change delay between pulses and energy. Second choice allows to vary properties of laser pulses in wide range (energy, delay, wavelength etc.) but more difficult in adjustment. A pulsed Nd:YAG laser can be used to obtain two or several pulses. Flashlamp pump pulse duration is usually several hundreds microseconds. If active Q-switch is opened with Marx bank two times during flash lamp pump than two pulses can be provided. First time it is opened at beginning of pumping pulse and after short period second trigger is used to generate second laser pulse. Delay between pulses must be greater 20 μs to achieve sufficient energy for second pulse.

7.2.2. Optical Systems

Focusing of laser beam on samples surface is usually made with lens. Typically focal plane is about 50 to 200 mm but for low energy pulses lens with shorter focal plane is used. In some cases it is useful to deliver laser beam by fiber optics. In such case laser ligh can be transferred by fiber and then focused after fiber output if power pulse can damage fiber. For fibers with high damage threshold laser light can be focused on fiber entrance and than output of fiber is directed on sample. Collection of plasma light is performed by optical system and transferred to detector. Several different systems are used to collect plasma light. In simplest case single lens can be used to collect plasma emission. Any single lens will suffer from chromatic aberration and to eliminate such problem a specially designed system of several lenses with differentially directed aberrations is used. Second way to escape chromatic aberration is to use a mirror that provides single focus for different wavelengths.

There are two schemes used for collecting, side-view and top-view scheme. For first case plasma radiation is projected by optical system (lens or telescope) side by side to detection system. This system allows to obtain signal from various plasma locations and to receive spatial resolved study of plasma. Alternative scheme top-view is also called backscattering scheme. This scheme uses same optical way to focus laser beam and to collect plasma radiation. For example laser pulse is passed through a pierced mirror and focused by lens on sample and same lens is used to collect plasma light and than parallel beam is reflected by pierced mirror. Second lens is used to collect parallel beam from plasma on spectrograph slit. In some cases a fiber optics cable is used to transfer plasma light to spectroscopy system.

Plasma radiation can be focused by lens to fiber input and then transferred to detector. In simplest case fiber tip can be directed few mm above plasma to collect radiation.

7.2.3. Spectra resolution

Spectra of laser plasma contain continuous emission and characteristics lines for plasma species (atoms and ions). These lines are used for analysis of sample composition and for temperature and electron density determination for plasma. Different elements have unique lines but laser plasma is a source with high temperatures and high particle density thus lines can be overlapped due to line broadening. Several spectral resolution methods are used in LIBS. Usually spectrograph system is used to obtain spectra but spectral filters also can be used to monitor selected lines from plasma irradiation.

Optical filters

Color filter transmit only interested spectral region and is used for plasma imaging or to decrease scattered light in spectral resolution system. Long pass edge filters transmit irradiation only for wavelengths greater or lower than edge wavelength and such filters are usually used to remove second order lines in spectra. Band pass filter pass radiation inside some region that can be used to select some lines. If spectral region of interest is near laser radiation than a notch filter can be used to deplete laser line and to detect nearby lines. For this type of filter a wide range of spectra is transmitted and small spectral range (1 - 5 nm) is reflected with several orders of reduction.

The acousto-optical tunable filter was used in several studies to select studied lines. This device contains a crystal transparent to studied radiation and a high frequency driver on one plane. This driver produces high frequency modulation of sound waves in crystal that formed a diffraction grating. Interested spectral region is reflected by this grating in spectral window about 5 nm in range 200 – 4000 nm. Efficiency is about 80% for reflected beam. Spectral range is chosen by changing high frequency of driver that changes a stationary sound wave in crystal and grating.

Spectrograph

To resolve irradiation in spectra a spectrograph is used in majority of cases. The spectra resolution element in most cases is a diffraction grating. Several systems of spectrograph are used: Czerny – Turner, Paschen - Runge, Littrow and Ebert - Fastie. The most popular scheme of spectrograph in LIBS is Cherny – Turner scheme. In this scheme light is passed through input slit and reaches mirror that collimated it and directed it to diffraction grating. In ideal case all area of diffraction grating will be covered with radiation to increase spectral resolution. Than diffraction grating resolves light in spectra and reflects it to second mirror that focus it at output plane. Detector is placed at output focal plane and is used to digitize spectra. Cherny – Turner spectrograph is popular due to high resolution in combination with compact size.

Convention spectrograph can be used to detect emission in spectral window determined by focal plane of spectrograph and size of array detector. Echelle spectrograph was suggested to detect spectra in wide range during single measurement [25]. The diffraction grating named echelle is a central part of this device. This grating has low lines on millimeter but works at high angle and forms a dispersed spectrum at high diffraction orders in short spectral ranges. All orders of grating are overlapped and a second dispersive element, like prism, is used to resolve this orders. As a result a complex picture called echelleogram is achieved at focal plane of spectrograph with one dimension for wavelength and second dimension for different orders. This picture is digitized by detector and than processed by software to obtain high resolution spectrum in wide window. Usually a spectrum in range 200 – 800 nm is detected for single measurement with high spectral resolution.

7.2.4. Detectors

Several types of detectors are used in LIBS depending on study purpose. Detector choice is depend on spectral range and needed time resolution. First studies of LIBS were performed with photomultiplier tubes. These devises are very sensitive and have ns time resolution but can be used only with monochromators. Different covering of photocathode allows to detect light from UV to far IR. The photomultiplier tubes were very popular but now are rarely used due inability to detect part of spectra during single measurement. Nowdays for spectra detection a different array type devices are used (photodiode arrays, CCD, CMOS etc.). Photodiode arrays consists of several hundreds photodiodes stacked in linear array and signal is digitized by sequential reading of diodes.
The most popular detectors for LIBS are CCD and ICCD. Coupled charge device (CCD) is a two dimension array of light sensitive elements that is called pixels. CCD diagonal can be from 1/4 to 4/3 inches and can be up to 6144x4608 pixels. Iradiation is absorbed by photosensitive area of pixel and generates charges in pixel. After signal aqusition all charges are read sequentially by following scheme. For first row these charges are shifted to special pixel line called shift register and than charge from each pixel is digitized by analog-to-digital converter sequentially. After all pixels are read from this row than next all charges in rows are shifted down seqentially in array and second row is reached shift register and digitized. All pixels will be digitized in this way and a two dimensional picture will be formed. If signal level is too low than pixel charges can be united in analog form before digitizing and that procedure is called pixel binning. This procedure is used to decrease digitizing noise and to reduce time needed to read all pixels. In simplest case pixels in column can be binned to produce higher signal and decrease time of reading. CCD has several advantages compared to PDA arrays. First advantage is to produce two dimensional picture that is used for studies along entrance slit and to detect picture for echelle spectrograph. However dynamic range of CCD is one order smaller than for PDA. CCD is suffered from effect called blooming when high intensity light produces too many charges that can flow to near pixel in row. In this case it is not possible to detect level of light for nearby pixels. Another type of two dimensional arrays is CMOS detector. This detector is differed by pixel design i.e. amplifier and digitizing devices are stored in each pixel. Consequently it is possible to read each pixel independently that is very attractive to detect low intensity light near pixel with high level of signal. CMOS sensor is characterized by smaller photosensitive area and it has lower sensitivity than CCD detector. Additionally linear dynamic range of CCD is better compared to CMOS. Readout time of CCD or CMOS is about several milliseconds and to perform time resolved studies in microsecond time range a fast shutter should be used. Optoeletronic device called multichanel plate is used to provide a fast shutter for time resolved studies. In combination with CCD

detector this device is called intensified CCD (ICCD). The device allows to achieve short exposure time from few nanoseconds to several seconds. Additionally this type of device has multiplication of 10^3 for incident light that can be used in low light measurements. Different detectors have been compared for LIBS analysis and it was found that despite differences in light sensetivity for CCD and ICCD analytical capabilities were better for CCD [26].

7.3. Reproducibility of LIBS Measurements

The performance of any atomic spectrometric method is characterized especially by both the precision of the analytical procedure and the accuracy of the results. Quantitative analysis in LIBS involves the emission intensity of an element in the plasma relating to the concentration of that element in the target (solid or liquid sample). The intensity of the analytical line for the determination of the compound of interest can be presented as follows:

$$I_i^X \sim f \cdot \eta_X \cdot c_X \cdot m_{abl} \cdot e^{-\frac{E_i}{kT_{exc}}}, \tag{1}$$

where f is the efficiency of emission collection, η the efficiency of atomization, m_{abl} the ablated mass, T_{exc} the excitation temperature, E the energy of upper level. Any random changes of these parameters can strongly affect reproducibility. The variation in size, shape and spatial position of laser torch from pulse to pulse influences the portion of collected light undoubtedly. Another problem arising from laser sampling process is the element fractionation, a situation in which the composition of the atomic vapor analyzed does not represent the bulk sample [27-29]. In addition, laser microanalysis is rather sensitive to inhomogeneity of the sample. In turn, the analytical signal intensity of an element in the plasma is related to the ejected sample mass depending on laser radiation parameters (laser energy and focusing). The laser plasma characteristics, including the variation of its temperature and the interaction of the plasma with the sample (some of the post-breakdown phenomena), also produce an influence on analytical signal [30]. Moreover, the variations of the sample matrix can affect the analytical signals even more as a consequence of the induced change of plasma characteristics. In this section experimental efforts and approaches to the production, measurement and processing of analytical signal in order to improve reproducibility in LIBS is considered. Firstly, influence of experimental conditions (parameters of laser radiation, atmosphere, set-up assembling etc) on reproducibility will be overviewed. Then the use of an internal standard originating from classical spectrochemical analysis will be discussed. Finally, the several approaches to data processing based on the selection only statistically significant data from initial measured dataset will be briefly described.

7.3.1. Experimental Arrangements for Improving Reproducibility

To evaluate all sources of errors in LIBS the processes of laser ablation will be considered in stages in short. The interaction of laser radiation with matter begins with the absorption of light by the sample surface. As a result, the electron temperature of the surface layer of matter reaches a value ~ 2500 K for the time in 10^{-14}-10^{-13} s. It is believed that the phonons of the crystal lattice of the matter does not absorb the energy of laser photons [31]. The absorbed energy in a time of 10^{-12}-10^{-11} s is thermalized and, as a result, the surface layer of the sample is strongly heated. Accordingly, the so-called *thermal model* of laser-induced solid heating and ablation adequately describes the interaction of matter with the most common nanosecond laser pulses [32]. Fast local heating of the material leads to a cascade of laser-induced processes: the emission of neutrals, ions and electrons, the melting and the evaporation of the material, the generation of acoustic waves, the thermoelastic deformations and the thermal destruction of the sample, the formation and expansion of the near-surface plasma, the interaction of laser radiation with the newly formed plasma and its heating up to temperatures of ~ 10^4 K [32-35]. The expulsion of melt from the crater, being accompanied by the injection of the drops into the plasma and the formation of the rim around the crater, occurs under the pressure of vapor when the irradiance on the metal surface reaches the values greater than 10^6 W/cm^2 [33, 34, 36]. With further increase of irradiance up to 10^9 W/cm^2 the explosive boiling with the generation of the particles of different sizes is observed [36].

The use of shorter (sub-picosecond) laser pulses causes the features of the ablation kinetics which cannot be described within the framework of conventional thermal model. Therefore, the *two-temperature model* of laser ablation was proposed to describe the transient phenomena in a nonequilibrium electron gas and lattice during subpicosecond laser exposure [31, 32, 37]In accordance with it, the lattice temperature at the beginning of the absorption is significantly different from the electron temperature. In other words, a few tens of femtoseconds in the surface layer of the sample thickness of 1-100 nm there is a "hot electron gas" and the "cold lattice", which leads to an explosive ejection of hot electrons with the simultaneous destruction and removal due to Coulomb forces "cold" material of the crystal lattice [32, 37]. In this case, the plasma is formed directly from a solid substance without melting the material. The absence of melting is confirmed by the lack of the rim around the crater [38, 39].

It is evident that many processes, within the thermal and two-temperature model both occur under nonequilibrium conditions in the lifetime 10^{-6} to 10^{-4} sec of the laser-induced plume [34]. This, apparently, causes the fact that existing models of laser ablation do not allow to predict such important analytical characteristics for the specific impulse as the mass of the evaporated material, the number of different kinds of particles in the laser plume and their relative portion, the temperature of laser-induced plasma. They are influenced by the parameters of laser radiation and its focusing, composition and pressure of the surrounding atmosphere, the material properties [32, 33]. Below the ways to reduce the influence of various processes on the reproducibility of the analysis will be considered.

Solidified slurries around the crater associated with the expulsion of molten material by gas-steam cloud [40]. Accordingly, the solidified melt are eliminated in the absence of fusion in the case of laser ablation by ultrashort pulses or refractory materials [41]. The cracks are formed as a result of thermoelastic deformations during ablation of glass or ceramics [42]. There are the traces of decomposition at evaporation of thermally unstable substances, such as polymers [43]. The creation of deeper and deeper craters by successive laser shots could induce errors in the quantitative depth profiling analysis by LIBS, especially in those analyses leading to the generation of significantly deep cavities [44]. Moreover, the trend of LIBS signal may be complicated as the decreasing power density is compensated by enhanced plasma confinement as the crater deepens. Therefore, the set of gate delay and gate width, properly adapted to the crater-depth dependent behavior of the emission intensity, is needed to produce the desired depth independent emission data [45]. Explosive boiling of molten metal forms the complex shape of the crater bottom with the irregularities up to 0.5 microns [33, 34, 36]. The use of a femtosecond laser to evaporate the sample in an inert atmosphere at low pressure yields craters regular and, more importantly, reproducible form [38]. Furthermore, the absence of heat transfer in the sample volume and laser plasma shielding provide a decrease in the fractionation of elements during evaporation and dramatically reduce the error of analysis. Displacement of the wavelength of evaporation laser in a vacuum UV range can reduce the depth of a crater, and the roughness of the crater bottom [46]. Direct comparison of the influence of IR (1064 nm) and UV (355 nm) lasers on the analytical performance demonstrated the best reproducibility with the use of 355 nm laser in single-pulse mode and the combination of the 532 nm radiation for the first pulse and the 355 nm radiation [47]. If the higher number of elements involved in the ablation process of the alloy (e.g. bronzes) a more complex behavior exists during ablation. The plume composition for ultraviolet light in bronzes is in any case closer to the one of the target with respect to the one obtained with IR, but always remains too rich in zinc content [48]. Thus, ultrashort UV laser ablation is the direct and most effective way to ensure the reproducibility of the size of the crater. Nevertheless, under these conditions the mass of the ejected matter is fluctuated strongly from pulse to pulse [38]. Obviously, the beam profile affects the process of ablation and crater formation. Usually the use of Gaussian beams or flat-top (collimated) beam profile (in the case of local microanalysis) is considered preferable to provide craters with the most accurate and reproducible form. Generally, these expectations were confirmed by the quantitative comparison of the analytical performance of different beam profiles. Gaussian laser beams produced more uniform craters than beams with hot spots in their profiles. The plume particles would be less heterogeneous in terms of temperature and velocity. The plume would disperse slower, fencing in the analytes to emit more brightly for a longer time. The resultant signal-to-noise ratio of the analyte signal was 2 to 3 times better for Gaussian beam [49]. It is preferable to use single mode Gaussian beam if only one shot at a sample surface is possible to achieve (stand-off analysis or analysis of movable objects) due to better precision can be achived. Although the lower LODs were obtained for the multimode beam, a degrading of analytical capabilities was observed with the energy equal to the Gaussian beam. Thus better sensitivity obtained for multimode beam was attributed only to higher energy of this beam [22].

Focused laser beam may overcome the threshold value of fluence of the breakdown in atmosphere before the focus position, especially if some aerosol particles are irradiated incidentally. This is explained easily by the depth-of-focus typical size which has the order of millimeters for the conventional lasers and focusing system. In many works the focal point is set below the target surface to eliminate the moving of the breakdown position. Obviously the optimal position of focus inside sample depends on focusing system and material (the threshold value of breakdown): 5-15 mm for steels [50, 51] and only 3 mm for coal [52]. Such focusing configuration was found to provide a stable breakdown, reducing the RSD value from 15-42 % to 1-4 %. If breakdown in air above is excluded by focusing condition the near-surface plasma torch expansion is in the direction perpendicular to the sample surface. The coaxial arrangement of emission collection system and laser torch simplifies the experimental set-up and improves reproducibility due to small changes in the plasma-to-lens distance when crater is deepened by consecutive shots at the same point on the sample surface or some roughness are presented on the tilting target. The plasma displacement has minimal influence on the LIBS signal, because of the depth-of-focus of the detection system is also typically in millimeter's range that is greater than crater depth. This configuration is convenient for performing on-line or remote analysis [53]. It is known known that laser produced plasmas are spatially and temporally inhomogeneous media [4, 30, 34]. The shot-to-shot variations in the atomic densities and temperature distribution inside the plasma could affect the reproducibility when only the central part of the plasma is probed. Chaleard et al. suggested to improve the reproducibility of the measurements by focusing the whole volume of the plasma onto the entrance slit of the spectrometer [23]. On the other hand for depth-profile analysis such configuration can strengthen the tailing affect and, consequently, reduce precision and reproducibility. The tailing effect can be defined as the contribution of the upper layer signal to the depth profile graph when only the lower layer signal should appear. Čtvrtníčková with colleagues [54] suggested a special optical device (pinhole) to improve the depth profiling capability of LIBS by reducing the contribution of light emission from the borders of the plume.

The ambient media influences the shape and the parameters of laser-induced-plasma and consequently their reproducibility greatly. Due to its relative low ionization potential argon atmosphere causes fast ionization and plasma ignition and, consequently, sufficient plasma shielding coupled with the increase of temperature and electron density, helium, conversely, produces the lower temperature, electron density and emission intensity. Lee with colleagues [55] demonstrates that with a reduction in the pressure under air or argon atmosphere, the position of maximum intensity of plasma emission moved away from the target surface. However, in the case of helium, the position of maximum intensity changed little from a reduction in the pressure, moving a little closer to the surface under the low pressure At the same time temperature does not change greatly with atmospheric pressure of argon and air and a reduction till 50 Torr. The temperature of plasma had a maximum in helium atmosphere at 200 Torr, and significantly decreased at low and high pressure. The use of helium is preferable to prevent accidental changes of the amount of light from laser plasma entering the spectrograph. However, in this case special attention should be pay to correct the possible temperature fluctuations, for example, with the use of an internal standard. Salle with collegues [56] demonstrated that the the RSD is lower for the minor elements (except for Sr) at atmospheric pressure in air than at 7 Torr of CO_2. This was related with greater number of each species ablated due to reduced plasma shielding, resulting also less shot-to-shot perturbations of the pulse energy reaching the surface. At the same time the RSD is lower for the major elements at low pressure. This may be related to reduced self-absorption and line broadening effects of major species at the lower pressure. The redeposition of ejected material may be the source of additional errors in the case of ablation in the same spot. It was demonstrated by Vadillo et al [57] that craters formed in a vacuum are completely free of material deposited on the rim, while at atmospheric pressure the deposited material at the top edges of the crater created a ring around the tip that was clearly visible. The same group found that ablation rate reaches a plateau at 200 mbar for every metal (Ti, Al, Fe) and stay stable till 0.3 bar, regardless of the fluence [58]. So it is evident that the reduced pressure can provide the preferable conditions for reproducible results.

Another source of noise from plasma is its interaction with an aerosol generated during laser ablation. The time-resolved measurements of laser-induced plasma emission revealed a general increase in localized plasma temperature with increasing delay time, which is attributed with an initial suppression of plasma temperature about the aerosol particles as plasma energy is required to vaporize and ionize the aerosol particle mass [59]. Clusters are formed in the laser torch as a result of condensation of the vaporized material [32, 60]. And with it, the evaporation of material by short-wavelength ultraviolet radiation decreases the average size of clusters and their total amount [62]. The average size of clusters produced by the femtosecond evaporation of bronzes in 2-3 times less and the amount of particles greater than 5-10 times those for the nanosecond evaporation [63]. This reduces the fractionation during the evaporation of matter from the droplets of the ejected material [64, 65]. It is obvious that the using femtosecond pulses or uv laser also diminishes the influence of fractionation on analysis result and the error will be smaller in this case.

Early stages of plasma evolution are characterized by the Bremsstrahlung continuum emission, and only several hundred nanoseconds later the characteristic spectra of atomic lines are revealed, which allows qualitative and quantitate analysis. Although the continuum emission hides the atomic lines and looks like a noise, it corresponds to a plasma emission and it cannot be separated with the statistical methods. Since the background emission decreases the contrast of observed atonic lines and thus worsens signal-to-background and signal-to noise ratios, it is necessary to minimize this kind of radiation by choosing appropriate delay after the laser pulse for spectra acquisition. However, depending on the ambient gas pressure, laser energy and other factors, the plasma lifetime may vary from 200-300 ns to more than 40 ms. The optimal gating parameters providing the best signal-to-noise ratio due to high line intensity and low background must be determined case by case. Additionally, matrix effect for aerosol particles is attributed primarily to perturbations in the localized plasma properties. These perturbations are minimized at longer plasma delay times; hence quantitative LIBS analysis of aerosol particles should be performed with careful attention given to the temporal plasma evolution [59].

7.3.2. Internal Standards

The majority of precise quantitative analytical techniques using emission spectra are based on some variant of the internal-standard principle first enounced in 1929 by German experimental physicist Walther Gerlach [66]. It was most important in elevating the spark optical emission method of spectrochemical analysis from semiquantitative to quantitative analysis status. The ratio of the intensity of the analytical spectral line to the intensity of a spectral line of another component of the material should be measured instead of considering the absolute intensity of the spectral line. Despite the change in measured values, the intensity ratio remained almost the same. Two types of internal standard lines can be used [67]: (i) a weak (nonsensitive) spectral line of the major component of the sample; (ii) a strong (sensitive) line of an element that is not initially present, but added to the sample. W. Gerlach was also the first to call the spectral line pairs above mentioned homologous if they worked well together to compensate for changes in excitation conditions. It is important to note that not all the line pairs of an analyte and matrix are homologous. What defines a pair of homologous spectral lines? Unfortunately no set of necessary and sufficient conditions has been formulated until now. It is important to state, however, that their excitation potentials must be similar (see Section 7.3.2.1) and their intensities differ no more than 10 times.

So in analytical chemistry a typical internal standard is considered as an element, a definite quantity of which is added to the analyzed sample (for powder and liquid samples). For solid samples an element of known and steady concentration can be used as an internal standard. Usually the matrix element is suitable in the case of direct analysis of solids. From practical point of view in spectrochemical analysis the ratio of line intensities of element sought and a reference element (internal standard) should depend only on the fractional concentration of analyte. There are many works on laser ablation applications in spectrochemical analysis in which a wide range of experimentally measured signals (total plasma emission, acoustic wave, background emission etc.) is considered "*as an internal standard*". To avoid any confusion we will use the term "internal standard" in its common meaning.

The internal standard method was applied to improve reproducibility even in the first publication on LIBS [68]. Emission signals of the line pairs Ni 3414.77 Å/Fe 3443.79 Å and Cr 3445.52 Å /Fe 3443.79 Å were measured and the average log intensity ratios plotted against the known corresponding nickel and chromium percentages in the analyzed stainless steel samples produced the analytical curves. To improve precision in LIBS, the method of internal standards is still widely used because the analytical signal is proportional to the population of the analyte excited states and therefore very sensitive to the temperature fluctuations (especially in laser micro plasma). Better precision through the routine use of internal standards was demonstrated in many papers on LIBS, e.g. the internal standardization method using a Fe emission line provided a better correlation between As concentration in soil samples and the double-pulsed LIBS signal than the use of any other element (R changed from 0.469 to 0.954 and soil matrix effects were minimized) [69]. The concept of internal standard or intensity ratios may be useful not only for improving of reproducibility but also for identification of explosives [70, 71] and elemental analysis of organics [72, 73].

7.3.2.1. Criteria for Selection of Reference Line

It is commonly assumed the presence of local thermal equilibrium (LTE) to describe laser plume in LIBS. In this case the material balances of Boltzmann, Saha and Maxwell are in equilibrium and atoms, ions and electrons have the same temperature. At this stage of equilibrium thermodynamic properties have to be specified locally and instantaneously. Although matter and radiation cannot be described with the same temperature, the excitation kinetics in LTE is determined by material particles and is specified by the equality of the temperature of materials particles and the “excitation temperature” [74]. In what follows it is implied when "the temperature" will be mentioned. For transition occurring between two levels of an atom or ion in a homogeneous optically thin media, the observed intensity is given by:

$$I = f\frac{hcngA}{4\pi\lambda Z}e^{-\frac{E}{kT_{exc}}}, \tag{2}$$

where n is the total quantity of species, A the transition probability for spontaneous emission, g the statistical weight of quantum state, λ line wavelength, Z the partition function. If the intensity ratio of a spectral line emitted by an analyte (A) to that emitted by the major component as reference one (B) is taken, then we have:

$$\frac{I^A}{I^B} = \frac{n^A g^A A^A \lambda^B Z^B}{n^B g^B A^B \lambda^A Z^A} e^{-\frac{\Delta E}{kT_{exc}}} \tag{3}$$

The intensity of the element selected as an internal standard is usually used to compensate pulse-to-pulse variations of analytical signal or matrix effects. It is clear from Eq. 3 that transitions with a similar upper state energy should be used to neglect the intensity ratio dependence on temperature. There are some additional limitations for using internal standard in LIBS. Variations of temperature and electronic density due to plasma inhomogeneity can make worse the normalization. So it is preferably to use emission signal from the same small region of plume and species in the same ionization stage [75]. The factor of selective evaporation of elements, obviously, needs to be considered. Practically, this is quite difficult to satisfy all above mentioned requirements, even to find an appropriate pairs of line with close upper levels due to small instrumental spectral window or saturation of measured matrix line intensities due to relatively small dynamic range of detector.

To overcome the influence of plasma temperature Loge [76] suggested the step-by-step normalization involving determination of the instantaneous plasma temperature where (i) the temperature is calculated, (ii) the relative population of the atomic level of internal standard is obtained at this temperature and (iii) finally the measured emission from the sought atomic species is normalized using the relative population of the atomic level that gave rise to the emission line intensity which is calculated from the temperature. The example of gases analysis using nitrogen as internal standard is given, although the method appropriates both gas and solids. Application of this approach can significantly simplify selection of reference and analytical line and improve effectiveness of internal standard. Similar methodology was suggested by Mukherjee et al. [77] to control stoichiometry of oxidative coating of aluminum nanoparticles. The plasma temperature was calculated by emission atomic lines of ambient gas (Ar), then the population densities of oxygen, aluminum and argon (internal standard) species were determined by Maxwell–Boltzmann equation. Ratios of these values were calculated to determine stoichiometry of oxidized nanoparticles.

To neglect electronic density influence on lines intensities ratio it is necessary to use the same ionization stage of the analyte and the internal standard. To consider the influence of both main plasma characteristics (excitation temperature and electronic density) in the calibration model Panne et al. [78] proposed a normalization procedure using Saha–Boltzmann equilibrium relationships. From general relation between line intensity and total amount of species the following equation for ratio of element densities in plasma can be derived:

$$\frac{n^A}{n^B} = \frac{A^B g^B}{A^A g^A}\left(\frac{\lambda^A}{\lambda^B}\right) e^{-\frac{E^A - E^B}{kT_{exc}}} \frac{{}^qZ^A(T_{exc})}{{}^qZ^B(T_{exc})} \frac{{}^qW^B(T_{exc}, n_e)}{{}^qW^A(T_{exc}, n_e)} \frac{I^A}{I^B}, \tag{4}$$

where n_e is the electron density in the plasma, q ionization stage, ${}^qW^X(T_{exc}; n_e)$ the probability for occurrence of the ionization stage q, ${}^qZ^X(T_{exc})$ the partition function of the quantum states of the particular state of ionization, I the lines intensity. The ratio between the ion and atom density of element in a multicomponent plasma was derived for a given ionization stage using the Saha–Eggert equation. Atoms and singly charged ions were considered because only these species usually can be found in laser plasma at the time of observation. The electron density was determined by measurement of the Stark-broadening of the Mg I 285.213 nm line. The temperature was determined *via* the Boltzmann plot for the Mg II line series. For compensation of changes in plasma temperature and electron density, all line ratios intensities were corrected according to Eq. 4. The pulse-to-pulse variation decreases from about 14% to approximately 5%. Linearity of calibration curve plotted for element ratio was significantly improved. (R^2 was increased for Mg II/Si I ratio from 0.63 to 0.94).

The utility of internal standard depends to a large extent on the selection of appropriate reference line in all its application and modifications. There is the general criterion (thermodynamic) of closeness of upper level energies above mentioned, but really it describes requisite condition resulting from excitation of atoms in laser plasma. Because of the transfer of species from condense phase into the plasma, an ambient atmosphere may influence analytical signal more than excitation conditions in plasma. Though the thermodynamic criterion is enough in many cases and provides a good result with the selected line of matrix element, e.g. in Ref. [79, 94] and many other works discussed above. Usually this statement refers to the industrial applications of LIBS, when close group of objects under investigation and internal standard are selected for each element of interest [14, 51, 80, 81, 92]. Using commercial prototype of an instrument for the rapid determination of magnesium stearate (MgSt) distribution within and between tablets, it was possible to perform accurate quantitative analysis of MgSt at the 0.5% level [82]. In some cases internal standard improves figures-of-merit even without fulfilling thermodynamic criteria [73, 83], while in others the final result is worsened [84]. Unfortunately there is a very lack of investigations on establishing criteria for internal standard selection. In Ref. [85, 86] the internal standard was chosen by comparing calibration curves of the same element obtained using different internal standards and emission lines. The pairs giving the best linearity of calibration curve were used for analytical measurements. However, this is very time-consuming procedure and the set of reference materials is needed.

Application of internal standard means that intensities of reference and analytical line are both related to the fluctuated parameters. To avoid strong definition of this relation based on the model of ablation process (that is practically impossible), the efficiency of the diminishing of the fluctuations influence on calibration can be studied by the correlation strength between intensities of analytical and selected reference lines. We will try to demonstrate the hypothetical cases of the use of two different lines as the reference ones with simulated data. These data presented in Figure 2 were obtained by random number generator, but it should be referred neither to any specific spectral lines nor to any experimental conditions. The raw calibration graph (Fig. 2,a) is not suitable for analysis. If the reference line is poorly correlated with analytical line intensity (R set to 0.79 by the simulation), we will get no calibration graph improvement (Fig. 2,b). And vice versa, if another reference line (R set to 0.97) is chosen, normalization by the intensity of this reference line provides good calibration graph (Fig. 2,c). There are several papers where correlation between signals was used for internal standardization. Boyain-Goitia et al. [87] suggested special procedure which based on the use of correlation plots for referencing the elemental line intensities to the carbon matrix of the sample represented by molecular bands. The intensities of selected molecular and atomic peak pairs were plotted against one another from each individual recorded single-pulse LIBS. The intensity correction factors were calculated from these curves to normalize spectral data (both molecular bands and atomic lines). This procedure allowed determination of the relative element concentration distributions in the different types of pollen (bioaerosols). Zhang et al. [88]selected the optimal analytical O I lines for calculation of the O I/N I ratios by direct study of the correlation between the spectrum and the N I line at 746.8 nm previously selected as a reference line. For each pixel of spectrum, the correlation value between the corresponding signal intensity and the N I line intensity was obtained through 1500 plasma spectra. The peaks with correlation coefficients larger than 0.95, were chosen (three O I lines). From the quantitative results, the accuracy of organic oxygen measurement in coal was estimated to be in the range of 1.15–1.37% while the average absolute relative error was 19.39%. Thus, the study of correlation between analytical and reference lines, molecular bands or any other signals can provide the easy and accurate choice of normalization strategy that gives assured improvement of the figures-of-merit. Another advantage of "correlation approach" is that no spectral line parameters (even wavelength) are needed and only experimental data are involved in the reference line selection.

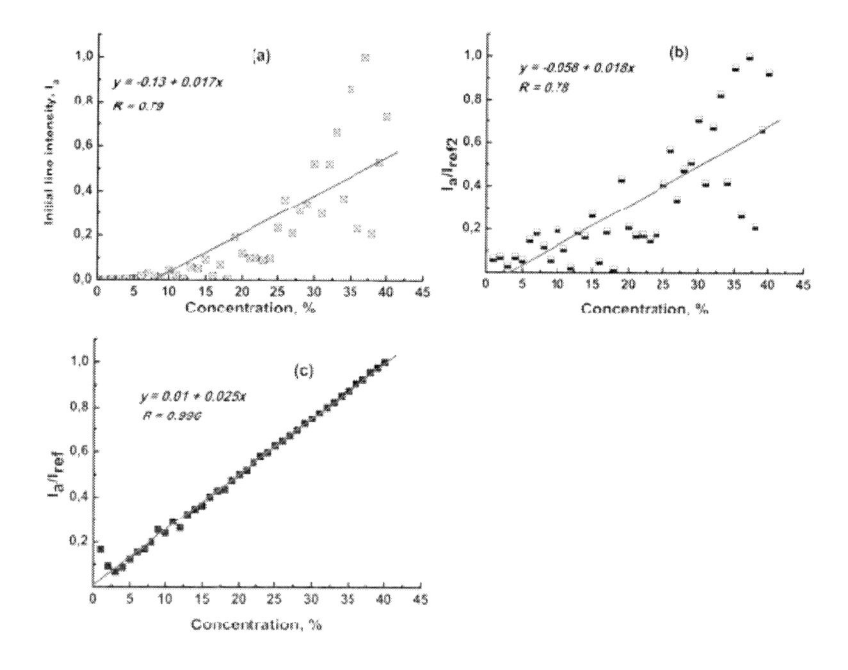

Figure 2. Hypothetical cases of the use of two different lines as the reference ones: a) raw calibration graph; b) calibration graph obtained by normalization on intensity of the reference line having a bad correlation with analytical line intensity; c) calibration graph obtained by normalization on intensity of the reference line having a good correlation with analytical line intensity.

7.3.2.2. Correction of Specific Processes Influence

The influence of the sample temperature on LIBS analysis was studied by López-Moreno et al. [89] by determination of the steel slag composition. A composition of the samples at high temperature was evaluated using a calibration curve performed at low temperature. The method uses a transfer function which avoids the need for building a calibration plot at high temperature. Boundary condition must be obeyed for the model to work properly: the emission response must remain equivalent both for the element of interest and the internal standard along the working temperature range. Dependencies of Si I 288.16 nm and Mg II 279.55 nm signals normalized to Ca I 300.68 nm exhibited a similar trend with a linear increase of the temperature up to approximately 850° C. It was found the excellent correlation of the Si /Ca and Mg/Ca concentration ratios measured by remote LIBS at 850° C and the nominal values obtained by XRF at room temperature equal to 0.9875 and 0.9351 respectively.

Selective evaporation (fractionation) during laser ablation may be a serious obstacle for internal standard normalization in LIBS, since number density ratio is influenced. This effect occurs quite often on sampling of brasses and bronzes due to the significant differences in vapor pressure of Zn and Cu providing difficulties for calibration [90]. Ko et al. in their early work [91] on internal standardization in LIBS measured the Zn/Cu ratio for brass analysis. After a study of the line intensity evolution it was shown the possibility of using this ratio with an appropriate time delay of spectra registration. It was also sufficient to choose the interval for measuring signals both for femtosecond and nanosecond LA to obtain a linear calibration by internal standardization of the Zn to the Cu line intensity [92-94]. It should be noted that the raw calibration was failed in this case. Fornarini et al. [27] have developed interesting approach of calibration analysis when preferable experimental conditions were selected after theoretical calculation of plasma state based on thermal evaporation model. Different kinds of internal standards were theoretically considered and it was suggested that Zn as a reference element is mainly responsible for the lack of plasma stochiometry. Actually, the best linearity of calibration curves with correlation coefficients 0.950-0.996 was obtained for Zn-rationed values. Pershin et al. [95] corrected the line intensities by taking into account the Prokhorov–Bunkin melt transparency wave. Special correction coefficient W_i was needed for each element:

$$W_i = (\rho c T_m + \lambda_m + \lambda_v)_i T_{mi}, \tag{5}$$

where ρ is the material density, c its specific heat, T_m the melting point, λ_m the latent heat of melting, λ_v the latent heat of evaporation. Initial LIBS data for quaternary bronze were not good correlated (especially in double-pulse mode) with certified values even by employing Cu I at 510.55 nm as internal standard. After correction procedure the intensities ratios were brought into close agreement with the respective values of concentration in solid sample. A serious limitation of this approach is the necessity to select spectral lines under certain criteria, for which the concentration of elements is proportional to the certain "constant of the process", which remains unchanged at the fixed experimental parameters and is inversely proportional to the laser pulse energy, and the intensity of the corrected lines. Special procedure of emission signals registration taking into account information about different "fractions" of ablated material was developed by Santagata et al [96, 97] for double-pulse LIBS. The resulting data for each sample were obtained by summing together the emission spectra belonging to the same spectral region and collected at definite inter-pulse delays in studied range (1–196 μs). Although a different grade of fractionation can occur, the fs/ns double-pulse LIBS can anyway provide accurate quantitative analysis of the minor constituents of copper-based-alloys (correlation coefficients of calibration curve were increased from 0.844-0.997 to 0.992-0.999). Similar results were obtained for ns-ablation. It was shown that the ratio of spectral intensities of various elements *does* depend on the time of measurement [98]. These findings point out that long integrated and nongated LIBS analyses may perform better, when the ratio of spectral lines is applied as an internal standard.

Internal standard itself implies the use of calibration strategy for quantitative analysis. A number of preferable experimental tunings and improvements of internal standard technique were discussed above to make its application more successful. However, there are several interesting approaches of enabling internal standard for other analytical tasks. Sun and Yu [99] proposed a simplified procedure for correcting self-absorption in calibration-free (CF) LIBS with reference lines. This procedure, named internal reference for self-absorption correction (IRSAC), was started by estimating the temperature from the Boltzmann plot without any correction and choosing an emission line for each species that was not self-absorbed ("internal reference"). Then the deviation of observed intensity ratios and values expected from Eq. 2 were corrected by coefficient f_λ^b defined as the self-absorption coefficient at the definite wavelength with the value between 0 and 1. Therefore, supposing self-absorption of reference line negligible ($f_A^b \approx 1$), an approximate measurement of the self-absorption coefficient can be obtained:

$$f_B^b = \frac{I^B g^A A^A}{I^A g^B A^B} e^{-\frac{\Delta E}{kT_{exc}}} . \tag{5}$$

The line intensities corrected by the previous temperature are used to calculate a new temperature. Finally, the optimal self-absorption coefficients were determined by an iterative procedure until the convergence of the correlation coefficients on the Boltzmann plot. Through the IRSAC method, the accuracy of material composition determination was improved greatly, e.g. Cu content in aluminum alloys was estimated to be 4.79 % with IRSAC (89.54 % usual CF-LIBS, 3.99% certified value). The bottleneck of the method is selection of emission line *a priori* free from self-absorption. It seems reasonable to supplement described approach with self-absorption test of reference line.

7.3.3. Data Processing

The shot-to-shot variability significantly influence the reproducibility of LIBS even in optimal experimental condition and with the use of internal standard. The variety of preliminary processing of spectral data usually based on some statistical approach was proposed to overcome or account the incidental variations of analytical signal. The simplest way of improving signal-to-noise ratio is averaging a number of pulses. Palanco and Laserna [100] was estimated the limit of reproducibility enhancement for the aluminium alloys ablation. The calculated dependence of SNR with the square root of the photons arriving the detector and the SNR of the Al I 396.15 nm line have been plotted against the number of averaged spectra. The experimental SNR increases with the number of averaged spectra like theoretical one until it hits the system noise floor at about 100 averaged spectra. At this point the SNR improved by approximately a tenfold. After this point, no substantial improvement is observed, which points to excess flicker noise becoming the limiting noise source. Opposite to photon noise which increases proportionally to the square root of the signal and enables SNR improvement through signal averaging, flicker noise is proportional to the signal itself and thus, prevents SNR enhancement from signal averaging. Lentjes with colleagues [101] suggested averaging of correlation coefficients instead of spectra. Linear correlation analysis was used as a technique for the identification of samples with a very similar chemical composition by LIBS. The spectrum of the "unknown" sample is correlated with a library of reference spectra. It was found out that the number of spectra to be averaged is equal to the number of correlation coefficients to be averaged. The value of 99.9% right identification was achieved for both variants. The benefit of using averaged spectra over averaging correlation coefficients is a faster calculation.

Position of laser breakdown in water is very unstable and influences reproducibility greatly. Lazic with colleagues [102, 103] developed simple data processing procedure for underwater LIBS to select automatically only the spectra characterized by similar plasma temperature, which are related to their continuum spectral distribution. Application of such a procedure improves the measurement accuracy and leads to linear calibration curves instead non-correlated ones for raw data. This opens a possibility for direct quantitative LIBS analyses of submerged sediments or other soft or rough surface materials underwater. The group of Hahn suggested several approaches for discarding bad measurements. A conditional analysis approach was proposed that separates single-shot particle hits from misses [104]. Single-shot spectra were then identified that had metal signals that were twenty times greater than the multi-shot averaged metal signals, thus showing significant SNR improvement. Also a filtering algorithm was implemented to deal with the significant signal noise associated with single-shot spectra and to eliminate irregular spectra in [105]. These irregular spectra are attributed to a variety of factors including weak ionization of the plasma, only a small amount of analyte being present in the plasma volume, and imperfect collection of the plasma emission. In a later work [106] the two methods of discrimination was compared with the use of the peak-to-base ratio and the SNR. The SNR method showed improved spectral discrimination.

7.4. Methods for Enhancement of LIBS Sensitivity

In analytical chemistry the term 'sensitivity' means the slope of a linear calibration curve or, in the case of a nonlinear curve, the first derivative of a calibration function on analyte content or its amount [107]. Often when discussing the sensitivity of an analytical method, a lower limit of determining content (or detection limit) is implied. Although both quantities are functionally related through the ratio for a limit of detection (LOD): LOD = $3\sigma/S$, where σ - standard deviation of the background emission or blank sample and S – sensitivity, nevertheless, LOD or its analogs better show the detection power of the method. Further the term of ‘sensitivity’ will be understood by us as exactly this parameter, i.e. LOD. Unfortunately, LIBS sensitivity for detection of some important elements, such as As, Se, P, Hg etc., is low (at the level 1-100 ppm) than provided by other spectrochemical techniques (LA-ICP-MS, ETAAS etc). This point is a major challenge to the wider analytical application of the method in field and/or laboratory conditions. Indeed, now a reliable knowledge about the content of rare-earth metals at sub-ppm levels should be obtained for geological *in situ* investigations of ores, rocks, stones, etc. The environmental monitoring of soils and waters requires even more lower LODs toxic traces. The short lifetime of laser-induced plasma and its small sizes are main reasons that why the low level of a signal can be obtained under analytical measurements.

Integrated intensity of emission lines in atomic-emission spectrometry is directly proportional to the number of particles N on the upper level (when self-absorption is assumed to be negligible) [108]:

$$I_\lambda(t) = hcAN(t),$$

where A - the probability of spontaneous emission for this transition (Einstein’s coefficient). Certainly, the number of particles N is a function of time due to the very short period of plasma emission (usually about 0.1-100 μs for pulsed ablation). A typical shape of the dependence $N(t)$ includes three parts: a fast growth at the range of 10-1000 ns, maximal values at a plateau about 0.5-2.5 μs and slow exponential decay up to line disappearance. On the other hand, the strong background interfering with analytical lines is observed during plasma expansion due to Bremsstrahlung radiation from moving electrons and recombination between electrons and ions. As a result, the standard deviation of background, σ, will be a function of time too and in extreme case at t>>10-100 μs, i.e. a plasma luminosity time, σ will be equal to the detector noises. So, the signal-to-noise ratio (SNR), being the most important analytical value to compare detection power of several techniques, $SNR = I_\lambda/\sigma$ has to depend on delay time. Usually the function has a maximum at 1-5 μs under atmospheric pressure. The task of improving LIBS sensitivity is reduced to increase the number of emitting particles and increase the luminosity time of an analytical line. It can be achieved by: 1) more strong ablation, which leads to a growth of evaporated mass and the number of all particles (atoms, ions, clusters etc.); 2) resonance excitation of atoms (due to additional light sources, for example, a radiation of dye laser); 3) thermal excitation of analyzed element atoms (due to particle collisions and effective conversion of a kinetic energy to internal one); and 4) selection of an appropriate time window (i.e. optimization of temporal conditions). Below, we consider all the currently known methods to improve sensitivity, except the last item on the choice of time window.

7.4.1. Electric Discharge Assisted Techniques

A first way for sensitivity improvement in laser spark spectrometers from Carl Zeiss, Jarrell-Ash, Lomo etc. was often used in 1960-1980 years. This technique, where the relative cool laser-induced vapors reached a pair of electrodes are cross-excited by a spark discharge, was first introduced by Brech and Cross [109]. This combination was virtually a kind of emission spectrometry with spark discharge, in which laser vaporization was attached to a spectrometer as a way of sample introduction into the discharge channel. The main advantages of this approach were: (i) the extension of the luminosity time of plasma to 10-1000 μs and, certainly, increase of the intensity; (ii) the reduction of the continuous background of laser plasma and its noises, i.e. significant improvement of sensitivity; and (iii) a flexibility of available spark discharge spectrometers to add laser ablation chamber [30]. A large RSD of 15% to 40% [110] being due to a large discharge capacity of several μF and several kV discharge voltages was often problem for analytical application of the method. Nevertheless, after the intensified CCD detectors appeared and became widespread, an interest to such combination significantly dropped. An attention to the technique has recently renewed due to search for inexpensive and sufficiently sensitive methods for laser microanalysis. For example, Nassef et al [111] has recently proposed the use of a spark discharge-laser induced breakdown spectroscopy (SD-LIBS) for analysis of conductive samples Al and Cu, and has demonstrated the possibility to enhance the signal intensity with better SNR. Higher electrical energy is deposited into the plasma, higher background emissions due to the continuous electron Bremsstrahlung emission and the emission from electrode material. Therefore, to provide the enhancement of SNR, the discharge voltage and currents should be kept at low levels, but this way will also decrease the intensities of atomic lines and increase discharge instabilities. Another technique, spark induced breakdown spectroscopy (SIBS) has been investigated by Hunter et al [112] to solve these problems. A high voltage (15–40 kV) low current arc is first used to initiate an ion channel between the electrodes. Then, once the channel is open, the charged capacity discharges between electrodes. However, Matsuta et al. [113] has reported that a laser pulse prior to high voltage pulse was able to control the discharge position regardless of the surface conditions and improve discharge stability.

Two configurations for spark discharge combined with laser ablation were previously developed. In the first scheme, the discharge occurs between two electrodes placed over the surface of analyzed sample at a distance of 2-5 mm. The arrangement was often applied to spectrochemical analysis before improvement of CCD detectors, and it is the most widely during last decade. The obvious advantage of this technique is the ability to analyze a non-conductive material such as soils, ceramics, plastics etc. However, as mentioned above, high voltage pulse produced an intense emission background is required to initiate a discharge. In the second scheme proposed by Chen et al [114], the sample is used itself as one of the electrodes. This option is suitable only to conducting materials (metals, alloys) or coatings on conducting material; nevertheless, the scheme has some advantages. Firstly, the size of a luminous area in discharge is larger due to larger surface of electrode-sample. Secondly, to initiate an ion channel near the sample surface by means of laser-induced charges, lower voltage can be applied to a gap between electrodes. One more advantage for both schemes is that lower laser energy (from one to second order of magnitude [111]) can be delivered to the sample for its ablation without a loss of detection power.

Many applications of LIBS combined with spark discharge were described in the literature especially before 1990 years. Therefore our considerations about their analytical possibilities will be limited here by a number of works published during a last decade. Hunter et al [115, 116] have developed a special probe to monitor of heavy metals (Pb, Hg, As, Cd, U etc) in aerosols and airborne particulates. LODs achieved by the technique were equal to 10 $\mu g/m^3$ for Pb and Hg. Enhancement factors of SNR up to 10 in comparison with conventional

LIBS were achieved by Hassef et al [111] who used Cu and Al targets. Unfortunately, this group investigated only emission lines of matrix elements without traces detection. Hunter et al [112] have reported that LODs of lead, chromium, mercury and cadmium in soils were decreased down to 25 ppm. Comparison between well known double-pulse LIBS often used as a more sensitive tool and SIBS performed by Belkov et al [117] was shown that there were no significant difference in sensitivity. Another attempt of such comparison made more recently by Li et al [118, 119] has shown that enhancement factors for SIBS were larger in 2-3 times for traces and in 10-100 times for matrix elements, as well as reproducibility of signals was significantly better (in 10-15 times). More recently, Chen et al [114] have proposed a technique to stabilize the discharge by low-energy laser pulse used for ignition prior to spark discharge instead of high-voltage arc used by Hunter et al [112] to ionize a gap between electrodes. Extremely low LOD, 2 ppb, for mercury traces in aqueous solution deposited on a copper substrate was achieved by means of this approach. Despite the cheapness of the high voltage block in comparison with the second laser source and better sensitivity, the use of spark discharge has some drawbacks. Contaminations introduced by electrode materials, as well as spectral interference of analytical lines of traces with lines of electrode material, significantly degrade the selectivity. Since the channel ionization occurs in the ambient gas, pressure and composition of this gas limit the application of these techniques. With decreasing pressure indeed it is necessary to apply more voltage to the gap. Remote analysis of dangerous objects such as nuclear materials with additional discharge spark becomes impossible. In other words, spark discharge has limited applications in analysis of not very complex sample (in particular the matrix and minor components) only in laboratory conditions in air.

In addition to techniques with spark discharge, glow discharge was also examined to enhance LIBS sensitivity in a series of papers from Laserna's group [120-122]. Due to low pressure and low voltage required for maintaining discharge, this approach provides better reproducibility and larger enhancement of SNR value (up to 80). Lower laser fluence down to subthreshold values (0.17 mJ/pulse) can be delivered on the surface without a significant loss of sensitivity similarly to a spark discharge. It leads to the shallower craters (at least in 5 times) and allows the depth profiling with a better resolution. Another nice feature of this approach is that both sources, i.e. laser and discharge, can be independently and gently adjusted to optimize the sample removal and its excitation. On the other hand, unlike the spark discharge, a special chamber for ablation at reduced pressure (~0.5 mbar) should be constructed to realize this technique. It strongly limits the possible analytical applications of such approach. Moreover, there is no estimation of enhancement factors for trace elements detection by means of LIBS with a glow discharge combination.

7.4.2. Effect of Electromagnetic Field

Because laser plasma contains the charged particles, ions and electrons, the external electromagnetic field can confine plasma expansion under several conditions. One of the first methods was proposed for the magnetic confinement of the plasma, in which laser plasma expanding in a steady magnetic field slowed down at a certain magnetic field strength. The physical aspect of the magnetic confined plasma was thoroughly studied by many investigators. Bhadra postulated that laser-induced plasmas would be stopped by a magnetic field B in a distance $r{\sim}B^{-2/3}$ [123]. Another estimation of the radius at which plasma stop its expansion was obtained by Huba et al [124] from the suggestion about the equality of the kinetic energy of plasma with that of magnetic energy at a certain radius, named as magnetic confinement radius of plasma. To compare the evolution of our plasma with and without magnetic field at certain strength B, parameter $\beta=8\pi NkT_e/B^2$, i.e. a ratio of kinetic energy to magnetic energy, was proposed by Rai with colleagues [125]. So, if we take plasma expansion velocity in the absence v_1 and in the presence v_2 of a magnetic field, ratio between them will be expressed as a function of β:

$$\frac{v_2}{v_1} = \left(1 - \frac{1}{\beta}\right)^{1/2}.$$

The equation shows that for a weak magnetic field (higher β) there are no changes in the velocity, while for a very strong magnetic field (low β) plasma confinement will be effective due to the only way of energy loss through a radiation. As the magnetic field does not interact with the electrons moving along the field lines or at a small angle to the lines, plasma will be transformed into an oblong shape [126].

Let's consider the evolution of the plasma interaction with field and the emission intensity resulting from plasma confinement. Parameter β will be a function a time due to the temporal variations of plasma density and its temperature. In extreme case, when $\beta = 1$ (i.e. the kinetic energy density of plasma will be equal to the energy density of the magnetic field), the field stops the plasma expansion (certainly talk about the electronic and ionic component of plasma). An increase in the number of collisions in plasma and, as a result, growth of the number of excited atoms and emission enhancement at this moment will occur. Therefore, evolution curves will pass by a time through a peak or two peaks corresponding to stopping electrons and ions [126]. Position and amplitude of this peak slightly depend on laser energy. It will be reasonable to assume that the peak will shift toward shorter delays with increasing magnetic field strength B. However, if the density of magnetic energy will be higher than the energy required for complete ionization of atoms, the atomic emission drops significantly due to the prevailing ionization and ion emission. Apart from the pure hydrodynamic resistance to the external magnetic field, the difference between mechanisms of excitation and deactivation of the upper states results in the peculiarity of field effect on these transitions. For example, Joshi et al [127] have recently found that the Li I lines at 670.8 and 610.3 nm responded to the external magnetic field in different ways: the resonance line increased with increasing field strength, while the 610.3 nm line strongly dropped. Conversely, both ionic lines at 548.4 and 478.8 nm increase in the field. So, a compromise between the rates of collisional and radiative processes should be considered in order to better choose an analytical line. On the other hand, Stoiljković et al [128] have recently studied a number of factors influenced on enhancement in magnetic field by means of the principal components analysis. In the case of atomic lines, they have found that there was the best correlation between the enhancements and first ionization energies as a negative correlation, while, in the case of ionic lines, the best correlation of the enhancements with the sum of the first ionization energies and oxide bond energies was obtained as a strong positive correlation. In other words, to choose analytical lines which would be expected as more sensitive to magnetic confinement some properties of transitions can be obtained only from experiments (e.g. rates of collisional and radiative process), while other properties can be took from known data (e.g. ionization potentials).

The cross-excitation of laser-induced vapor in spark discharge and transversal magnetic field was first demonstrated by Petrakiev et al [129] for the analysis of minerals and alloys. For a number of elements enhancement factors between 1.1 and 11.5 were achieved. Mason with colleagues [130-132] have proposed a peculiar approach to intensify the emission which consisted in stopping the expansion of plasma by a pulsed magnetic field. Enhancement factors achieved varied from 2 to 9, depending on the point of observation. It was found that the factors obtained for singly and doubly ionized species are larger in 2-3 times than for neutral atoms of the same element. From an analytical point of view, the pulsed magnetic field for plasma confinement leads to a significant worsening of accuracy due to the strong broadening for neutral atom lines and favors the self-reversal phenomenon. Several works published by Rai with colleagues [133-135] dealt with the use of a magnetic field as a way to increase LIBS signal. These authors have reported an enhancement of 1.5–2 times of LIBS sensitivity for Mg, Ti, Cr and Mn determination in aqueous solutions and solid targets in a steady magnetic field (with ~5 kG). LODs achieved were reduced only in 2 times for Mn (0.83 ppm) and Mg (0.23 ppm). Recently, Shen et al [126], studying laser plasma expansion after ablation of aluminum, copper and cobalt targets, have found that an increase of emission lines of aluminum and copper was observed in external magnetic field (B = 0.8 T) with enhancement factors varied between 2-8, while a quenching of emission was observed for cobalt lines. The one reason for such a strong matrix effect is that there is the compromise between the magnetic force playing an important role in magnetically confined LIBS and the Lorentz force for ferromagnetic materials, such as Co targets. Another possible reason for the reduction is that the lifetime of the laser-induced Co plasma was short, and the number density of the excited Co atoms decreases rapidly. Note, no trace elements were studied by these authors.

In spite of a well-established theoretical description of a physical nature for magnetic confinement of laser plasma, an interest in such technique in order to enhance LIBS emission is negligible for several reasons. Firstly, Zeeman broadening of spectral lines and possible line shifts due to multiplet splitting can lead to the significant accuracy errors. A growth of field intensity, which should enhance the magnetic stopping of plasma, will also increase the accuracy errors due to a larger splitting of multiplets. In addition to line broadening and line shifting, a strong heterogeneity [136] of the magnetic field applied to a small plasma should be taken into account to reduce a contribution of magnetic field variations to the changes of LIBS signal from pulse-to-pulse. Secondly, LODs of trace elements achieved by the approach slightly differed from LODs obtained by means of conventional LIBS technique; and an enhancement of analytical signal does not able to compensate for the loss of analytical accuracy. Thirdly, new nodes, such as magnets, synchronization block and power supply for magnets, needed to adapt conventional LIBS system for magnetic confinement, seem unsuitable for a work at field conditions and especially for remote analysis.

7.4.3. Double Pulse Method for Laser Induced Breakdown Spectroscopy

Analytical capabilities of LIBS are not so impressive compared to LA –ICP –AES and LA –ICP-MS due to poor sensitivity and precision. In order to improve sensitivity it was suggested to use additional laser pulse with μs delay after first pulse [137]. The application of two or multiple sequential pulses resulted in intensity enhancement, better signal - to – nose ration and longer plasma emission. Double pulse method represents a very effective approach to improve analytical performance of LIBS. Pair of laser pulses allows to increase intensity of atomic and ionic lines for up to 200 times. Improvement of analytical capabilites with simple addition of second laser pulse resulted in double pulse method popularity growth in LIBS community. Double pulse method uses additional laser pulse that can be used as an instrument for laser matter interaction study. All these achievements resulted in growth of double pulse method publications during last decade. To date, double pulse procedure became a well-known and very popular method in LIBS community. Double pulse studies have been recently summarized [11, 24]

Double pulse method is a technique when two consecutive laser pulses with short delay between them are used for laser sampling. Typical delay between pulses is in 100 ns – 200 μs range. For longer delays enhancement of intensity was not observed. For shorter delays signal enlargement was small. Several terms are used in literature for describing laser ablation with sequential laser pulses: double pulse, spark pairs, dual pulse methods. Term 'spark pairs' has been introduced by Cremers et. al. [137] to described laser ablation by two laser pulses with microsecond delay. Currently two alternative terms are widely used in LIBS community: double pulse and dual pulse methods. According to Oxford dictionary a meaning of adjective 'double' can be expressed as 'consisting of two equal, identical, or similar parts or things'. This definition is slightly different from that given by 'dual': 'consisting of two parts, elements'. Consequently, it should be recommended to use term 'double' pulse method in case of nearly equal laser pulses characteristics (the same wavelength, duration of pulse etc.). Second term 'dual' pulse should be used in case of different laser pulses properties (wavelength, duration, etc.). However one can find in literature that both terms are used interchangeably. If multiple consecutive laser pulses are used for sampling than method is called multiple pulse method.

First experiments for laser ablation with double pulses were carried out in the begin of 1970-s by Piepmeier et. al. [138]. They have observed small enhancement of atomic lines intensity in spectra and explained it by absorption of laser irradiation in plasma. The reproducibility of effect was poor due to laser system imperfection at that time and no paper has been followed. "Renaissance of double pulse" method was started by Cremers, Radziemski and Loree [137] where they suggested to use second laser pulse for bulk water solutions analysis. The analysis of bulk water samples is always a problem due to strong depletion of atoms excitation during plasma expansion in water. Consequently plasma in water is characterized by low intensity spectra and short lifetime. It was suggested to use first laser pulse to form vapor bubble in water and then to use second laser pulse to create laser plasma inside this bubble. This scheme allowed to register spectra lines with high excitation energy like B I 249.8 and Be I 249.5 lines and substantially increased spectra intensity for all elements under study. This procedure resulted in improved sensitivity of analysis and allowed to achieve low ppm limits of detection for bulk liquids in LIBS. Surprisingly, this study became unnoticed by scientists despite remarkable improvements achieved simply by adding second laser pulse, like several orders spectra intensity increase, improvement of signal - to - noise ratio and longer plasma lifetime. For solid samples in air first experiments with double pulse method have been carried out by Pershin et. al. [139-141]. Pulsed solid state Nd:YAG laser with passive Q-switching was used for laser ablation of aluminum alloys. Laser system was able to generate up to 6 pulses with delay 20 - 40 μs. When two pulses with 25 μs delay have been used for sampling than signal - to - noise ratio increase for up to 10 times for Al I and Al II lines. Authors have proposed possible mechanism of intensity improvement with major factors attributed to increased ablated mass and changes of condition for plasma formation [142, 143]. Unfortunately these results were published in USSR and were unknown for LIBS community. Two years after Uebbing et. al. [144] suggested to use second laser pulse to reheat plasma and to increase spectra intensity. Authors have detected higher spectra intensity (factor of 10) while temperature was slightly lower after reheating pulse. Sattmann et. al. [145] used multiple laser pulses for steel analysis and observed improvements of analytical results. Same group uses double pulse method for several impressive implementations of LIBS systems in industry [146-149]. Several groups have been focused on double pulse method fundamental studies and a number of studies have been followed. Stratis et. al. [150, 151] proposed pre-ablation double pulse method when first laser pulse was used to produce air spark above sample and second pulse was used for ablation. Followed experiments included double pulse studies with laser pulses of ns and fs durations [152, 153] Same group carried out comprehensive studies for different schemes and different wavelengths for double pulses to determine influence of experiment parameters on observed signal improvement [154-156]. Semerok et. al. studied ultrashort double pulse sampling [157, 158]. A comparison of single and double pulse procedures was studied with ultrashort pulses by several groups [159-162]. Different schemes of double pulse method were used by Palleschi and co-workers to study influence of geometry and conditions [163-166]. Several papers were devoted to explanation of double pulse effect [169, 170]. Same group [171] has presented study of double pulse efect dependence on sample compositio. Other groups were also interested in explanation of spectra intensity increase for ablation with double pulses. Different schemes and conditions were used to understand influence of laser parameters for double pulse effect [172-175]. St-Onge et. al. extensively studied two pulse ablation with UV and IR laser pulses [176, 177]. Colao et. al. [178-181] has studied influence of different laser parameters on signal enhancement. Russo's group carried out systematic study of plasma expansion, temperature and electron density for double pulse plasma [47, 182]. An application of multipulse method for compact economy LIBS system has been proposed by Galbacs et al. [183-185]. Double pulse procedure for aerosol analysis was performed by Windom et. al. [186, 187]. Several numerical models describing double pulse procedure have been suggested recently [188, 191].

Double pulse sampling was successfully used in other atomic spectroscopy methods with laser ablation. Laser ablation with double pulse method allowed to obtained more finer aerosol and to improve analytical performance of laser ablation inductively plasma mass spectrometry [61, 192]. Pair of laser pulses with short delay was successfully used in a number of applications: thin film synthesis [193], nanoparticles and nanostructures synthesis [194-197], laser machining [198, 200].

7.4.3.1. Double Pulse Method Configurations

Double pulse method realizations have been made with different configurations: different geometry of laser beams, different collecting optics, different laser wavelength and pulse duration, different conditions of ambient (air, water) etc. Two different configurations for double pulse procedure are used referring to relative position of laser beams. In collinear configuration two laser beams are coaxial or almost coaxial and focused at the same spot on sample. If one laser beam is a normal to sample surface and second is parallel to surface than this scheme is called orthogonal configuration. In such case one pulse is used for ablation while other can promote favorable conditions for ablation (pre-pulse) or additional excitation of atoms in plasma (re-heating).

Collinear scheme can be carried out with single laser or with two laser systems. Single laser system is more compact and easily adjusted but has several limitations. It is not generally possible to obtain two pulses with delay shorter than 20 μs and to independently change energy for first and second pulses with single laser. However single laser was used for pulses generation with 6 μs delay but this setup needed two amplifiers to produce laser pulses with enough energy [146]. Double pulse procedure with two laser systems is more difficult in optics adjusting and synchronization but more flexible in experiments.

Optical schemes used for double pulse experiment (Figure 3):

1. collinear beams:

1.1. collinear beams for two pulses within the same flashlamp pumping (a) [140, 145, 201]

1.2. collinear beams with two pulses from different lasers focused on the same spot (b) [163]

2. orthogonal beams:

2.1. first pulse is fired above surface and after short delay second is used for sampling – pre-ablation (c) [150]

2.2. first pulse is used for sampling and after delay second pulse is used to reheat plasma – reheating scheme (d) [144]

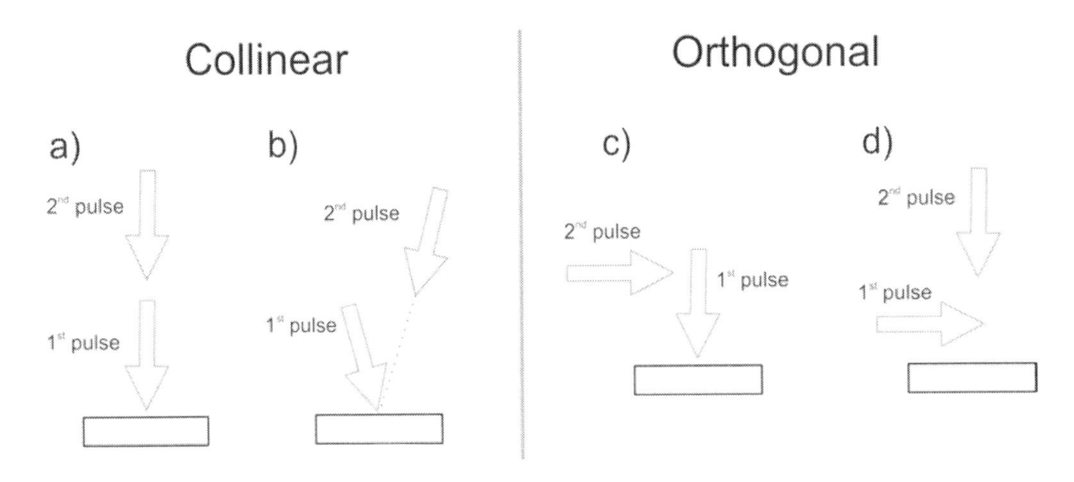

Figure 3. Double pulse configurations for LIBS. Collinear scheme with two pulses with single laser (a) and with different lasers (b). Orthogonal scheme with pre-pulse (c) and re-heating (d).

First variant of coaxial scheme was proposed in [138]. First experiment utilizing orthogonal scheme was suggested by Cremers et.al. in [137] for water samples analysis. Uebbing et al. [144] suggested to use reheating laser pulse to increase signal for laser ablation in air. They used two different lasers to produce two laser pulses. First laser pulse was used for ablation and second for plasma re-heating. In spite of 10 times increase of line intensity double pulse effect was not impressive due to fact that re-heating pulse energy was also about

10 times higher than energy of first pulse. Temperature was about 10000 K after second pulse while 12000 K was detected after first pulse. Consequently, excitation with reheating laser pulse had low efficiency. Stratis et al. [150, 151] suggested orthogonal scheme with pre-ablation spark. First laser beam was parallel to target surface and formed a laser spark in air. Second laser beam was focused on sample surface and after several μs delay were used for sampling.

7.4.3.2. Double vs Single Pulse

When laser ablation is carried out with two sequential pulses than usually first subject to discuss is an intensity increase. This intensity enhancement is an important characteristic of double pulse procedure since better sensitivity and lower limits of detection. But there are several other plasma properties that are different for double pulse method and that can be effectively used in different applications. To describe double pulse effect in details a comparison of double and single pulse ablation is made.

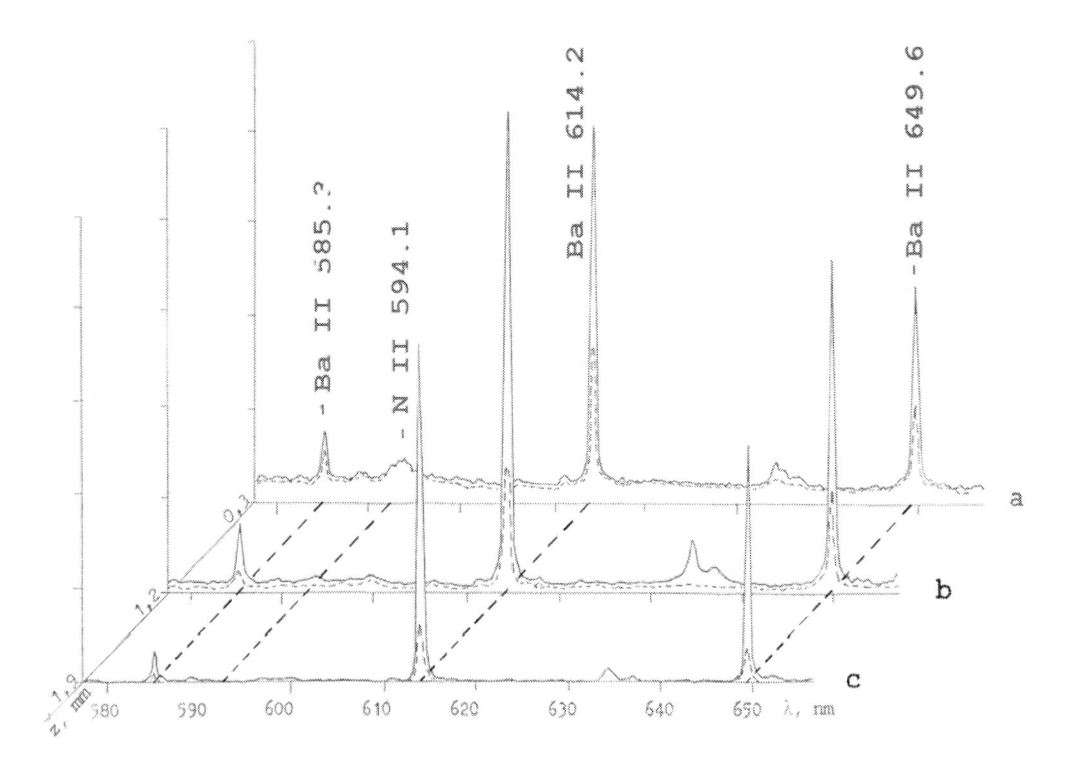

Figure 4. Double pulse spectra of glass sample (time integrated spectra, collinear scheme, 1064 nm, 10 ns, delay 25 μs) Spectra were detected at different distance from sample surface 0.3 mm (a), 1.2 mm (b), 1.9 mm (c).

Intensity increase of atomic and ionic lines in spectra has been observed for most double pulse studies [11, 24, 137, 139, 146, 147, 150, 155, 158, 166, 173]. For double pulse method increase of signal was varied from 2 to 240 depending on used laser pulses charachteristics and samples. For collinear scheme with two pulses it was observed by Pershin et. al. [140, 139] that signal enhancement for Al I lines was about 10 times. Benedetti et al. have observed up to 50 times higher signals for double pulse method compared to single pulse [165]. Colao et. al [178] observed intensity increase for Al I and Al II lines and longer duration of plasma emission. Stratis et. al. [150, 151] implement pre-ablation variant of double pulse method for steel samples and detected improved intensity for all elements of impurities and major components. It was observed that signal enhancement is matrix independent but higher ablation is observed for samples with low melting point [152, 171]. Matrix dependence of double pulse signal was observed for steels, rocks and ceramics [174]. Gautier et. al. [173] studied dependence of atomic an ionic signals enhancement on excitation energy of transition. They determined that the greater excitation energy the higher signal enhancement. It should be noted that for double pulse procedure background intensity was also amplified with increase of lines in spectrum. Consequently, signal-to-noise ratio should be used for double pulse improvement discussion instead of increased intensity.

Differences of signal enhancement by various scientific groups in comparable conditions could be explained by differences in plasma size and dynamics compared to single pulse study. It was observed by Pershin et. al. [139] that signal improvement was depended on distance from sample surface. Cristoforetti et al. [167] carry out detailed study of plasma dynamics by imaging. They have observed that double pulse plasma shape and dynamics substantially changed compared to single pulse plasma. It was shown that signal enhancement could be easily overestimated for up to several times. They have determined that if intensity is compared for most bright locations for double pulse and single pulse plasmas than factor of signal enhancement was about 2 times. But if intensity from local point out of the center was used for single pulse plasma than enhancement factor was about 12 times due to greater plasma size for double pulses. It was suggested that signal enhancement should be estimated by space integrated emission from all plasma regions.

For laser ablation of liquids by double pulse method a impressive improvement of signal was obtained [137]. Cremers et. al. could detect Be I and B I lines for double pulse plasma in bulk water while they weren't able to observed selected lines for single pulse. Signal enhancement of 8 times was obtained for delay 20 μs between two pulses. Kumar et. al. [217, 218] have observed increase of intensity for Mg I and Cr I lines for 40 and 10 times correspondingly. This improvement of intensity substantially increased sensitivity of analysis and allowed to achieve ppb limits of detection for Mg and Cr.

Size and shape of laser plasma changed significantly in case of double pulse technique [47, 137, 167, 182]. Several studies with plasma imaging have been shown that for double pulse ablation plasma size was always greater compared to single pulse plasma. Mao et al. [47, 182] have studied double pulse plasma in collinear scheme with two Nd:YAG lasers. They have observed that plasma dimension was increased two times for double pulse. Plasma shape had changed dramatically after second laser pulse and can be described with mushroom shape. Cristoforetti et. al. [167] observed increase of plasma volume for 5 times for double pulses. Authors explained shape and size changes for second pulse by first plasma shockwave influence on second plasma propagation. Detailed study of plsma imaging has been carried out by Noll et. al. [149] with streak camera and fast photography. They determined that plasma volume was 10 times greater for double pulse sampling. It was observed that second plasma expansion was significantly different from single pulse plasma evolution.

Increased ablation for different samples was detected for double pulses [145, 150, 157]. Sattmann et. al. [145] have observed 8 times greater ablated mass for double pulse sampling. They have found that increase of ablated mass was the dominant reason for signal increase [149]. Stratis et. al. [150, 154] observed increased crater dimensions for sampling with double pulses. They have discussed that signal increase should be explained by increased ablation mass. Sturm et. al. [177] studied double pulse method with IR and UV pulses and they could control crater volume with changing delay between pulses. Balzer et. al. [202] used double pulse method to control coating layer in industry in wide range of thickness due to bigger crater volume. For ultrashort laser pulse sampling with two pulses deeper craters and increased volume of craters was obtained [157].

Plasma temperature comparison for double and single pulses has been studied in many papers [146, 150, 172-174, 176]. For single and double pulse cases it was observed that temperature was comparable. Sturm et. al. [146] observed higher temperature for 1000 K in case of double pulse plasma. Stratis et. al. [150] detected that single pulse plasma had lower temperature for 20% compared to single pulse for orthogonal scheme. St-Onge et. al. [176] observed 30% temperature increase for two pulse sampling with UV and IR laser pulses. In other conditions double pulse plasma has nearly the same temperature [177]. At the same time Gautier et. al. [172-174] observed tiny decrease of temperatures for double pulse plasma.

Electron density was observed to be near the same for double and single pulse plasma. Benedetti et. al. [165] studied influence of pulse energies on plasma characteristics and haven't observed any differences of electron density in wide range of densities. St-Onge et. al. [177] has observed lower electron density for dual pulse plasma for short delays after pulse. However electron density of dual pulse plasma changed slowly compared to that for single pulse plasma and after 500 ns electron density for plasma exceeded that for single pulse plasma. Colao et. al. [178] studied electron density changes for double pulse plasma of Al sample. Similar to previous group results they have observed same tendency for electron density. Increased electron density for plasma near surface was observed by Burakov et. al. [203] for aluminum samples.

Plasma lifetime was increased substantially due to enhanced signal and it slower decay. Double pulses were used for steel samples analysis by Sturm et. al [146] and they observed 1.5-fold increase of plasma emission duration. Colao et. al. have observed increased emission duration for double pulse plasma [178] and explained by more effective plasma excitation by second laser pulse.

Ablation threshold was decreased for double pulse experiments. For double pulse experiments with collinear scheme ablation threshold decreased for 20 times [178]. Smaller ablation threshold was shown to depend on delay between two pulses and for delay above 60 μs ablation threshold was almost the same as for single pulse study [179]. For double pulses with femtosecond duration it was detected that ablation threshold was decreased if delays shorter 500 fs were used [204].

7.4.3.3. Effect of Interpulse Delay in Double Pulse Method

Double pulse effect was observed at different conditions for delay in range 1 ps – 100 μs. Depending on interested plasma characteristic optimal delay will be different for laser pulses with different properties (wavelength, duration, energy). Cristoforetti et. al. [163, 164] studied delay influence on signal increase for ns pulses in range 0 to 20 μs with collinear scheme.

They used bronze samples and found that for atomic lines Cu I signal increase had maximum for 0.7 μs delay while for ionic line Cu II greater delay with 2 μs and later was preferred. For Al samples slightly longer delay times were determined with 1 μs for atomic lines and 4 μs for ionic lines. Gautier et. al. [173] have detected optimal delay for collinear and orthogonal schemes. Signal was maximized for atomic lines at delay 0.5 μs while for ionic lines it should be greater 1 μs in collinear scheme. For orthogonal scheme with re-heating setup [172, 174] signal for atomic lines was increased for delay 200 ns. St-Onge et. al. [176] have used two laser pulses with collinear and orthogonal schemes and found that for atomic lines optimal delay should be less 0.5 μs while for ionic lines it should be longer than 3 μs. Stratis et. al. [150, 151] carried out detailed study for orthogonal geometry with pre-spark and re-heating schemes to determined best conditions for signal increase. They observed signal improvement for 33 times for atomic lines of copper sample if second pulse delayed for 2 μs. In case of lead sample signal was 11 times greater for double pulse method with 1 μs delay. Nevertheless signal enhancement was independent for wide range of delays from 1 to 100 μs in pre-spark scheme. For short delays below 0.5 μs a decrease of signal was observed that was explained it by air plasma shielding for second laser pulse. Mao et. al [182] studied influence of delay between pulses on evaporated mass, spectra intensity, plasma temperature and electron density. Intensity increased abruptly for delays greater 200 ns. Temperature and electron density were higher for double pulse plasma if two pulses with 200 ns or longer delay were used for ablation. In all studies presented optimal delay for atomic lines is in range 0.5 – 1 μs while for ionic line a bigger 2 μs delay was detected.

For fs double pulses it was found that shorter delays were more effective compared to ns pulses. Short pulses studies were implemented for two points of view. First goal was to carry out double pulse sampling with fs pulses that is known for absence of fractionation and lower ablation threshold. Second idea was to use second fs pulse as a tool for laser plasma study. Scaffidi et. al. [152, 154] studied intensity improvement and material removal with fs and ns pulses. They have observed that laser ablation with fs pulse and re-heating is more effective. A significant increase of mass removal was observed for aluminum and copper samples. For copper sample intensity of Cu I lines in spectra increased 3 times with optimal delay 5 μs between pulses but for lead shorter delay 2.5 μs was applied. Roberts et. al. [205] studied mass removal dependence on delay between fs pulses. Increased ablated mass was observed for delays less 10 ps, for larger delays this parameter was nearly the same as for single pulse and for delays greater 200 ps a decrease of ablated mass observed. Semerok et. al. [157] studied ablation with double pulses with re-heating scheme. For delays < 1 ps crater was formed as for single pulse procedure. For delay 1 – 10 ps a decrease of crater depth was detected that were attributed to partial plasma shielding. If second pulse arrives after 100 ps delay than plasma completely shields laser pulse. Based on this results it can be suggested that plasma was formed for 1 ps after laser pulse. Consequently, to increase intensity of atomic lines authors used 100 – 200 ps delay between pulses to additionally re-heat the plasma.

Significant intensity improvement was obtained with two fs pulses in [162]. For delays 0 – 50 ps signal was increasing with delay increasing. For delays longer 50 ps intensity was maximal and remains constant. Improvement factor up to 10 times was observed for different samples. For laser ablation of silicon with fs pulses in collinear scheme only 2 times higher intensity of atomic line was achieved for delays 31 and 80 ps [206].

7.4.3.4. Influence of Laser Parameter on Double Pulse Method

Wavelength

Application of laser pulses with different wavelengths was suggested in [177]. Laser system with 266 nm and 1064 nm were used to ablate aluminum sample. UV pulse is less shielded by laser plasma and more effectively vaporize and atomize sample while IR pulse is effectively absorbed by plasma. Idea was to use first UV pulse for effective ablation and then IR pulse for re-heating. It was found that such sequence allowed to improve atomic and ionic lines intensity compared to combination of two IR pulses in collinear scheme. For Si I 288.16 improving factor was about 30 times and about 100 - fold for Al II 281.62 line. They have carry out detailed study of craters volumes for different delays. Combination of two pulses resulted in 8 – fold deeper crater and no dependence on delay was observed for delay greater 0.5 μs. It is interesting to mention that if UV and IR pulses were fired simultaneously than crater volume was less compared to UV single pulse sampling. It was explained that UV pulse easily generates plasma that is effectively absorbed IR pulse. IR pulse effectively absorbed by plasma thus increased plasma temperature and electron density and that is resulted in improvement of absorption of UV pulse. In such case all laser pulses energy will be effectively shielded by plasma and less material will be ablated. Consequently, combination of UV and IR pulses allows to control ablated mass simply by adjusting delay.

Another study with combination of 1064, 532, 355 nm pulses were presented by Russo [47]. Study was devoted to analytical performance estimation with different laser pulses. For all possible arrangements of different wavelengths signal was increased for collinear scheme. The combination of 532 nm for first pulse and 1064 nm for second pulse provided best sensitivity and worst reproducibility. The optimal precision was obtained with combination of 532 nm first pulse and second 355 nm pulse but sensitivity was poor. This combination provided matrix independent ablation that allowed to obtain unite calibration curve for two types of copper and zinc alloys. Cahoon et. al. [207] used UV pulse as a first and then IR pulse to reheat plasma for glass samples analysis. Combination of two pulses increased in ablation and improved analysis of LIBS. The combination of 1064 nm pulse and 10.6 μm pulses were applied in [208]. The enhancement factor was line depended and changes from 25 to 300 times. Two laser pulses should overlapped in time for more than 1 μs to observe enhancement.

Energy

Influence of energy for first and second pulses was studied for aluminum alloys at [165]. For collinear scheme it was suggested to keep second pulse energy unchanged while first pulse energy was varied. Signal improvement was greatest if first pulse energy was one third part from second pulse energy. Gautier et. al. [175] studied influence of energy in orthogonal scheme for both pulses. For pre-spark scheme they have observed that maximum signal 20% was obtained if energy of first pulse was a half of second pulse energy. But for re-heating scheme signal enhancement was 1.5 times if energy of first pulse was larger. Double pulse method with single laser was used for analysis of steel samples in water [201]. Single laser generated two laser pulses during fashlamp pulses. It was observed that for maximum atomic line intensity it is preferable to use 120 mJ for first pulse and 280 mJ for second. Such energy ratio was explained by bobble dynamics that first pulse should have enough energy to form bobble but it is preferable to use maximum energy for second pulse for higher sample ablation and plasma excitation. In case of double pulse method for liquids analysis with two different lasers another best ratio of pulse energies was observed [218]. For both pulses nearly equal energy was the best choice. If first pulse energy was increasing than first laser plasma effectively shielded second laser pulse. If second pulse energy was increasing than spectra intensity showed a saturation effect. Authors suggested to optimize energy ratio and pulse delay for every setup to achieve best results.

Pulse Duration

Scaffidi et. al. [152] studied dual pulse method with ns and fs pulses and different combination of laser pulses. Dual pulse schemes with pre-spark and re-heating were used. Intensity was increased substantially for re-heating scheme with 30-fold intensity for Cu I line and 80 – fold intensity for Al I lines. It was found that for all conditions studied ns pulses is more effective as second re-heating pulse rather than first sampling pulse. Influence of delay between pre-spark fs pulse and ablation ns pulse on ablation mass and spectra intensity was studied in [156]. Ablated mass for dual pulses was increased for 10 times for aluminum and 8 times for copper with only 4 and 3 times greater intensity of spectra.

When two fs pulses were used for laser ablation than shorter delays between pulses were used in order obtain effect of double pulses [157]. For delays shorter than 10 ps a strong plasma shielding was detected that decrease signal. To increase signal a delays of 100 ps were used due to absence of shielding by plasma. Spectroscopy study of double pulse fs plasma was performed in [206]. For delays 31 and 80 intensity enhancement of Si I line was 2 times. Authors didn't observe differences between plasma dynamics for double and single pulse procedures. Detailed study of delay influence for two fs pulses was made [206]. Delay between pulses was changed and for delays greater 50 ps signal increased 10 times for different lines.

7.4.3.5. Laser Ablation with Multiple Pulses

First experiments with multiple pulses were carried out in 1969 by Piepmeier et al. [138]. Later Sattmann et al. [145] used solid state laser with active Q-switch that can be opened several times during single flash lamp. This system generated up to 6 pulses and it was possible to vary delay between pulses in 2 - 80 μs range in order to achieve better analytical performance. Authors detected increased intensity of atomic lines, increased electron temperature and density, increased ablation and plasma size for multiple pulses ablation. In next article [146] of the same group steel samples were analysed and LOD were better than 10 ppm for C, P, S, Cr and Ni. Further paper was focused on effect of ambient gases (argon, oxygen and nitrogen) on ablation rate for multipulse sampling [209]. Comparative study [210] of single pulse method and double and triple pulses method was performed. Longer lifetime and greater plasma dimensions were observed for multipulse method. Plasma electron temperature and electron density were observed to be species depended. In case of three or two laser pulses higher temperature and electron density were determined for atoms but for ions dependence was opposite.

Galbacs et. al suggested to use 11 consecutive laser pulses for sampling and achieve significant improvement of analytical capabilities [183]. They have used economy laser system with passive Q-switch and control the number of pulses in train by flashlamp pumping. Signal enhancements were improved more than two orders of magnitude. In most recent paper by Galbacs et. al. [184] two systems with different lasers (Nd:YAG with active Q-switch and Nd:GGG with passive Q-switch) and different detectors (ICCD and CCD) have been used to study effect of pulse number on intensity increase and plasma properties. Intensity increased for up to 10 times for multiple pulse sampling. It was observed that ablated mass was increasing if number of pulse in train was amplifying. Temperature of plasma was detected to be higher in case of additional pulses sampling. The maximum signal enhancement was observed for trains with 5 laser pulses.

Laser sources with high repetition rate [14, 16] could be suggested for multiple pulse applications. Microchip laser with high repetition rate up to 200 kHz can be used for this. Quantitative analysis of low-alloy steel was made with microchip laser system [14] (1 kHz, 50 μJ). Improved sensitivity was detected but effect of consecutive pulses didn't observed. Cristoforetti et. al. [16] used laser with higher repetition rate and bigger pulse energy for sampling (8 kHz, 80 μJ). Time of atomic line emission was < 1 μs and no influence of consecutive pulses was detected. Diode pumped laser [17] system with high energy (0.8 mJ, 20 ns, 1.8 kHz) was used for analysis. Improved intensity was explained by greater quantity of laser plasma with no influence of pulses. Fiber laser [18] with high repetition rate and powerful pulses (200 kHz, 0.8 mJ) was used for analysis It was observed that sequential pulses effect was absent despite short delay between pulses and high energy.

7.4.3.6. Double Pulse Mechanisms and Theoretical Modeling

Starting from first experiments by double pulse method several mechanisms were proposed to explain signal improvement. Most studies were focused on explanation of observe nonlinear improvement for double pulse intensity. Recently numerical modeling was suggested [188-190].

Experimental studies

Pershin et. al. [139, 142] used collinear scheme for double pulse method and suggested mechanisms to explain detected 10-fold intensity increase. They observed improved intensity of ionic aluminum lines but intensity of continuous background was same as for single pulse. Mass removal was increased for double pulse method but on the other hand reduction of atmospheric gases emission was observed. They suggested two mechanisms to explain these results. First, plasma effectively absorbed laser pulse in case of single pulse but for double pulses second laser pulse didn't absorbed by plasma and more energy is transferred to sample thus ablation is enhance and plasma was more effectively excite by second laser pulse. Second, laser plasma from first laser pulse formed a region with air rarefaction and second plasma expands in such rarefied gas with decreased density that provided favorable conditions for plasma excitation.

For double pulse procedure it was supposed by [150] that signal enhancement should be attributed to increased ablated mass and greater size of plasma. For double pulse plasma radiant power was higher. Temperature was slightly higher for double pulse plasma but this increase couldn't explain intensity improvement. Signal increase was explained by greater ablated mass. Bigger craters for double pulses were attributed to more effective coupling of second laser pulse to sample. Noll et. al. [149] suggested that key effect of intensity increase for collinear scheme was the decrease of particle density above samples due to first pulse. Second laser pulse interacts with sample surface without or by substantially plasma shielding.

It was observed that re-heating of first plasma was not observed and second laser pulse more efficiently used for sampling. Second plasma expands more rapidly in reduced particle density atmosphere after first pulse that resulted in greater plasma size. For double pulse and single pulse plasmas same electron density and temperature were detected. Improved signal was explained by increased ablation and effective excitation of plasma in rarefied media. Colao et. al. [179] observed decrease of emission of atmospheric gases and explained it by reduction of air molecules above surface after first laser pulse plasma.

Cristoforetti et. al [163] performed a detailed study of double pulse effect at different pressures. Double pulse intensity increase was maximized for pressure 100 torr for two and single pulses. For single pulse it was observed signal decrease with pressure increase but for double pulses intensity of atomic line in spectra didn't depend on pressure. Intensity enhancement decreased for double pulses if pressure was smaller than 100 torr. Such dependence was described by plasma shielding. First laser pulse formed plasma and after plasma recombination a rarefied area was formed near sample surface. Second pulse generate plasma in rarefied atmosphere thus plasma shielding was smaller that resulted in increased ablated mass and greater plasma dimensions. For double pulse with orthogonal configuration it was observed [166] that same mechanisms should be used for explanation of increased intensity. Plasma shielding was studied for double and single pulse ablation of aluminum samples in detailed paper [170]. Strong influence of plasma shielding on plasma formation was observed for single pulse while shielding was small for plasma formed by second pulse.

It was observed [209] that double pulse procedure was influenced by gas properties above sample. Several noble gases and oxygen were used for study ablation process. If gas density was increasing than double pulse effect for ablated mass was decreasing. It was observed that oxygen have low influence on dependence of evaporated mass. For double pulse sampling in Ar a 8-fold increase of line intensities was detected and was optimal for signal improvement.

Stratis et. al. [150-153] studied double pulse setup with pre-pulse scheme. A substantial increase with 10 to 30 times was observed for atomic lines of copper and lead compared to laser ablation without pre-pulse. The increased signal was explained by larger ablated mass as was detected by crater study. It was shown that signal increase didn't correlate with thermal conductivity of sample or melting point. To describe observed changes for double pulse procedure several mechanisms were suggested. It was stated that first plasma could heat the sample surface that changed optical properties of surface and second laser pulse effectively vaporized sample. Second mechanism is a formation of rarefied environment by pre-spark shockwave that produce favorable conditions for plasma excitation by second pulse. These processes produced increased mass removal and larger plasma volume that suggested to be core factors.

Gautier et. al. [173, 175] studied dependence of double pulse plasma properties on different parameters. The aim of study was to compare pre-spark and re-heating schemes. The optimal delay for pre-spark was about 20 μs and about 200 ns for re-heating scheme. For orthogonal configuration intensity enhancement was higher for re-heating scheme compared to pre-spark. Their next study [174] was focused on collinear scheme and dependence of signal increase for this scheme. It was observed that optimal delay between pulses for ionic greater 2 μs and for atomic less 1 μs. Results were explained by decrease of particle density of atmospheric gases that promotes lower shielding. Hohreiter et. al. [211] carried out double pulse study with pre-spark for analysis of Mg and Si in glass. They haven't detected changes in energy transmission for double pulses. It was suggested that initial breakdown, laser pulse and plasma interaction and plasma dynamics didn't change for double pulse method. To explained increase of Mg and Si neutral lines intensity it was proposed that shockwave formed with pre-spark change conditions for laser sampling with second pulse.

Based on previous studies mechanisms of double pulse signal enhancement is suggested:

1) first plasma additional excitation by second laser pulse;
2) sample heating by first plasma that resulted in increased ablated mass for second pulse;
3) reduction of atmospheric particle density after shockwave front formed by first plasma that resulted in decreased shielding of second laser pulse, increased ablation and provided favorable conditions for second plasma formation

In case of multiple pulse sampling it was suggested that there are three mechanisms of intensity increase in collinear multipulse method [184]: plasma reheating by sequential laser pulses, increased mass ablation due to first pulse preconditioning and additional breakdown with multiple pulses on droplets from previous pulse. Increased ablated mass due to surface heating and air rarefaction were supposed to be main factors of signal enhancement for multiple pulses.

Theoretical studies

Theoretical model have been suggested in [189] to explain double pulse effect. Model was studied laser ablation of solid sample in air. Because model was limited to describe plasma dynamics for first 100 ns than two pulses with delays 10 – 100 ns were studied. It was found that surface temperature is lower for double pulse than for single pulse. For double pulse sampling surface stayed in molten phase and increased evaporation was attributed to melt splash by second pulse. Laser shielding was insignificant for second pulse consequently pulse energy were effectively transferred to sample. According presented models temperature is lower for double pulse method for first tens ns but temperature was decreasing slower and after 200 ns it was higher for double pulses.

Wu et. al. [191] proposed a model to describe double pulse sampling with delay up to 100 ns. Greater values of electron density and temperature were obtained for two UV pulses. Second laser pulse was not directly interacts with sample but shielded by first plasma. Rai et. al. [188] suggested simple model to describe processes and laser plasma dynamics for double pulse procedure. Model was able to describe plasma dynamics for 0 to 30 ms delays between pulses. According this calculations optimal time between pulses was detected to be 4 μs that coincide with experiment. For double pulse plasma temperature was the same as for single pulse plasma. A significant signal improvement was explained by increase of ablated mass for double pulses.

7.4.3.7. Double Pulse Studies for Liquid Samples

The analytical capabilities of LIBS of liquid samples is often poor compared to such method as traditional atomic emission and atomic absorbtion techiques [212, 213]. But the strongest point of LIBS is a possibility of online and remote analysis of samples under hard conditions or hazard environment [4]. The major disadvantage of bulk liquids samples analysis by LIBS is the significantly reduced plasma emission intensity and time duration in comparison with that obtained from ablation on solids in air. Consequently, to overcome problem several procedures were used: laser ablation of water surface (low reproducibility), laser ablation of jets and surface excitation scheme for laminar flows [214], and double pulse technique. Double pulse method proposed by Cremers et. al. [137] allows to analyse bulk liquids with improved sensitivity and reproducibility. Other sampling procedures (liquid jets or laminar flow) were advanced with double pulse method [218].

The idea of two sequential pulses suggested by Cremers et. al. was to use first laser pulse to form bobble with vapor and than second pulse is used to obtain plasma inside bobble. Authors [137] suggested that if laser plasma will be formed inside bobble than conditions will near the same as for laser sampling in air. Two different lasers were used in collinear and orthogonal schemes. It was suggested that collinear scheme was preferred to orthogonal due to better reproducibility of bobble formation. Optimal delay between pulses was estimated 18 μs with up to 8 times enhancement of signal. Double pulses allowed to detect B I 249.68 and Be I 249.45 lines. Improvement of LOD for B I line was more than 15 times with 80 ppm for water bulk samples. Followed paper by the same authors with very detailed experiments has been published only after 12 years [201]. In this study proposed double pulses were used for solid samples analysis in liquid media. Previous supposition was proved for importance of bobble dynamics on increase of lines intensity for double pulse mode. For laser ablation of samples in water a single laser with successive pulses were used. Double pulse method, called repetitive spark pairs in original paper, allowed to detect Al I and Fe I lines that couldn't be detect by single pulse sampling. Shadowgraph studies in combination with plasma imaging have been carried out. According these studies it was suggested that best delay between pulses should be slightly less than the time needed for bobble to reach it maximum size. Optimal ratio for first - to - second pulse has been determined on the basis of bobble dynamics. It was observed that it was preferable to use only small part of energy for first laser pulse since it was needed to produce bobble. Major part of energy should be used for second laser pulse to increase evaporated mass and to excite plasma. Electron temperature was not significantly higher in case of double pulse method in water compared to laser sampling in air: 7000 K for air and 8000 K for water. However electron density was 8 times greater for double pulse sampling in water. Effect of double pulse intensity enhancement was explained by increase of ablated mass and by more proper conditions for plasma excitation. Increase of sensitivity for double pulse method was element depended with up to 20-fold improvements. The idea of bobble formation in order to improve sensitivity was used by Nyga and Neu [215]. Two fibers were used for analysis, first fiber were used for laser pulses delivery while second was used to collect plasma emission. First laser pulse was used to form a bobble on the surface of submerged sample and when bobble radius reached second fiber tip than a second laser pulse was used to form plasma.

Nakamura et. al. [216] have successfully used double pulse method for iron detection in water to detect pipes corrosion degree in water heating systems. Collinear scheme with two laser systems were used to ablate water in nozzle. A signal enhancement of 5 times was achieved for Fe lines. Detection limit was improved more than 40 times with best value of 16 ppb for double pulse method. Kumar et. al. studied double pulse method for laser ablation of liquid jet [217, 218]. For optimal conditions improvement of LOD for Mg was about 40 times and 10 times for Cr. The optimal delay between pulses was detected to be between 2 and 3 μsec.

Bulk liquids analysis with double pulse method was made by Kuwako et al. [219]. They have used both single and double pulse method for sodium solution analysis. For optimal conditions detection limit for sodium was about 0.1 ppb. Pearman et al. [220] analysed water solutions of Ca, Zn, Cr in wide range of laser energy, delay between pulses. Authors used orthogonal scheme and achieved higher increase of signal - to - noise ratio compared to collinear geometry. They have observed 250 intensity increase for best conditions and could achieve 41 ppb for Ca. Double pulse method was also efficient for analysis of Cr in water and allows to improve sensitivity for 9 times [221].

Double pulse method was tested as a perspective technique for water analysis under high pressure [222, 223]. Double pulse effect for signal was decreasing with pressure increasing and disappears above 120 bar pressure. Authors explained such dependence by bobble dynamics in liquid at high pressure. In the next studies analytical capabilities of LIBS have been determined for deep ocean water analysis [224, 225]. It should be noted that optimal delay between pulses was shorter compared to normal pressure experiments and was estimated to be 200 ns according signal optimization. Despite poor detection limits (50 ppm for Ca and Na) it was recognized that double pulse method is a perspective analytical technique for deep ocean study due to ability for online study.

Double pulse method has been successfully used in several applications of laser induced breakdown spectroscopy for analysis of solid samples submerged in sea. Cremers et. al. [201] used single laser for two laser pulses to perform analysis of solids in liquids. Observed increased signal allowed to detect additives in steels with several hundreds ppm. Extensive study of samples analysis under water was made by De Giacomo's group [226, 227]. They use collinear scheme with single laser to produce two pulses for sampling. Strong confinement of plasma in double pulse mode was detected which led to higher temperature and electron density in plasma. Detailed study for estimation of analytical capabilities was performed [228]. Double pulse studies was performed with collinear scheme for analysis of copper samples in water solution. Optimal delay was detected to be 100 μs for atomic lines and 50 μs for ionic lines.

For double pulse improvements of samples under water and in bulk liquids all groups suggested that bobble dynamics played a key role. Cremers et. al. [201] indicated that double pulse mode is depending on bobble formation and evolution. However only in few studies this process was studied theoretically.

De Giacomo et. al [190] have studied theoretically and experimentally the double pulse laser ablation of titan sample in water. They have studied bobble formation and collapse in details and have shown that bobble can be exist for 300 μs. It was observed under studied conditions that bobble reached maximum value at 150 μs and pressure inside bobble decreased to 0.2 atmosphere. However to improve spectra intensity best delay between laser pulses was observed for period of starting bobble growth when plasma was expanding in increasing bobble. Experimental best delay was found to be 80 μs for studied scheme. Theoretical and experiment studies showed that plasma formation is strongly affected by chemical reactions for double pulse plasma

7.4.3.8. Analytical Capabilities of Double Pulse LIBS

A detailed discussion of double pulse method and improved signals was made in previous pages thus this section will discuss most remarkable applications with double pulse LIBS. Special consideration will be paid for important features that was applied in study with double pulses. In most studies for double pulse method it was shown that analytical capabilities showed significant improvement of signal - to - noise ratio resulted in greater sensitivity. In some cases double pulses was used as a unique procedure to implement analysis in situ (e.g. submerged samples). Analytical capabilities of double pulse method are summarized in Table 2.

A great number of impressive double pulse applications were made for analysis of bulk liquids and solids in water by a number of research group but it was discussed above and this section will be focused on analysis of solids in gases. Cristoforetti et. al. [169] used double pulse method to improve analytical capabilities for steel and ceramic analysis. It was observed that intensity enhancement allowed to improve limits of detection more than 10 times. Bazler et. al. [202] applied double pulse method for online measurements of thickness of Zn coating on steel sheets. Double pulses were used to increase ablated mass to be able to measure layer thickness up to 12 μm at production line. Noll et. al. [146] used double pulse method to improve signal for low alloy additives in steel samples and achieved detection limits for carbon, sulfur and phosphor below 9 ppm. Double pulse method was used for control of technological lines in steel industry [147]. It was suggested to use double pulses to increase sensitivity used for analysis of high allow steels. Improved precision of analysis was achieved additionally to better signal for laser sampling with two pulses. Detection limits for 9 elements were below 10 ppm.

Improvement of ablated mass was used for soil analysis with double pulse LIBS [168]. Increased intensity for atomic lines allowed to detect lines of impurities and to carry out calibration-free LIBS. Kwak et. al. [69] studied double pulses for analysis of arsenic in soils. Improved intensity and precision was observed for double pulses that resulted in 2-fold lower detection limits. Study of application of double pulses for analysis of steel samples was made in [299]. Multiple pulses were used as effective and simple way to remove scale layer of samples and to improve analytical capabilities. Galbacz et. al. [183, 184] suggested to use multiple pulses from laser to increase sensitivity of analysis. For multiple pulses signal increased up to 120 times. Economy CCD detector in combination with developed laser system allowed to improve sensitivity and enhance reproducibility of analysis by LIBS. Better analytical capabilities for multiple pulses sampling were achieve for gold alloys [185]. Colao et. al. [230] used double pulse method to control cleaning of copper samples. Double pulse method was used for profiling analysis of ceramics samples [231]. For double pulses scheme with re-heating an improved resolution of depth analysis was observed to be 2.5 μm.

Double pulse method was used [232] for analysis of fossil snake vertebrae and it was proved that this technique could be powerful tool for analysis of fossils. Double pulse technique was used for stand-off analysis for range 20 m [233]. Improved selectivity and higher sensitivity allowed to detect hazardous samples.

Table 2. Double pulse method analytical performance

Element	Line	Matrix	conditions[1]	LOD, ppm SP [2]	LOD, ppm DP	Reference
Solid						
Al I	396.14	steel	collnr, air	30	20	[169]
C I	193.09	steel	collnr, Ar	-	7	[146]
Cr I	425.43	Al	collnr, air	100	10	[169]
Cr I	425.43	steel	collnr, air	70	50	[169]
Cr II	267.72	steel	collnr, Ar	-	7	[146]
Cu I	324.75	steel	collnr, air	25	5	[169]
Cu I	324.75	Al	collnr, air	150	80	[169]
Cu I	327.29	Al	orthgnl, air	2	2	[173]
Fe I	371.99	Al	collnr, air	400	50	[169]
Fe II	259.94	Al	orthgnl, air	6	3	[173]
K I	766.51	glass	orthgnl, air	5.9	3.6	[207]
Mg I	285.21	Al	collnr, air	30	4	[169]
Mg II	280.27	Al	orthgnl, air	0.5	0.2	[173]
Mn I	294.92	Al	collnr, air	100	90	[169]
Mn I	482.34	steel	collnr, air	300	120	[169]
Mn II	294.91	Al	orthgnl, air	8	3	[173]
Ni I	341.47	Al	collnr, air	600	100	[169]
Ni II	231.60	steel	collnr, Ar	-	6	[146]
P I	178.28	steel	collnr, Ar	-	9	[146]
S I	180.73	steel	collnr, Ar	-	8	[146]
Si I	288.16	Al	collnr, air	18	8	[172]
Si I	288.16	steel	collnr, air	100	4	[169]
Ti I	334.9	Al	collnr, air	100	10	[169]
Ti I	308.80	steel	collnr, air	50	25	[169]
Ti II	323.45	Al	orthgnl, air	3	4	[173]
Liquid						
B I	249.45	water / ethanol	collnr	1200	80	[137]
Ca II	422.7	water	orthgnl	-	0.04	[220]
Ca II	422.7	water high pressure	collnr	-	50	[225]
Cr I	425.4	water	orthgnl	-	1	[220]
Fe I	259.96	water	collnr	0.6	0.016	[216]
Na I	589.3	water high pressure	collnr	-	50	[225]
Na I	589.3	water	collnr	0.01	0.0001	[219]
Mg II	279.5	water jet	collnr	0.23	0.06	[217]
Zn I	472.2	water	orthgnl	-	18	[220]
Submerged						
Cu I	324.75	steel in water	collnr	-	520	[201]
Cr I	425.44	steel in water	collnr	-	360	[201]
Mn I	403.45	steel in water	collnr	-	1200	[201]
Si I	288.16	steel in water	collnr	-	1200	[201]

[1] Conditions of double pulse study: double pulse scheme and environmental conditions. Collinear scheme – collnr, orthogonal scheme – ortthngl, atmospheric air - air, other gases - Ar, etc.

[2] Limits of detection for single pulse (SP) method and double pulse method (DP)

7.4.4. Resonant Excitation

Resonance excitation of atoms in plasma by additional light source, such as dye lasers, diode lasers or OPO with non-linear crystals, allows the complete transfer of atoms from ground state to required level under saturation. Two approaches to achieve a positive gain in

LIBS emission were previously developed. First way is based on the resonant excitation of analyte atoms; and then laser-induced fluorescence (LIF) of these atoms is recorded simultaneously with an emission spectrum of laser plasma. In another way atoms of matrix elements being resonantly excited by light source give their excess internal energy to other atoms in plasma (including analyte atoms) through collisions. After thermal redistribution of energy, analyte atoms and, certainly, other collisionally excited atoms are emitted photons. Both ways will be thoroughly discussed below.

The methodology of the first approach and instrumental arrangement is similar to LIF technique combined with laser ablation which thoroughly discussed previously in several excellent reviews [234-236]. Because spectral lines are often overlapping especially in spectra with numerous lines, originating from the main matrix elements in e.g. steels, LIF can provide more selectivity of analysis. Another problem occurring in spectrochemical emission methods is that many analytical lines are only fully resolved at very high spectral resolution. In combination with a limitation of useful area of detector (CCD or photodiode array), this may severely restrict rapid multi-element analysis. Fluorescence technique with LIBS seems to be advantageous for sensitive analysis free from strong spectra interferences due to element-specific excitation in LIF. Since the additional excitation allows the transfer of analyte atoms from their ground level to the desired excited state so that the intensity of the line corresponding to this transition in an emission spectrum increases manyfold. As a result of such transfer, other lines of analyte can be enhanced too. An excitation of Al I line at 396.15 nm observed by Telle with colleagues [237] can be mentioned here as an example when Al I line at 394.40 nm was excited by a tunable laser. Apart from filling of upper level, sometimes the collisional deactivation of upper level observed in dense plasma resulted in appearance of emission lines which have upper level with lower energy than originating transition. So, many lines of Fe I species were obtained to be enhanced in 50-100 times [237].

However, the combination of LIF with LIBS has several restrictions due to physical nature of fluorescence. Resonant excitation can only be used for "cold" plasma, when the number of atoms in the ground state is large and enough to yield the maximum effect of the extra stimulation. The recording of signal is delayed in this case in 2-5 μs after ablation pulse, and the majority of the strongest emission lines is reduced or even disappeared. Another feature of this approach is the fact that only one selected element can be further excited, which leads to the physical impossibility of rapid multi-element analysis of traces. Gobernado-Mitre et al [238] have shown the use of LIF in enhancing weak Al I lines at limestone ablation. Then Hilbk-Kortenbruck et al [239] have demonstrated that the coupling LIBS and LIF determination of cadmium and thallium were more sensitive than LIBS measurements alone (about 20–50 times). Koch et al [240] have recently used resonance excitation to determine indium in liquids. Lui et al [241] have observed resonantly enhanced strong Pb lines. A further loss of simplicity in instrumental arrangement is also the cost of using additional laser sources for LIF to excite selected atoms in the LIBS plasma and, therefore this kind of scheme appears unsuitable to field applications. In contrast to the non-resonant multipulse excitation of plasma, resonance excitation of atoms of matrix elements leads to the transfer of energy through collisions by matrix particles to all other particles in plasma, including the analyte. This approach allows more efficient "pumping" of the plasma by energy and, consequently, increases the intensity of analytical lines by 100 times [242]. However, to implement it requires a tunable laser source, such as dye lasers, diode laser or parametric oscillator.

7.4.5. Plasma Confinement

Another way for compressing laser plasma consists of a spatial confinement of plasma front propagation. It is well known that the formation and expansion of laser-induced plasma in a gas is accompanied by the formation of a shock wave. Actually, if a small obstacle (for example, a plate) is placed across a shock wave path, the wave is back reflected to the plasma front by the obstacle. This back reflection leads to an increase of the number of collisions among particles in the plasma, which in turn originates an increase of the number of atoms in high-energy states and hence an enhancement of emission intensity. This model foresees an enhancement factor for single-pulse LIBS and requires a simpler instrumental arrangement than what is needed in other mentioned approaches. Corsi et al. [44] and Zeng et al. [243] have studied the confinement in the craters or small cavities characterized by a very small size (much smaller than laser plume size). An increase of the emission intensity from the plasma confined in the cavity has been obtained together with a strong broadening of silicon lines. Alternatively, Shen et al. [244]have used a cylindrical pipe with diameters of 11 mm, 14 mm and 20 mm (much larger than plasma size), which confined the laser plume only from two sides. The maximal intensification factor achieved was not exceeding the value of 3. Concerning the mechanical confinement so far the enhancement effect was obtained only for single matrix elements, such as aluminium or silicon. In contrast, Budi with colleagues [245] showed that there is no an increase in the intensity of copper lines in the plasma at reduced pressure (from 2 to 30 Torr). Note, that in their work the distance between the glass plates was about 6 mm, which is approximately equal to the size of the plasma at reduced pressure. Recently, Popov et al. [246] reported the observation of significant, at about 10 times, increasing the ionic lines of iron and atomic lines of arsenic evaporation of standard soil samples. This was achieved by developing a special miniature chamber 4 × 4 mm, having a cylindrical symmetry, with a small hole for the incoming laser beam and a hole for observing the plasma emission. The use of the chamber allows the decrease LODs of several toxic elements (such as arsenic, mercury, lead etc.) in strongly polluted soils. [247] More recently, Guo with colleagues [248] proposed the use of an aluminum hemispherical cavity (diameter = 11.1 mm) to confine plasmas in air from a steel target with a low concentration manganese. A significant enhancement (factor up to 12) in the emission intensity of Mn lines was observed at a laser fluence of 7.8 J /cm^2 when the plasma was confined by the hemispherical cavity, leading to an increase in plasma temperature about 3600 K. It should be noted here that a variety of possible mechanisms responsible for enhancement by using the small cavity is proposed. For example, several investigators supposed that more efficient collection of plasma radiation lead to an increase of emission intensity due to chamber walls.

Acknowledgments

This work is supported by the Russian Foundation for Basic Research, grants No 11-02-01202, 11-03-01187 and by the Ministry of Education and Science of the Russian Federation project 2011-1.2.2-220-010_156.

LITERATURE

[1] R.E. Russo, X. Mao, O.V. Borisov, Laser ablation sampling, *Trends Anal. Chem.* 17 (1998) 461-469.

[2] J.D. Winefordner, I.B. Gornushkin, T. Correll, E. Gibb, B.W. Smith, N. Omenetto, Comparing several atomic spectrometric methods to the super stars: special emphasis on laser induced breakdown spectrometry, LIBS, a future super star, *J. Anal. At. Spectrom.* 19 (2004) 1061-1083.

[3] Cremers D.A., Radziemski L.J. *Handbook of Laser-Induced Breakdown Spectroscopy,* New York, 2006, 283 p.

[4] *Laser Induced Breakdown Spectroscopy (LIBS): Fundamentals and applications* / Ed. by A.W. Miziolek, V. Palleschi and I. Schechter. Cambridge, 2006, 640 p.

[5] *Laser-Induced Breakdown Spectroscopy* / Ed. by J.P. Singh and S.N. Thakur. Elsevier Science, 2007, 429 p.

[6] V. Zorba, X. Mao, R.E. Russo, Optical far- and near-field femtosecond laser ablation of Si for nanoscale chemical analysis, *Anal. Bioanal. Chem.* 396 (2010) 173-180.

[7] J. Markowe, LIBS instrumentation for fully automated industrial applications, 5th Conference on Laser-Induced Breakdown Spectroscopy "LIBS 2008", *Book of Abstract*, Berlin, 2008, p. 178.

[8] Mars Technology Program of NASA, *http://marstech.jpl.nasa.gov/index.cfm.*

[9] Svelto O., *Principles of Lasers*, Springer, New York, 2009, 620 p.

[10] C.J. Lorenzen, C. Carlhoff, U. Hahn, M. Jogwich, Applications of Laser-induced Emission Spectral Analysis for Industrial Process and Quality Control, *J. Anal. At. Spectrom.* 7 (1992) 1029-1035.

[11] J. Scaffidi, S.M. Angel, D.A. Cremers, Emission Enhancement Mechanisms in Dual-Pulse LIBS, *Anal. Chem.* 78 (2006) 24-32.

[12] K.L. Eland, D.N. Stratis, T. Lai, M.A. Berg, S.R. Goode, S.M. Angel, Some Comparisons of LIBS Measurements Using Nanosecond and Picosecond Laser Pulses, *Appl. Spectrosc.* 55 (2001) 279-285.

[13] K.L. Eland, D.N. Stratis, D.M. Gold, S.R. Goode, S.M. Angel, Energy Dependence of Emission Intensity and Temperature in a LIBS Plasma Using Femtosecond Excitation, *Appl. Spectrosc.* 55 (2001) 286-291.

[14] C. Lopez-Moreno, K. Amponsah-Manager, B.W. Smith, I.B. Gornushkin, N. Omenetto, S. Palanco, J.J. Laserna, J.D. Winefordner, Quantitative analysis of low-alloy steel by microchip laser induced breakdown spectroscopy, *J. Anal. At. Spectrom.* 20 (2005) 552-556.

[15] K. Amponsah-Manager, N. Omenetto, B.W. Smith, I.B. Gornushkin, J.D. Winefordner, Microchip laser ablation of metals: investigation of the ablation process in view of its application to laser-induced breakdown spectroscopy, *J. Anal. At. Spectrom.* 20 (2005) 544-551.

[16] G. Cristoforetti, S. Legnaioli, V. Palleschi, A. Salvetti, E. Tognoni, P.A. Benedetti, F. Brioschi, F. Ferrario, Quantitative analysis of aluminium alloys by low-energy, high-repetition rate laser-induced breakdown spectroscopy, *J. Anal. At. Spectrom.* 21 (2006) 697-702.

[17] M. Hoehse, I. Gornushkin, S. Merk, U. Panne, Assessment of suitability of diode pumped solid state lasers for laser induced breakdown and Raman spectroscopy, *J. Anal. At. Spectrom.* 26 (2011) 414-424.

[18] M. Baudelet, C.C.C. Willis, L. Shah, M. Richardson, Laser-induced breakdown spectroscopy of copper with a 2 mu m thulium fiber laser, *Opt. Express* 18 (2010) 7905-7910.

[19] J.-F. Y. Gravel, F.R. Doucet, P. Bouchard, M. Sabsabi, Evaluation of a compact high power pulsed fiber laser source for laser-induced breakdown spectroscopy, *J. Anal. At. Spectrom.*, 25 (2011) doi: 10.1039/C0JA00228C.

[20] C. Barnett, E. Cahoon, J.R. Almirall, Wavelength dependence on the elemental analysis of glass by Laser Induced Breakdown Spectroscopy, *Spectrochim. Acta Part B* 63 (2008) 1016-1023.

[21] ISO Standard 11146, *"Lasers and laser-related equipment – Test methods for laser beam widths, divergence angles and beam propagation ratios"*, 2005

[22] V. Lednev, S.M. Pershin, A.F. Bunkin, Laser beam profile influence on LIBS analytical capabilities: single vs. multimode beam, *J. Anal. At. Spectrom.* 25 (2010) 1745-1757.

[23] C. Chaleard, P. Mauchien, N. Andre, J. Uebbing, J.L. Lacour, C. Geertsen, Correction of matrix effects in quantitative elemental analysis with laser ablation optical emission spectrometry, *J. Anal. At. Spectrom.* 12 (1997) 183–188.

[24] V.I. Babushok, F.C. DeLucia Jr., J.L. Gottfried, C.A. Munson, A.W. Miziolek, Double pulse laser ablation and plasma: Laser induced breakdown spectroscopy signal enhancement, *Spectrochim. Acta Part B* 61 (2006) 999-1014.

[25] H.E. Bauer, F. Leis, K. Niemax, Laser induced breakdown spectrometry with an echelle spectrometer and intensified charge coupled device detection, *Spectrochim. Acta Part B* 53 (1998) 1815-1825.

[26] M. Muller, I. B. Gornushkin, S. Florek, D. Mor, U. Panne, Approach to Detection in Laser-Induced Breakdown *Spectroscopy, Anal. Chem.* 79 (2007) 4419-4426.

[27] L. Fornarini, F. Colao, R. Fantoni, V. Lazic, V. Spizzicchino, Calibration analysis of bronze samples by nanosecond laser induced breakdown spectroscopy: A theoretical and experimental approach, *Spectrochim. Acta Part B* 60 (2005) 1186-1201.

[28] C.C. Garcia, H. Lindner, A. von Bohlen, C. Vadla, K. Niemax, Elemental fractionation and stoichiometric sampling in femtosecond laser ablation, *J. Anal. At. Spectrom.* 23 (2008) 470-478.

[29] V.N. Lednev, S.M. Pershin, Plasma stoichiometry correction method in laser-induced breakdown spectroscopy, *Laser Phys.* 18 (2008) 850–854.

[30] Moenke-Blankenburg L., *Laser Microanalysis*, John Wiley & Sons, New York, 1989, 288 p.

[31] L.A. Falkovsky, E.G. Mishchenko, Electron-lattice kinetics of metals heated by ultrashort laser pulses, *J. Exp. Theor. Phys.* 88 (1999) 84-88.

[32] S.I. Anisimov, B.S. Luk'yanchuk, Selected problems of laser ablation theory, *Phys.-Uspekhi* 45 (2002) 293-324.

[33] R.E. Russo, X. Mao, H. Liu et al., Laser ablation in analytical chemistry - a review, *Talanta.* 57 (2002) 425-451.

[34] Delone N.B., *Basics of Interaction of Laser Radiation with Matter*, Editions 1478 Frontieres, Cedex, France, 1993, 401 p.

[35] S.A. Darke, J.F. Tyson, Interaction of laser radiation with solid materials and its significance to analytical spectrometry, *J. Anal. Atom. Spectrom*. 8 (1993) 145-209.
[36] J.M. Fishburn, M.J. Withford, D.W. Coutts, J.A. Piper, Study of the fluence dependent interplay between laser induced material removal mechanisms in metals: Vaporization, melt displacement and melt ejection, *Appl. Surf. Sci.* 252(2006) 5182-5188.
[37] S.-S. Wellershoff, J. Hohlfeld, J. Güdde, E. Matthias, The role of electron–phonon coupling in femtosecond laser damage of metals, *Appl. Phys. Part A* 69 (1999) S99–S107.
[38] X. Zeng, X.L Mao., R. Greif, R.E. Russo, Experimental investigation of ablation efficiency and plasma expansion during femtosecond and nanosecond laser ablation of silicon, *Appl. Phys. Part A* 80 (2005) 237-241.
[39] G. Seifert, M. Kaempfe, F. Syrowatka, C. Harnagea, D. Hesse, H. Graener, Self-organized structure formation on the bottom of femtosecond laser ablation craters in glass, *Appl. Phys. Part* 81 (2005) 799-803.
[40] J.H. Yoo, S.H. Jeong, R. Greif, R.E. Russo, Explosive change in crater properties during high power nanosecond laser ablation of silicon, *J. Appl. Phys*. 88 (2000) 1638-1649.
[41] M. Guillong, I. Horn, D. Günther, Capabilities of a homogenized 266 nm Nd:YAG laser ablation system for LA-ICP-MS, *J. Anal. At. Spectrom*. 17 (2002) 8-14.
[42] Y. Hirayama, H. Yabe, M. Obara, Selective ablation of AlN ceramic using femtosecond, nanosecond, and microsecond pulsed laser, *J. Appl. Phys*. 89 (2001) 2943-2949.
[43] F. Raimondi, S. Abolhassani, R. Brütsch, F. Geiger, T. Lippert, J. Wambach, Quantification of polyimide carbonization after laser ablation, *J. Appl. Phys*. 88 (2000) 3659-3666.
[44] M. Corsi, G. Cristoforetti, M. Hidalgo, D. Iriarte, S. Legnaioli, V. Palleschi, A. Salvetti, E. Tognoni, Effect of laser-induced crater depth in laser-induced breakdown spectroscopy emission features, *Appl. Spectrosc*. 59 (2005) 853-860.
[45] M. Pardede, T.J. Lie, K.H. Kurniawan, H. Niki, K. Fukumoto, T. Maruyama, K. Kagawa, M.O. Tjia, Crater effects on H and D emission from laser induced low-pressure helium plasma, *J. Appl. Phys*. 106 (2009) 063303.
[46] J. Gonzalez, X.L. Mao, J. Roy, S.S. Mao, R.E. Russo, Comparison of 193, 213 and 266 nm laser ablation ICP-MS, *J. Anal. At. Spectrom*. 17 (2002) 1108-1113.
[47] V. Piscitelli S, M.A. Martínez L., A.J. Fernández C., J.J. González, X.L. Mao, R.E. Russo, Double pulse laser induced breakdown spectroscopy: Experimental study of lead emission intensity dependence on the wavelengths and sample matrix, *Spectrochim. Acta Part B* 64 (2009) 147–154.
[48] L. Fornarini, V. Spizzichino, F. Colao, R. Fantoni, V. Lazic, Influence of laser wavelength on LIBS diagnostics applied to the analysis of ancient bronzes, *Anal. Bioanal. Chem*. 385 (2006) 272–280.
[49] W.L. Yip, N.H. Cheung, Analysis of aluminum alloys by resonance-enhanced laser-induced breakdown spectroscopy: How the beam profile of the ablation laser and the energy of the dye laser affect analytical performance, *Spectrochim. Acta Part B* 64 (2009) 315–322.

[50] C. Aragon, J.A. Aguilera, F. Penalba, Improvements in quantitative analysis of steel composition by laserinduced breakdown spectroscopy at atmospheric pressure using an infrared Nd: YAG laser, *Appl. Spectrosc.* 53 (1999) 1259 – 1267.

[51] I. Bassiotis, A. Diamantopoulou, A. Giannoudakos, F. Roubani-Kalantzopoulou, M. Kompitsas, Effects of experimental parameters in quantitative analysis of steel alloy by laser-induced breakdown spectroscopy, *Spectrochim. Acta Part B* 56 (2001) 671-683.

[52] J. Li, J. Lu, Z. Lin, S. Gong, C. Xie, L. Chang, L. Yang, P. Li, Effects of experimental parameters on elemental analysis of coal by laser-induced breakdown spectroscopy, *Opt. Laser Technol.* 41 (2009) 907–913.

[53] S. Palanco, J. Laserna, Remote sensing instrument for solid samples based on open-path atomic emission spectrometry, *Rev. Sci. Instrum.* 75 (2004) 2068-2074.

[54] T. Čtvrtníčková, F.J. Fortes, L.M. Cabalín, J.J. Laserna, Optical Restriction of Plasma Emission Light for Nanometric Sampling Depth and Depth Profiling of Multilayered Metal Samples, *Appl. Spectrosc.* 61 (2007) 719-724.

[55] Y.-I. Lee, T. L. Thiem, G.-H. Kim, Y.-Y. Teng, J. Sneddon, Interaction of an Excimer-Laser Beam with Metals. Part III: The Effect of a Controlled Atmosphere in Laser-Ablated Plasma Emission, *Appl. Spectrosc.* 46 (1992) 1597 - 1604

[56] B. Salle, D.A. Cremers, S. Maurice, R.C. Wiens Laser-induced breakdown spectroscopy for space exploration applications: Influence of the ambient pressure on the calibration curves prepared from soil and clay samples, *Spectrochim. Acta Part B* 60 (2005) 479-490.

[57] J.M. Vadillo, J.M.F. Romero, C. Rodriguez, J.J. Laserna, Depth-resolved analysis by laser-induced breakdown spectrometry at reduced pressure, *Surf. Interface Anal.* 26 (1998) 995-1000.

[58] J.M. Vadillo, J.M.F. Romero, C. Rodriguez, J.J. Laserna, Effect of plasma shielding on laser ablation rate of pure metals at reduced pressure, *Surf. Interface Anal.* 27 (1999) 1009-1015.

[59] P.K. Diwakar, P.B. Jackson, D.W. Hahn, The effect of multi-component aerosol particles on quantitative laser-induced breakdown spectroscopy: Consideration of localized matrix effects, *Spectrochim. Acta Part B* 62 (2007) 1466 – 1474.

[60] H.-R. Kuhn, J. Koch, R. Hergenröder, K. Niemax, M. Kalberer, D. Günther, Evaluation of different techniques for particle size distribution measurements on laser-generated aerosols, *J. Anal. At. Spectrom.* 20 (2005) 894-900.

[61] H. Lindner, J. Koch, K. Niemax, Production of ultrafine particles by nanosecond laser sampling using orthogonal pre-pulse laser breakdown, *Anal. Chem.* 77 (2005) 7528-7533.

[62] V. Kanicky, V. Otruba, ·J.-M. Mermet, Characterization of acoustic signals produced by ultraviolet LA-ICP-AES, Fresenius *J. Anal. Chem.* 363 (1999) 339–346.

[63] C. Liu, X.L. Mao, S.S. Mao, X. Zeng, R. Greif, R.E. Russo, Nanosecond and femtosecond laser ablation of brass: particulate and ICPMS measurements, *Anal. Chem.* 76 (2004) 379-383.

[64] J. Koch, M. Wälle, J. Pisonero, D. Günther, Performance characteristics of ultra-violet femtosecond laser ablation inductively coupled plasma mass spectrometry at 265 and 200 nm, *J. Anal. At. Spectrom.* 21 (2006) 932-940.

[65] J. Koch, D. Günther, Femtosecond laser ablation inductively coupled plasma mass spectrometry: achievements and remaining problems, *Anal. Bioanal. Chem.* 387 (2007) 149-153.

[66] W. Gerlach, E. Schweitzer, Die Chemische Emissions Spektralanalyse, 1 , L.Voss, Leipzig, 1929.

[67] V. Thomsen, Walther Gerlach and the Foundations of Modern Spectrochemical Analysis, *Spectroscopy* 17 (2002) 117-120.

[68] E.F. Runge, R.W. Minck, F.R. Brian, Spectrochemical analysis using a pulsed laser source, *Spectrochim. Acta Part B* 20 (1964) 733-736.

[69] J.H. Kwak, C. Lenth, C. Salb, E.J. Ko, K.W. Kim, K. Park, Quantitative analysis of arsenic in mine tailing soils using double pulse-laser induced breakdown spectroscopy, *Spectrochim. Acta Part B* 64 (2009) 1105 – 1110.

[70] F.C. De Lucia, Jr., R.S. Harmon, K.L. McNesby, R.J. Winkel, Jr., A.W. Miziolek, Laser-induced breakdown spectroscopy analysis of energetic materials, *Appl. Optics* 42 (2003) 6148-6152.

[71] S. Rai, A.K. Rai, S.N. Thakur, Identification of nitro-compounds with LIBS, *Appl. Phys. Part B* 91 (2008) 645–650.

[72] M. Tran, Q. Sun, B.W. Smith, J.D. Winefordner, Determination of C:H:O:N ratios in solid organic compounds by laser-induced plasma spectroscopy, *J. Anal. At. Spectrom.* 16 (2001) 628-632.

[73] S. Kaski, H Häkkänen, J. Korppi-Tommola, Determination of Cl/C and Br/C ratios in pure organic solids using laser-induced plasma spectroscopy in near vacuum ultraviolet, *J. Anal. At. Spectrom.* 19 (2004) 474–478.

[74] J. A.M. van der Mullen, Excitation equilibria in plasmas; a classification, *Phys. Rep.* 191 (1990) 109-220.

[75] R.S. Adrain, J. Watson, Laser microspectral analysis: a review of principles and applications, *J. Phys. Part D* 17 (1984) 1915-1940.

[76] G.W. Loge, *Method for determining the concentration of atomic species in gases and solids,* Patent no. WO97/15811 (1997).

[77] D. Mukherjee, A. Rai, M.R. Zachariah, Quantitative laser-induced breakdown spectroscopy for aerosols via internal calibration: Application to the oxidative coating of aluminum nanoparticles, *Aerosol Sci.* 37 (2006) 677–695.

[78] U. Panne, C. Haisch, M. Clara, R. Niessner, Analysis of glass and glass melts during the vitrification process of fly and bottom ashes by laser-induced plasma spectroscopy. Part I: Normalization and plasma diagnostics, *Spectrochim. Acta Part B* 53 (1998) 1957–1968.

[79] A.S. Eppler, D.A. Cremers, D.D. Hickmott, M.J. Ferris, .A.C. Koskelo, Matrix Effects in the Detection of Pb and Ba in Soils Using Laser-Induced Breakdown Spectroscopy, *Appl. Spectrosc.* 50 (1996) 1175-1181.

[80] L.M. Cabalín, D. Romero, C.C. García, J.M. Baena,·J.J. Laserna, Time-resolved laser-induced plasma spectrometry for determination of minor elements in steelmaking process samples, *Anal. Bioanal. Chem.* 372 (2002) 352–359.

[81] P. Lucena, J.J. Laserna, Three-dimensional distribution analysis of platinum, palladium and rhodium in auto catalytic converters using imaging-mode laser-induced breakdown spectrometry, *Spectrochim. Acta Part B* 56 (2001) 177-185.

[82] L. St-Onge, J.F. Archambault, E. Kwong, M. Sabsabi, E.B. Vadas, Rapid quantitative analysis of magnesium stearate in tablets using laser induced breakdown spectroscopy, *J. Pharm. Pharmaceut. Sci.* 8 (2005) 272-288.

[83] H. Fink, U. Panne, R. Niessner, Process analysis of recycled thermoplasts from consumer electronics by laser-induced plasma spectroscopy, *Anal. Chem.* 74 (2002) 4334- 4342.

[84] B. Sallé, J.L. Lacour, P. Mauchien, P. Fichet, S. Maurice, G. Manhès, Comparative study of different methodologies for quantitative rock analysis by Laser-Induced Breakdown Spectroscopy in a simulated Martian atmosphere, *Spectrochim. Acta Part B* 61 (2006) 301-313.

[85] M.A. Ismail, H. Imam, A. Elhassan, W.T. Youniss, M.A. Harith, LIBS limit of detection and plasma parameters of some elements in two different metallic matrices, *J. Anal. At. Spectrom.* 19 (2004) 489-494.

[86] S. Koch, R. Court, W. Garen, W. Neu, R. Reuter, Detection of manganese in solution in cavitation bubbles using laser induced breakdown spectroscopy, *Spectrochimica Acta Part B* 60 (2005) 1230–1235.

[87] A.R. Boyain-Goitia, D.C.S. Beddows, B.C. Griffiths, H.H. Telle, Single-pollen analysis by laser-induced breakdown spectroscopy and Raman microscopy, *Appl. Optics 42* (2003) 6119-6132.

[88] L. Zhang, L. Dong, H. Dou, W. Yin, S. Jia, Laser-Induced Breakdown Spectroscopy for Determination of the Organic Oxygen Content in Anthracite Coal Under Atmospheric Conditions, *Appl. Spectrosc.* 62 (2008) 458-463.

[89] C. López-Moreno, S. Palanco, J.J. Laserna, Quantitative analysis of samples at high temperature with remote laser-induced breakdown spectrometry using a room-temperature calibration plot, *Spectrochim. Acta Part B* 60 (2005) 1034 – 1039.

[90] O.V. Borisov, X.L. Mao, A. Fernandez, M. Caetano, R.E. Russo, Inductively couplet plasma mass spectrometric study of non-linear calibration behaviour during laser ablation of binary Cu–Zn alloys, *Spectrochim. Acta Part B* 54 (1999) 1351–1365.

[91] J.B. Ko, W. Sdorra, K. Niemax, On the internal standardization in optical emission spectrometry of microplasmas produced by laser ablation of solid samples, Fresenius *J. Anal. Chem.* 335 (1989) 648- 651.

[92] L. St-Onge, M. Sabsabi, P. Cielo, Quantitative analysis of additives in solid zinc alloys by laser-induced plasma spectrometry, *J. Anal. At. Spectrom.* 12 (1997) 997–1004.

[93] V. Margetic, A. Pakulev, A. Stockhaus, M. Bolshov, K. Niemax, R. Hergenröder, A comparison of nanosecond and femtosecond laser-induced plasma spectroscopy of brass samples, *Spectrochim. Acta Part B* 55 (2000) 1771-1785.

[94] A. Elhassan, A. Giakoumaki, D. Anglos, G.M. Ingo, L. Robbiola, M.A. Harith, Nanosecond and femtosecond laser induced breakdown spectroscopic analysis of bronze alloys, *Spectrochim. Acta Part B* 63 (2008) 504–511.

[95] S.M. Pershin, F. Colao, V. Spizzichino, Quantitative analysis of bronze samples by laser-induced breakdown spectroscopy (LIBS): a new approach, model, and experiment, *Laser Phys.* 16 (2006) 455–467.

[96] A. Santagata, R. Teghil, G. Albano, D. Spera, P. Villani, A. De Bonis, G.P. Parisi, A. Galasso, fs/ns dual-pulse LIBS analytic survey for copper-based alloys, *Appl. Surf. Sci.* 254 (2007) 863–867.

[97] R. Teghil, A. Santagata, A. De Bonis, G. Albano, P. Villani, D. Spera, G. P. Parisi and A. Galasso, Applications of ultra-short pulsed laser ablation: thin films deposition and fs/ns dual-pulse laser-induced breakdown spectroscopy, *Phys. Scripta* 78 (2008) 058113.

[98] V. Bulatov, R. Krasniker, I. Schechter, Converting spatial to pseudotemporal resolution in laser plasma analysis by simultaneous multifiber spectroscopy, *Anal. Chem.* 72 (2000) 2987–2994.

[99] L. Sun, H. Yu, Correction of self-absorption effect in calibration-free laser-induced breakdown spectroscopy by an internal reference method, *Talanta* 79 (2009) 388–395.

[100] S. Palanco, J, Laserna, Remote sensing instrument for solid samples based on open-path atomic emission spectrometry, *Rev. Sci. Instrum.*, 75 (2004) 2068-2074.

[101] M. Lentjes, K. Dickmann, J. Meijer, Calculation and optimization of sample identification by laser induced breakdown spectroscopy via correlation analysis, *Spectrochim. Acta Part B* 62 (2007) 56–62.

[102] V. Lazic, F. Colao, R. Fantoni, V. Spizzichino, Laser induced breakdown spectroscopy in water: improvement of the detection threshold by signal processing, *Spectrochim. Acta Part B* 60 (2005) 1002–1013.

[103] V. Lazic, F. Colao, R. Fantoni, V. Spizzichino, S. Jovićević, Underwater sediment analyses by laser induced breakdown spectroscopy and calibration procedure for fluctuating plasma parameters, *Spectrochim. Acta Part B* 62 (2007) 30-39.

[104] D.W. Hahn, W.L. Flower, K. Henken, Discrete particle detection and metal emissions monitoring using laser-induced breakdown spectroscopy, *Appl. Spectrosc.* 51 (1997) 1836–1844.

[105] J.E. Carranza, D.W. Hahn, Sampling statistics and considerations for single-shot analysis using laser-induced breakdown spectroscopy, *Spectrochim. Acta Part B* 57 (2002) 779–790.

[106] J.E. Carranza, K. Lida, D.W. Hahn, Conditional data processing for single-shot spectral analysis by use of laser-induced breakdown spectroscopy, *Appl. Opt.* 42 (2003) 6022–6028.

[107] Compendium of Analytical Nomenclature (Orange Book of IUPAC). *Definitive Rules* 1997. Part 18.4.3.2.

[108] *Demtröder W.*, Atoms, Molecules and Photons. *An Introduction to Atomic-, Molecular and Quantum-Physics*, Springer, 2010, 589 p.

[109] F. Brech, L. Cross, Optical Microemission Stimulated by a Ruby MASER, *Appl. Spectrosc.* 16 (1962) 59.

[110] S.D. Rasberry, B.F. Scribner, M. Margoshes, Laser probe excitation in spectrchemical analysis. I Investigation of quantitative aspects, *Appl. Optics* 6 (1967) 87–93.

[111] O.A. Nassef, H.E. Elsayed-Ali, Spark discharge assisted laser induced breakdown spectroscopy, *Spectrochim. Acta Part B* 60 (2005) 1564–1572.

[112] A.J.R. Hunter, R.T. Wainner, L.G. Piper, S.J. Davis, Rapid field screening of soils for heavy metals with spark-induced breakdown spectroscopy, *Appl. Optics* 42 (2003) 2102–2109.

[113] H. Matsuta, K. Kitagawa, K. Wagatsuma, Stabilization of the Spark-discharge Point on a Sample Surface by Laser Irradiation for Steel Analysis, *Anal. Sci.*, 22 (2006) 1275-1277.

[114] Y. Chen, Q. Zhang, G. Li, R. Li, J. Zhou, Laser ignition assisted spark-induced breakdown spectroscopy for the ultra-sensitive detection of trace mercury ions in aqueous solutions, *J. Anal. At. Spectrom.* 25 (2010) 1969-1973.

[115] A.J.R. Hunter, S.J. Davis, L.G. Piper, K.W. Holtzclaw, M.E. Fraser, Spark-Induced Breakdown Spectroscopy: A New Technique for Monitoring Heavy Metals, *Appl. Spectrosc.* 54 (2000) 575-582.

[116] A.J.R. Hunter, J.R. Morency, C.L. Senior, S.J. Davis, M.E. Fraser, Continuous emissions monitoring using spark-induced breakdown spectroscopy, *J. Air & Waste Manage. Assoc.* 50 (2000) 111-117.

[117] M.V. Belkov, V.S. Burakov, A. De Giacomo, V.V. Kiris, S.N. Raikov, N.V. Tarasenko, Comparison of two laser-induced breakdown spectroscopy techniques for total carbon measurement in soils, *Spectrochim. Acta Part B* 64 (2009) 899-904.

[118] K. Li, W. Zhou, Q. Shen, Z. Ren, B. Peng, Laser ablation assisted spark induced breakdown spectroscopy on soil samples, *J. Anal. At. Spectrom.* 25 (2010) 1475-1481.

[119] K. Li, W. Zhou, Q. Shen, J. Shao, H. Qian, Signal enhancement of lead and arsenic in soil using laser ablation combined with fast electric discharge, *Spectrochim. Acta Part B,* 65 (2010) 420-424.

[120] K.A. Tereszchuk, J.M. Vadillo, J.J. Laserna, Energy assistance in laser induced plasma spectrometry (LIPS) by a synchronized microsecond-pulsed glow discharge secondary excitation, *J. Anal. At. Spectrom.* 22 (2007) 183-186.

[121] K.A. Tereszchuk, J.M. Vadillo, J.J. Laserna, Glow-Discharge-Assisted Laser-Induced Breakdown Spectroscopy: Increased Sensitivity in Solid Analysis, *Appl. Spectrosc.* 62 (2008) 1262-1267.

[122] K.A. Tereszchuk, J.M. Vadillo, J.J. Laserna, Depth profile analysis of layered samples using glow discharge assisted Laser-induced Breakdown Spectrometry (GD-LIBS), *Spectrochim. Acta Part B* 64 (2009) 378-383.

[123] D.K. Bhadra, Expansion of a Resistive Plasmoid in a Magnetic Field, *Phys. Fluids* 11 (1968) 234-239.

[124] J.D. Huba, A.B. Hassam, D. Winske, Stability of sub - Alfvénic plasma expansions, *Phys. Fluids Part B* 2 (1990) 1676-1697.

[125] V.N. Rai, M. Shukla, H.C. Pant, An x-ray biplanar photodiode and the x-ray emission from magnetically confined laser produced plasma, *Pramana J. Phys.* 52 (1999) 49–65.

[126] X.K. Shen, Y.F. Lu, T. Gebre, H. Ling, Y.X. Han, Optical emission in magnetically confined laser-induced breakdown spectroscopy, *J. Appl. Phys.* 100 (2006) 053303.

[127] H.C. Joshi, Ajai Kumar, R.K. Singh, V. Prahlad, Effect of a transverse magnetic field on the plume emission in laser-produced plasma: An atomic analysis, *Spectrochim. Acta Part B* 65 (2010) 415-419.

[128] M.M. Stoiljković, I.A. Pašti, M.D. Momčilović, J.J. Savović, M.S. Pavlović, Principal component analysis of the main factors of line intensity enhancements observed in oscillating direct current plasma, *Spectrochim. Acta Part B* 65 (2010) 927-934.

[129] A. Petrakiev, G. Dimitrov, L. Georgieva, Laser microspectral analysis in a magnetic field, *Spectrosc. Lett.* 2 (1969) 97-106.

[130] K.J. Mason, J.M. Goldberg, Production and Initial Characterization of a Laser-Induced Plasma in a Pulsed Magnetic Field for Atomic Spectrometry, *Anal. Chem.* 59 (1987) 1250-1255.

[131] K.J. Mason, J.M. Goldberg, Characterization of a Laser Plasma in a Pulsed Magnetic Field. Part II: Time-Resolved Emission and Absorption Studies, *Appl. Spectrosc.* 45 (1991) 1444-1455.

[132] K.J. Mason, J.M. Goldberg, Characterization of a Laser Plasma in a Pulsed Magnetic Field. Part I: Spatially Resolved Emission Studies, *Appl. Spectrosc.* 45 (1991) 370-379.

[133] V.N. Rai, H. Zhang, F.Y. Yueh, J.P. Singh, A. Kumar, Effect of steady magnetic field on laser-induced breakdown spectroscopy, *Appl. Optics* 42 (2003) 3662-3669.

[134] V.N. Rai, A.K. Rai, F.-Y. Yueh, J.P. Singh, Optical emission from laser-induced breakdown plasma of solid and liquid samples in the presence of a magnetic field, *Appl. Optics* 42 (2003) 2085-2093.

[135] V.N. Rai, J.P. Singh, F.Y. Yueh, R.L. Cook, Study of optical emission from laser-produced plasma expanding across an external magnetic field, *Laser Part. Beams 21* (2003) 65-71.

[136] A. Neogi, R.K. Thareja, Instabilities in laser-produced carbon plasma expanding in a nonuniform magnetic field, *Appl. Phys. Part B* 72 (2001) 231-235.

[137] D.A. Cremers, L.J. Radziemski, T.R. Loree, Spectrochemical Analysis of Liquids Using the Laser Spark, *Appl. Spectrosc.* 38 (1984) 721-728.

[138] E.H. Piepmeier, H.V. Malmstadt, Q-Switched Laser Energy Absorption in Plume of an Aluminum Alloy, *Anal. Chem.* 41 (1969) 700–707.

[139] S.M. Pershin, Transformation of the optical spectrum of a laser plasma when a surface is irradiated with a double pulse, *Sov. J. Quantum Electron.* 19 (1989) 215-218.

[140] G.P. Arumov, A.Yu. Bukharov, O.V. Kamenskaya, S.Yu. Kotyanin, V.A. Krivoschekov, A.N. Lyash, V.A. Nekhaenko, S.M. Pershin, A.V. Yuzgin, Effect of surface irradiation regime on emission spectrum of a laser pulse, *Sov. Tech. Phys. Lett.* 13 (1987) 362–363.

[141] G.P. Arumov, A.Yu. Bukharov, V.A. Nekhaenko, S.M. Pershin, Double-pulse YAG:Nd^{3+} laser with a controllable delay in the 20 – 100 ns range, *Sov. J. Quantum Electron.* 18 (1988) 1085-1088.

[142] S.M. Pershin, Physical mechanism of suppression of the emission of radiation by atmospheric gases in a plasma formed as a result of two-pulse irradiation of the surface, *Sov. J Quantum. Electron.* 19 (1989) 1618-1619.

[143] S.M. Pershin, A.Yu. Bukharov, Enhancement of the contrast of laser plasma emission spectra accompanying two-pulse irradiation of a surface by neodymium laser radiation, *Sov. J. Quantum. Electron.* 22 (1992) 405-407.

[144] J. Uebbing, J. Brust, W. Sdorra, F. Leis, K. Niemax, Reheating of Laser-Produced Plasma by Second Pulse Laser, *Appl. Spectrosc.* 45 (1991) 1419-1423.

[145] R. Sattmann, V. Sturm, R. Noll, Laser-Induced Breakdown Spectroscopy of Steel Samples Using Multiple Q-Switch Nd-YAG Laser-Pulses, *J. Phys. Part D* 28 (1995) 2181-2187.

[146] V. Sturm, L. Peter, R. Noll, Steel analysis with laser-induced breakdown spectrometry in the vacuum ultraviolet, *Appl. Spectrosc.* 54 (2000) 1275-1278.

[147] L. Peter, V. Sturm, R. Noll, Liquid steel analysis with laser-induced breakdown spectrometry in the vacuum ultraviolet, *Appl. Optics* 42 (2003) 6199-6204.

[148] R. Noll, H. Bette, A. Brysch, M. Kraushaar, I. Monch, L. Peter, V. Sturm, Laser-induced breakdown spectrometry - applications for production control and quality assurance in the steel industry, *Spectrochim. Acta Part B* 56 (2001) 637-649.

[149] R. Noll, R. Sattmann, V. Sturm, S. Winkelmann, Space- and time-resolved dynamics of plasmas generated by laser double pulses interacting with metallic samples, *J. Anal. At. Spectrom.* 19 (2004) 419-428.

[150] D.N. Stratis, K.L. Eland, S.M. Angel, Dual-Pulse LIBS Using a Pre-ablation Spark for Enhanced Ablation and Emission, *Appl. Spectrosc.* 54 (2000) 1270-1274.

[151] D.N. Stratis, K.L. Eland, S.M. Angel, Enhancement of aluminum, titanium, and iron in glass using pre-ablation spark dual-pulse LIBS, *Appl. Spectrosc.* 54 (2000) 1719-1726.

[152] J. Scaffidi, J. Pender, W. Pearman, S.R. Goode, B.W. Colston, Jr., J.C. Carter, S.M. Angel, Dual-pulse laser-induced breakdown spectroscopy with combinations of femtosecond and nanosecond laser pulses, *Appl. Optics* 42 (2003) 6099-6106.

[153] D.N. Stratis, K.L. Eland, S.M. Angel, Effect of Pulse Delay Time on a Pre-ablation Dual-Pulse LIBS Plasma, *Appl. Spectrosc.* 55 (2001) 1297-1303.

[154] J. Scaffidi, W. Pearman, M. Lawrence, J.C. Carter, B.W. Colston Jr., S.M. Angel, Spatial and temporal dependence of interspark interactions in femtosecond–nanosecond dual-pulse laser-induced breakdown spectroscopy, *Appl. Optics* 43 (2004) 5243-5250.

[155] S.M. Angel, D.N. Stratis, K.L. Eland, T. Lai, M.A. Berg, D.M. Gold, LIBS using dual- and ultra-short laser pulses, *Fresenius J. Anal. Chem.* 369 (2001) 320-327.

[156] J. Scaffidi, W. Pearman, J.C. Carter, B.W. Colston, S.M. Angel, Temporal Dependence of the Enhancement of Material Removal in Femtosecond-Nanosecond Dual-Pulse Laser-Induced Breakdown Spectroscopy, *Appl. Optics* 43 (2004) 6492-6499.

[157] A. Semerok, C. Dutouquet, Ultrashort double pulse laser ablation of metals, *Thin Solid Films* 501 (2004) 453–454.

[158] P. Mukherjee, S. Chen, S. Witanachchi, Effect of initial plasma geometry and temperature on dynamic plume expansion in dual-laser ablation, *Appl. Phys. Lett.* 74 (1999) 1546-1548.

[159] M. Oba, Y. Maruyama, K. Akaoka, M. Miyabe, I. Wakaida, Double-pulse LIBS of gadolinium oxide ablated by femto- and nano-second laser pulses, *Appl. Phys. Part A* 101 (2010) 545-549.

[160] A. Santagata, D. Spera, G. Albano, R. Teghil, G. P. Parisi, A. De Bonis, P. Villani, Orthogonal fs/ns double-pulse libs for copper-based-alloy analysis, *Appl. Phys. Part A* 93 (2008) 929-934.

[161] T.Y. Choi, D.J. Hwang, C.P. Grigoropoulos, Femtosecond laser induced ablation of crystalline silicon upon double beam irradiation, *Appl. Surf. Sci.* 197–198 (2002) 720-725.

[162] V. Pinon, C. Fotakis, G. Nicolas, D. Anglos, Double pulse laser-induced breakdown spectroscopy with femtosecond laser pulses, *Spectrochim. Acta Part B* 63 (2008) 1006-1010.

[163] G. Cristoforetti, S. Legnaioli, V. Palleschi, A. Salvetti, E. Tognoni, Influence of ambient pressure on laser-induced breakdown spectroscopy in the parallel double - pulse configuration, *Spectrochim. Acta Part B* 59 (2004) 1907-1917.

[164] M. Corsi, G. Cristoforetti, M. Giuffrida, M. Hidalgo, S. Legnaioli, V. Palleschi, A. Salvetti, E. Tognoni, C. Vallebona, Three-dimensional analysis of laser induced plasmas in single and double pulse configuration, *Spectrochim. Acta Part B* 59 (2004) 723-735.

[165] P.A. Benedetti, G. Cristoforetti, S. Legnaioli, V. Palleschi, L. Pardini, A. Salvetti, E. Tognoni, Effect of laser pulse energies in laser induced breakdown spectroscopy in double-pulse configuration, *Spectrochim. Acta Part B* 60 (2005) 1392-1401.

[166] G. Cristoforetti, S. Legnaioli, L. Pardini, V. Palleschi, A. Salvetti, E. Tognoni, Spectroscopic and shadowgraphic analysis of laser induced plasmas in orthogonal double pulse pre-ablation configuration, *Spectrochim. Acta Part B* 61 (2006) 340-350.

[167] G. Cristoforetti, S. Legnaioli, V. Palleschi, A. Salvetti, E. Tognoni, Characterization of a collinear double pulse laser-induced plasma at several ambient gas pressures by spectrally- and time-resolved imaging, *Appl. Phys. Part B* 80 (2005) 559-568.

[168] M. Corsi, G. Cristoforetti, M. Hidalgo, S. Legnaioli, V. Palleschi, A. Salvetti, E. Tognoni, C. Vallebona, Double pulse calibration-free laser-induced breakdown spectroscopy: a new tool for in-situ analysis of polluted soils, *Appl. Geochem.* 21 (2006) 748-755.

[169] M.A. Ismail, G, Cristoforetti, S, Legnaioli, L. Pardini, V. Palleschi, A. Salvetti, E. Tognoni, M.A. Harith, Comparison of detection limits, for two metallic matrices, of laser-induced breakdown spectroscopy in the single and double-pulse configurations, *Anal. Bioanal. Chem.* 385 (2006) 316-325.

[170] G. Cristoforetti, G. Lorenzetti, P.A. Benedetti, E. Tognoni, S. Legnaioli, V. Palleschi, Effect of laser parameters on plasma shielding in single and double pulse configurations during the ablation of an aluminium target, *J. Phys. Part D* 42 (2009) 225207.

[171] G. Cristoforetti, S. Legnaioli, V. Palleschi, A. Salvetti, E. Tognoni, Effect of target composition on the emission enhancement observed in Double-Pulse Laser-Induced Breakdown Spectroscopy, *Spectrochim. Acta Part B* 63 (2008) 312-323.

[172] C. Gautier, P. Fichet, D. Menut, J.-L. Lacour, D. L'Hermite, J. Dubessy, Main parameters influencing the double-pulse laser-induced breakdown spectroscopy in collinear beam geometry, *Spectrochim. Acta Part B* 60 (2005) 792-804.

[173] C. Gautier, P. Fichet, D. Menut, J.-L. Lacour, D. L'Hermite, J. Dubessy, Study of double-pulse setup with an orthogonal beam geometry for laser-induced breakdown spectroscopy, *Spectrochim. Acta Part B* 59 (2004) 975-986.

[174] C. Gautier, P. Fichet, D. Menut, J. Dubessy, Applications of the double-pulse laser-induced breakdown spectroscopy (LIBS) in the collinear beam geometry to the elemental analysis of different materials, *Spectrochim. Acta Part B* 61 (2006) 210-219.

[175] C. Gautier, P. Fichet, D. Menut, J.-L. Lacour, D. L'Hermite, J. Dubessy, Quantification of the intensity enhancements for the double-pulse laser-induced breakdown spectroscopy in the orthogonal beam geometry, *Spectrochim. Acta Part B* 60 (2005) 265-276.

[176] L. St-Onge, V. Detalle, M. Sabsabi, Enhanced laser-induced breakdown spectroscopy using the combination of fourth-harmonic and fundamental Nd:YAG laser pulses, *Spectrochim. Acta Part B* 57 (2002) 121-135.

[177] L. St-Onge, M. Sabsabi, P. Cielo, Analysis of solids using laser-induced plasma spectroscopy in double-pulse mode, *Spectrochim. Acta Part B* 53 (1998) 407–415.

[178] F. Colao, V. Lazic, R. Fantoni, S. Pershin, A comparison of single and double pulse laser-induced breakdown spectroscopy of aluminum samples, *Spectrochim. Acta Part B* 57 (2002) 1167-1179.

[179] F. Colao, S. Pershin, V. Lazic, R. Fantoni, Investigation of the mechanisms involved in formation and decay of laser-produced plasmas, *Appl. Surf. Sci.* 197-198 (2002) 207-212.

[180] L. Caneve, F. Colao, R. Fantoni, V. Spizzichino, Laser ablation of copper based alloys by single and double pulse laser induced breakdown spectroscopy, *Appl. Phys. Part A* 85 (2006) 151-157.

[181] V. Lazic, F. Colao, R. Fantoni, V. Spizzicchino, Recognition of archeological materials underwater by laser induced breakdown spectroscopy, *Spectrochim. Acta Part B* 60 (2005) 1014-1024.

[182] X. Mao, X. Zeng, S.-B. Wen, R.E. Russo, Time-resolved plasma properties for double pulsed laser-induced breakdown spectroscopy of silicon, *Spectrochim. Acta Part B* 60 (2005) 960-967.

[183] G. Galbacs, V. Budavari, Z. Geretovszky, Multi-pulse laser-induced plasma spectroscopy using a single laser source and a compact spectrometer, *J. Anal. At. Spectrom.* 20 (2005) 974-980.

[184] G. Galbacs, N. Jedlinszki, K. Herrera, N. Omenetto, B.W. Smith, J.D. Winefordner, Study of Ablation, Spatial, and Temporal Characteristics of Laser-Induced Plasmas Generated by Multiple Collinear Pulses, *Appl. Spectrosc.* 64 (2010) 161-172.

[185] G. Galbacs, N. Jedlinszki, G. Cseh, Z. Galbacs, L. Turi, Accurate quantitative analysis of gold alloys using multi-pulse laser induced breakdown spectroscopy and a correlation-based calibration method, *Spectrochim. Acta Part B* 63 (2008) 591-597.

[186] B.C. Windom, P.K. Diwakar, D.W. Hahn, Dual-pulse Laser Induced Breakdown Spectroscopy for analysis of gaseousand aerosol systems: Plasma-analyte interactions, *Spectrochim. Acta Part B* 61 (2006) 788-796.

[187] M.E. Asgill, M.S. Brown, K. Frische, W.M. Roquemore, D.W. Hahn, Double-pulse and single-pulse laser-induced breakdown spectroscopy for distinguishing between gaseous and particulate phase analytes, *Appl. Optics* 49 (2010) C110-C119.

[188] V.N. Rai, F.Y. Yueh, J.P. Singh, Theoretical model for double pulse laser-induced breakdown spectroscopy, *Appl. Optics* 47 (2008) G30-G37.

[189] A. Bogaerts, Z. Chen, D. Autrique, Double pulse laser ablation and laser induced breakdown spectroscopy: a modeling investigation, *Spectrochim. Acta Part B* 63 (2008) 746-754.

[190] A. De Giacomo, M. Dell'Aglio, D. Bruno, R. Gaudiuso, O. De Pascale, Experimental and theoretical comparison of single-pulse and double-pulse laser induced breakdown spectroscopy on metallic samples, *Spectrochim. Acta Part B* 60 (2005) 975–985.

[191] B. Wu, Y. Zhou, A. Forsman, Study of laser-plasma interaction using a physics-based model for understanding the physical mechanism of double-pulse effect in nanosecond laser ablation, *Appl. Phys. Lett.* 95 (2009) 251109.

[192] M. Miclea, C.C. Garcia, I. Exius, H. Lindner, K. Niemax, Emission spectroscopic monitoring of particle composition, size and transport in laser ablation inductively coupled plasma spectrometry, *Spectrochim. Acta Part B* 61 (2006) 361-367.

[193] N. Jegenyes, Z. Toth, B. Hopp, J. Klebniczki, Z. Bor, C. Fotakis, Femtosecond pulsed laser deposition of diamond-like carbon films: The effect of double laser pulses, *Appl. Surf. Sci.* 252 (2006) 4667-4671.

[194] M. Sivakumar, B. Tan, K. Venkatakrishnan, Enhancement of silicon nanostructures generation using dual wavelength double pulse femtosecond laser under ambient condition, *J. Appl. Phys*. 107 (2010) 044307.

[195] S. Suzuki, R. Sen, T. Tamaki, H. Kataura, Y. Achiba, Single-walled carbon nanotube formation with double laser vaporization technique, *Eur. Phys. J. Part D* 24 (2003) 401-404.

[196] V.S. Burakov, A.V. Butsen, N.V. Tarasenko, Laser-induced plasmas in liquids for nanoparticle synthesis, *J. Appl. Spectrosc.* 77 (2010) 386-393.

[197] P. Mukherjee, S. Chen, J.B. Cuff, P. Sakthivel, S. Witanachchi, Evidence for the physical basis and universality of the elimination of particulates using dual-laser ablation. 1. Dynamic time-resolved target melt studies, and film growth of Y2O3 and ZnO, *J. Appl. Phys*. 91 (2002) 1828-1836.

[198] S.M. Pershin, Nonlinear increase in the interaction efficiency of a second pulse with a target upon excitation of a plasma by a train of pulses from a Nd:YAG laser, *Quantum Electron.* 39 (2009) 63-67.

[199] G. Cristoforetti, S. Legnaioli, V. Palleschi, E. Tognoni, P.A. Benedetti, Crater drilling enhancement obtained in parallel non-collinear double-pulse laser ablation, *Appl. Phys. Part A* 98 (2010) 219-225.

[200] S.M. Klimentov, S.V. Garnov, T.V. Kononenko, V.I. Konov, P.A. Pivovarov, F. Dausinger, High rate deep channel ablative formation by picosecond-nanosecond combined laser pulses, *Appl. Phys. Part A* 69 (1999) S633-S636.

[201] A.E. Pichahchy, D.A. Cremers, M.J. Ferris, Elemental analysis of metals under water using laser-induced breakdown spectroscopy, *Spectrochim. Acta Part B* 52 (1997) 25-39.

[202] H. Balzer, M. Hoehne, V. Sturm, R. Noll, Online coating thickness measurement and depth profiling of zinc coated sheet steel by laser-induced breakdown spectroscopy, *Spectrochim. Acta Part B* 60 (2005) 1172-1178.

[203] V.S. Burakov, A.F. Bokhonov, M.I. Nedelko, N.V. Tarasenko, Change in the ionisation state of a near-surface laser-produced aluminium plasma in double-pulse ablation modes, *Quantum Electron.* 33 (2003) 1065-1071.

[204] Y.P. Deng, X.H. Xie, H. Xiong, Y.X. Leng, C.F. Cheng, H.H. Lu, R.X. Li, Z.Z. Xu, Optical breakdown for silica and silicon with double femtosecond laser pulses, *Opt. Express* 13 (2005) 3096-3103.

[205] D.E. Roberts, A. du Plessis, L.R. Botha, Femtosecond laser ablation of silver foil with single and double pulses, *Appl. Surf. Sci.* 256 (2010) 1784-1792.

[206] J.T. Schiffern, D.W. Doerr, D.R. Alexander, Optimization of collinear double-pulse femtosecond laser-induced breakdown spectroscopy of silicon, *Spectrochim. Acta Part B* 62 (2007) 1412-1418.

[207] E.M. Cahoon, J.R. Almirall, Wavelength dependence on the forensic analysis of glass by nanosecond 266nm and 1064nm laser induced breakdown spectroscopy, *Appl. Optics* 49 (2010) C49-C57.

[208] D.K. Killinger, S.D. Allen, R.D. Waterbury, C. Stefano, E.L. Dottery, Enhancement of Nd:YAG LIBS emission of a remote target using a simultaneous CO_2 laser pulse, *Opt. Express* 15 (2007) 12905-12915.

[209] A. Löbe, J. Vrenegor, R. Fleige, V. Sturm, R. Noll, Laser-induced ablation of a steel sample in different ambient gases by use of collinear multiple laser pulses, *Anal. Bioanal. Chem.* 385 (2006) 326-332.

[210] C. Hartmann, A. Gillner, U. Aydin, R. Noll, T. Fehr, C. Gehlen, R. Poprawe, Investigation on laser micro ablation of metals using ns-multipulses, *J. Phys. Conf. Series* 59 (2007) 440-444.

[211] V. Hohreiter, D.W. Hahn, Dual-pulse laser induced breakdown spectroscopy: Time-resolved transmission and spectral measurements, *Spectrochim. Acta Part B* 60 (2005) 968-974.

[212] A. De Giacomo, M. Dell'Aglio, O. De Pascale, Single Pulse-Laser Induced Breakdown Spectroscopy in aqueous solution, *Appl. Phys. Part A* 79 (2004) 1035-1038.

[213] P.K. Kennedy, D.X. Hammer, B.A. Rockwell, Laser induced plasma in aqueous media, *Prog. Quantum Electron.* 21 (1997) 155-248.

[214] J.R. Watcher, D.A. Cremers, Determination of Uranium in Solution Using Laser-Induced Breakdown Spectroscopy, *Appl. Spectrosc.* 41 (1987) 1042-1048.

[215] R. Nyga, W. Neu, Double-pulse technique for optical emission spectroscopy of ablation plasmas of samples in liquids, *Opt. Lett.* 18 (1993) 747-749.

[216] S. Nakamura, Y. Ito, K. Sone, H. Hiraga, K.-I. Kaneko, Determination of Iron Suspension in Water by Laser-Induced Breakdown Spectroscopy with Two Sequential Laser Pulses, *Anal. Chem.* 68 (1996) 2981-2986.

[217] A. Kumar, F.Y. Yueh, J.P. Singh, Double - pulse laser-induced breakdown spectroscopy of liquid jets with different thickness, *Appl. Optics* 42 (2003) 6047-6051.

[218] V.N. Rai, F.-Y. Yueh, J.P. Singh, Study of Laser-Induced Breakdown Emission from Liquid under Double-Pulse Excitation, *Appl. Optics* 42 (2003) 2094-2101.

[219] A. Kuwako, Y. Uchida, K. Maeda, Supersensitive Detection of Sodium in Water with Use of Dual-Pulse Laser-Induced Breakdown Spectroscopy, *Appl. Optics* 42 (2003) 6052-6056.

[220] W. Pearman, J. Scaffidi, S.M. Angel, Dual-Pulse Laser-Induced Breakdown Spectroscopy in Bulk Aqueous Solution with an Orthogonal Beam Geometry, *Appl. Optics* 42 (2003) 6085-6093.

[221] V.N. Rai, F.Y. Yueh, J.P. Singh, Time-dependent single and double pulse laser-induced breakdown spectroscopy of chromium in liquid, *Appl. Optics* 47 (2008) G21-G29.

[222] M. Lawrence-Snyder, J. Scaffidi, S.M. Angel, A.P.M. Michel, A.D. Chave, Sequential-Pulse Laser-Induced Breakdown Spectroscopy of High-Pressure Bulk Aqueous Solutions, *Appl. Spectrosc.* 61 (2007) 171-176.

[223] M. Lawrence-Snyder, J. Scaffidi, S.M. Angel, A.P.M. Michel, A.D. Chave, Laser-Induced Breakdown Spectroscopy of High-Pressure Bulk Aqueous Solutions, *Appl. Spectrosc.* 60 (2006) 786-790.

[224] A.P.M. Michel, A.D. Chave, Double pulse laser-induced breakdown spectroscopy of bulk aqueous solutions at oceanic pressures: interrelationship of gate delay, pulse energies, interpulse delay, and pressure, *Appl. Optics* 47 (2008) G131-G143.

[225] A.P.M. Michel, A.D. Chave, Single pulse laser-induced breakdown spectroscopy of bulk aqueous solutions at oceanic pressures: interrelationship of gate delay and pulse energy, *Appl. Optics* 47 (2008) G122-G130.

[226] A. De Giacomo, M. Dell'Aglio, F. Colao, R. Fantoni, V. Lazic, Double-pulse LIBS in bulk water and on submerged bronze samples, *Appl. Surf. Sci.* 247 (2005) 157-162.

[227] A. De Giacomo, M. Dell'Aglio, F. Colao, R. Fantoni, Double pulse laser produced plasma on metallic target in seawater: basic aspects and analytical approach, *Spectrochim. Acta Part B* 59 (2004) 1431-1438.

[228] A. De Giacomo, M. Dell'Aglio, A. Casavola, G. Colonna, O. De Pascale, M. Capitelli, Elemental chemical analysis of submerged targets by double-pulse laser-induced breakdown spectroscopy, *Anal. Bioanal. Chem.* 385 (2006) 303-311.

[229] V. Sturm, J. Vrenegor, R. Noll, M. Hemmerlin, Bulk analysis of steel samples with surface scale layers by enhanced laser ablation and LIBS analysis of C, P, S, Al, Cr, Cu, Mn and Mo, *J. Anal. At. Spectrom.* 19 (2004) 451-456.

[230] F. Colao, R. Fantoni, V. Lazic, L. Caneve, A. Giardini, V. Spizzichino, LIBS as a diagnostic tool during the laser cleaning of copper based alloys: experimental results, *J.Anal. At. Spectrom.* 19 (2004) 502-504.

[231] T. Čtvrtníčková, F.J. Fortes, L.M. Cabalín, V. Kanický, J.J. Laserna, Depth profiles of ceramic tiles by using orthogonal double-pulse laser induced breakdown spectrometry, *Surf. Interf. Anal.* 41 (2009) 714-719.

[232] M. Galiová, J. Kaiser, K. Novotný, M. Ivanov, M. Nývltová Fišáková, L. Mancini, G. Tromba, T. Vaculovič, M. Liška and V. Kanický, Investigation of the osteitis deformans phases in snake vertebrae by double-pulse laser-induced breakdown spectroscopy, *Anal. Bioanal. Chem.* 398 (2010) 1095-1107.

[233] J.L. Gottfried, F.C. De Lucia, C.A. Munson, A.W. Miziolek, Double-pulse standoff laser-induced breakdown spectroscopy for versatile hazardous materials detection, *Spectrochim. Acta Part B* 62 (2007) 1405-1411.

[234] J.D. Winefordner, I.B. Gornushkin, D. Pappas, O.I. Matveev, B.W. Smith, Novel uses of lasers in atomic spectroscopy. Plenary Lecture, *J. Anal. At. Spectrom.* 15 (2000) 1161-1189.

[235] P. Stchur, K.X. Yang, X.D. Hou, T. Sun, R.G .Michel, Laser excited atomic fluorescence spectrometry - a review, *Spectrochim. Acta Part B* 56 (2001) 1565-1592.

[236] D.J. Butcher, Lasers as Light Sources for Analytical Atomic Spectrometry, *Appl. Spectrosc. Rev.* 42 (2007) 543-562.

[237] H.H. Telle, D.C.S. Beddows, G.W. Morris, O. Samek, Sensitive and selective spectrochemical analysis of metallic samples: the combination of laser-induced breakdown spectroscopy and laser-induced fluorescence spectroscopy, *Spectrochim. Acta Part B* 56 (2001) 947-960.

[238] I. Gobernado-Mitre, A.C. Prieto, V. Zafiropulos, Y. Spetsidou, C. Fotakis, On-Line Monitoring of Laser Cleaning of Limestone by Laser-Induced Breakdown Spectroscopy and Laser-Induced Fluorescence, *Appl. Spectrosc.* 51 (1997) 1125-1129.

[239] [1] F. Hilbk-Kortenbruck, R. Noll, P. Wintjens, H. Falk, C. Becker, Analysis of heavy metals in soils using laser-induced breakdown spectrometry combined with laser-induced fluorescence, *Spectrochim. Acta Part B* 56 (2001) 933–945.

[240] S. Koch, W. Garen, W. Neu, R. Reuter, Resonance fluorescence spectroscopy in laser-induced cavitation bubbles, *Anal. Bioanal. Chem.* 385 (2006) 312-315.

[241] S.L. Lui, Y. Godwal, M.T. Taschuk, Y.Y. Tsui, R. Fedosejevs, Detection of lead in water using Laser-Induced Breakdown Spectroscopy and Laser-Induced Fluorescence, *Anal. Chem.* 80 (2008) 1995-2000.

[242] C. Goueguel, S. Laville, F. Vidal, M. Sabsabi, M. Chaker, Investigation of resonance-enhanced laser-induced breakdown spectroscopy for analysis of aluminium alloys, *J. Anal. At. Spectrom.* 25 (2010) 635-644.

[243] X. Zeng, X. Mao, S.S. Mao, J.H. Yoo, R. Greif, R.E. Russo, Laser–plasma interactions in fused silica cavities, *J. Appl. Phys.* 95 (2004) 816-822.

[244] X.K. Shen, J. Sun, H. Ling, Y.F. Lua, Spatial confinement effects in laser-induced breakdown spectroscopy, *Appl. Phys. Lett.* 91 (2007) 081501.

[245] W.S. Budi, H. Suyanto, H. Kurniawan, M.O. Tjia, K. Kagawa, Shock Excitation and Cooling Stage in the Laser Plasma Induced by a Q-Switched Nd:YAG Laser at Low Pressures, *Appl. Spectrosc.* 53 (1999) 719-730.

[246] A.M. Popov, F. Colao, R. Fantoni, Enhancement of LIBS signal by spatially confining the laser-induced plasma, *J. Anal. At. Spectrom.* 24 (2009) 602-604.

[247] A.M. Popov, F. Colao, R. Fantoni, Spatial confinement of laser-induced plasma to enhance LIBS sensitivity for trace elements determination in soils, *J. Anal. At. Spectrom.* 25 (2010) 837-848.

[248] L.B. Guo, C.M. Li, W. Hu, Y.S. Zhou, B.Y. Zhang, Z.X. Cai, X.Y. Zeng, Y.F. Lu, Plasma confinement by hemispherical cavity in laser-induced breakdown spectroscopy, *Appl. Phys. Lett.* 98 (2011) 131501.

In: Laser-Induced Plasmas
Editor: Ethan J. Hemsworth, pp. 205-230

ISBN 978-1-61324-851-5

Chapter 8

PARAMETRIC INSTABILITIES OF ULTRAINTENSE LASER PULSES PROPAGATING IN PLASMAS

***Meenu Varshney Asthana*[1], *Sonu Sen*[2,3] *and Dinesh Varshney*[*3]**
[1]Department of Physics, M. B. Khalsa Cllege, Indore, India
[2]Department of Engineering Physics, Rishiraj Institute of Technology, Indore, India
[3]School of Physics, Vigyan Bhawan, Devi Ahilya University, Indore, India

ABSTRACT

Laser light propagation in plasma is known to be subject to parametric instabilities, where the laser energy decays into scattered light and plasma modes giving rise to stimulated Raman (Brillouin) scattering and related phenomena. The presence of instabilities such as SRS, SBS and self-focusing in laser plasma interaction can result in significant losses of the incident laser energy leading to poor laser plasma coupling. For the ultra-short laser pulses, the most important instabilities are electron plasma instabilities. In nonlinear regime the relativistic self-focusing has affect on SRS and SBS, hereby becomes important to investigate and understand them. In the present work we investigate nonlinear interaction of a high power intense Gaussian electromagnetic beam with the electron and ion waves in an unmagnetized plasma. The resulting stimulated Raman and Brillouin scattering phenomena are studied analytically and numerically. Based on WKB and paraxial theory the propagation characteristics and regimes of an intense laser pulse is completely determined by the degree of diffraction, nonlinear defocusing and self-focusing suffered by the beam as it traverses through the plasma. When the laser power exceeds the critical power, the laser beam can undergo periodic self-focusing due to relativistic nonlinearity. The effect of nonlinear coupling between the pump laser and scattered laser beam has been incorporated. The effect of finite laser beam size and that of scattered beam and relativistic self-focusing of the pump laser beam on SRS and SBS back-reflectivity have been illustrated. Numerical calculations are made for typical parameters of relativistic laser-plasma interaction processes applicable for wide range of arbitrary pump strength and background plasma density.

---

* Corresponding author Tel.: +91-731-2467028; Telefax: +91-731-2465689, Email: vdinesh33@rediffmail.com

1. INTRODUCTION

Parametric instabilities such as stimulated Brillouin scattering (SBS), stimulated Raman scattering (SRS), self-focusing and filamentation have long been of concern in inertial confinement thermonuclear fusion research, because high levels of scattered light can reduce energy coupling and modify the distribution, affecting the symmetry of capsule implosion [1]. The interaction of high intensity laser pulses with plasma has been a subject of much theoretical and experimental interest. Relativistic laser plasma interaction has been reviewed elegantly [2, 3].

Research into this area has been possible only within the past few years, with the development of lasers capable of delivering ultra-intense pulses with field strengths sufficient to drive electrons to relativistic velocities. In this regime, the propagation of laser beam, in underdense plasma having frequency smaller than the laser frequency, the fast igniter concept, especially, requires understanding of laser plasma interaction at relativistic laser intensities. Most of the studies related to laser plasma interactions are limited to the cases where the electron nonlinearity due to relativistic mass variation is accounted for. However, in many situations the laser can create different types of nonlinearities at different time scales according to the inequalities (a) $\tau < \tau_{pe}$ or (b) $\tau_{pe} < \tau < \tau_{pi}$; hence, one can have different time regimes. Here, τ is laser pulse duration, τ_{pi} is ion plasma period, and τ_{pe} is the electron plasma period. In case (a), i.e., $\tau < \tau_{pe}$, the relativistic nonlinearity is set up. This nonlinearity is set up almost instantaneously. In case (b), relativistic and ponderomotive nonlinearities are operative. In this case electrons are expelled from the channel due to electron ponderomotive force, but ions are not expelled due to their inertia.

The coupling of intense laser beams with plasmas has important implications for inertial confinement fusion [4, 5]. Of particular relevance are scattering instabilities, namely, the stimulated Raman and Brillouin scattering, filamentation, and self-focusing of the laser light. These instabilities can reduce the laser-plasma coupling efficiency and produce energetic electrons, which can preheat the fusion fuel and reduce the compression rate. They can also modify the intensity distribution, affecting the uniformity of energy deposition. For the ultra-short laser pulses, the most important instabilities are electron plasma instabilities. The properties of SRS, SBS, self-focusing and filamentation have earlier been investigated, ignoring interaction between them. In nonlinear regime the self-focusing influence the SRS and SBS, hereby becomes important to investigate and understand them. In SRS, an incident laser beam decays into a scattered wave and electron plasma wave satisfying the frequency and phase matching condition.

In the relativistic regime, some of the properties of the Raman instability are modified. Like it can occur at any density, provided that the laser can propagate, *i.e.*, $n_e < \gamma n_c$, while it was limited to $n_e < (1/4)n_c$ at moderate intensities. Moreover, instabilities previously known is the moderate intensity regime as two-plasmon instability and relativistic filamentation are in fact unified with the Raman instability in the relativistic regime. The growth rate of the instability can be extremely large of the order to a time scale of a femtosecond for a 1.06 μm laser light. It results in a rapid destruction of the wave front of the light, and a strong electron heating. Once the plasma is heated to a temperature in the MeV range, the growth rate and the domain of wave vectors for which instabilities occurs are reduced [6].

As a result of the instability, the front part of the pulse is strongly depleted, leading to strong increase of the initially cold electron temperature. When the electrons have reached MeV temperature, the instability is less efficient and the back part of the pulse is able to propagate. Thus, in SRS the electron plasma wave produces superathermal electrons that penetrate and preheat the target core and the scattered wave represents a substantial amount of wasted energy, *i.e.*, the energy that would otherwise get coupled to the target. Hereby, to ensure the amount of useful and dissipated energy in laser plasma coupling, Raman reflectivity becomes an important parameter to decide which we intend to study here.

The reflection of laser light arising from parametric instabilities can be an important process in laser fusion plasmas. In particular, the Raman backscattering of laser light in an underdense region can prevent the light from arriving at the critical density where enhanced absorption occurs. Stimulated Raman scattering (SRS) is a major instability, which plays a very important role in laser-plasma interaction. The SRS instability is the resonant decay of the incident laser wave into a scattered electromagnetic (EM) wave and an electron plasma wave. This electron plasma wave can have a very high phase velocity (of the order of the velocity of light), and so can produce very energetic electrons when it damps [1]. Such electrons can preheat the fuel in laser fusion applications. The Raman instability is of particularly significant concern because of large reflectivity and high-energy electrons. Control of laser-plasma instabilities such as SRS is very important for the success of laser fusion. Much experimental and theoretical work has been devoted to the study of the SRS instability [7-15]. In spite of the great deal of interest in SRS, the nonlinear saturation of the Raman process is not yet fully understood.

Stimulated Raman scattering driven by intense laser pulses has been studied by many authors. Modena [16] has observed experimentally Raman forward scattering (RFS) in the relativistic regime and presented various results. Among these results was the direct observation of RFS from the spectrum of the transmitted beam and the rapid broadening of the RFS signal above a certain density. Rousseaux *et al.* [17] have investigated experimentally forward and backward Raman scattering driven by intense laser pulse through moderately underdense plasmas. Backward SRS is strongly driven at 10^{17} W/cm^2 and gradually decreases at higher intensities. In contrast, forward SRS (F-SRS) continuously increases with the laser intensity. The accelerated relativistic electron in the forward direction appears to be correlated with the F-SRS and the intense electron heating is likely to play a major role in the temporal growth or inhibition of the instabilities.

Miyakoshi *et al.* [18] have studied experimentally stimulated Raman backscattering from millimeter-scale inhomogeneous plasma irradiated with ultraintense laser pulse. The backward SRS spectral intensities are increased with scattered light up to $1.2\lambda_0$ (wavelength of the ultraintense laser) and saturate at longer wavelength. Nonlinear phase mismatch was found to cause the saturation due to the relativistic mass increase of the electron. Guerin *et al.* [19] have studied Raman instabilities in the relativistic regime. They have derived the dispersion relation in one-dimensional overdense plasma and showed analytically and numerically that a relativistic electromagnetic wave is unstable in plasma, with a variety of regimes depending on the electron density, the wave intensity, and the initial spreading of the longitudinal electron distribution function.

Another parametric instability, which plays an important role in laser plasma interaction and may produce a significant amount of backscattered light is SBS, which is again a serious concern for inertial confinement fusion research. The loss of light due to reflectivity instabilities continues to be the most worrisome plasma effect in laser driven fusion. In solid target experiments, energy reflectivities attributed to SBS have varied from 0 to 50%. Little agreement between theory and experiments has been reported so far inspite of intensive studies of SBS during the last two decades. One of the main challenges in current research is a theoretical explanation of the relatively low level of reflectivity observed in large-scale laser fusion experiments [20] and in well-controlled interaction experiments with preformed plasma [21-23]. Simple estimates based on linear gain theory predict a SBS reflectivity much higher than that observed in these experiments. It is worth to and mention that many features of the experimental results in laser plasmas are still not completely understood.

The transverse intensity gradient of the laser beam (pump) generates a ponderomotive force, which modifies the background plasma density profile in a transverse direction to the laser beam axis. This modification in the density may lead to the steady state self-focusing of the laser beam and focusing of the ion-acoustic and the backscattered laser beam; consequently, the SBS is affected. A more comprehensive study of the effect of laser hot spots and the breakdown of linear instability theory with application to SBS has been conducted by considering the dynamic response of background density and taking the ion-acoustic wave in the heavy damping limit. The effect of transient self-focusing of the main laser beam, results in the high intensity periodical bursts of SBS. A reflecting rear boundary can also cause the backscattering to show non-stationary behaviour [24].

With the advent of the very high-power lasers, the electron velocity in plasma may become quite large, comparable to the free space velocity of the light. So, the effect of relativistic electron mass variation is important. Stimulated Brillouin backscattering (SBBS) has been studied in a regime of relatively low intensities in the past. Interesting new phenomena arises if one operates in the so-called strong-coupling regime. This regime is characterized by low plasma temperatures, a few 100 eV, and high laser intensities. The SBBS effect can be better seen at plasma densities above the quarter critical density in order to avoid the simultaneous excitation of stimulated Raman backscattering.

For the sake of completeness and to highlight the relativistic self-focusing effect with relativistic mass and ponderomotive nonlinearity on stimulated Raman backscattering (SRBS) and stimulated Brillouin backscattering (SBBS) processes, our motivation of the present work is, firstly to seek the role of pump (laser) beam propagation in a self-created plasma channel. It is seen that, the pump laser beam can undergo a relativistic self-focusing process and its intensity is affected during the process of propagation. The scattered wave and the laser pump exert a ponderomotive force on the electrons (ions), driving the acoustic (plasma) waves. The channel provides the radial localization of the pump and sideband waves. Secondly, the stimulated Raman and Brillouin backscattering process are studied for chosen laser beam and plasma parameters in self-trapped mode.

Based on paraxial theory and mathematical formulation the ponderomotive self-channeling, is presented in Section 2. The characteristic beam propagation equation along with self-trapping condition is derived in Section 3. The coupled mode equations for SRS and SBS back-reflectivity are respectively derived in Section 4 and 5. Numerical computations are performed for typical relativistic laser-plasma parameters to see the focusing behaviour of pump laser and scattered beam, and also the SRS/SBS reflectivity with gain. Results and discussion are made in Section 6.

2. Ponderomotive Self-Channeling

Consider the propagation of an intense Gaussian laser beam through cold plasma of electron density n_0, propagating along z-direction. The amplitude of the electric vector **E** in cylindrical coordinate can be expressed as

$$\mathbf{E} = \hat{x} A(r,z,t)\exp[-i(\omega t - kz)] \tag{1}$$

where,

$$A^2 = A_{00}^2 g(t)\exp\left(\frac{-r^2}{r_0^2}\right) \tag{2}$$

$$k(z) = \left(\frac{\omega}{c}\right)\varepsilon_0^{'1/2}$$

$$\omega_p(z) = \left(\frac{4\pi n_0(z)e^2}{m}\right)^{1/2} \tag{3}$$

is the electron plasma frequency, ε'_0 is the plasma dielectric constant and choosing step-pulse function $g(t) = 1$ for $t > 0$ and $g(t) = 0$ otherwise, 'e' and 'm' are the electronic charge and mass respectively and 'c' is the velocity of light.

For the present work, consider a plasma channel being formed due to subpicosecond laser pulse in a helium-gas jet plasma. The formation of a plasma channel can be most easily ascribed to the effects of ponderomotive force associated with the intense laser pulse as it propagates through the plasma. Initially, the laser pulse exerts a ponderomotive force on the plasma electrons and expels them radially. This sets up a large space charge force which propels the ions outward from the axis of the propagation. After the passage of the laser pulse, the ions continue to drift radially creating a plasma channel. For optical guiding of laser beam in plasmas, the radial profile of the index of refraction, must have a maximum on axis, causing the wavefront to curve inward and the laser beam to converge. When this focusing force is strong enough to counteract the diffraction of the beam, the laser pulse can propagate over a long distance and maintain a small cross-section (laser channel).

The index of refraction or dielectric function ε can be written as

$$\varepsilon(r) = \left[1 - \left(\frac{\omega_p^2}{\omega^2}\right)\left(\frac{n_e(r)}{n_0\,\gamma(r)}\right)\right] \tag{4}$$

here, $n_e(r)$ is the radial distribution of electron density, $\gamma(r)$ is the relativistic factor associated with the electron motion transverse to the laser propagation and 'ω' is the laser frequency. The dielectric function 'ε' has dependence on irradiance EE^* of a Gaussian beam and hence

'ε' is the function of r^2, therefore in the paraxial approximation 'ε' can be expressed in powers of r^2. It will be seen from this that an on-axis maximum can be created through modification of the radial profile of γ and/or n_e.

Considering the intensity dependence of electron mass, the relativistic ponderomotive force produced due to laser beam is

$$\mathbf{F}_p = e\nabla\phi_p \tag{5}$$

where,

$$\phi_p = -\left(\frac{mc^2}{e}\right)(\gamma - 1) \tag{6}$$

is the ponderomotive potential with relativistic factor $\gamma = \left[1 + a^2/2\right]^{1/2}$ and $a = \left(e|A|/m\omega c\right)$. At the front of the laser pulse ponderomotive force has axial and radial components. The radial ponderomotive force pushes the electrons radially outward, on the time scale of a radial space charge field $\mathbf{E}_s = (-\nabla\phi_s)$. Using Poisson's equation

$$\nabla^2\phi_s = 4\pi e(n_e - n_0) \tag{7}$$

For $\omega_p\tau > 1$, assuming a quasi-steady-state with $\phi_s = -\phi_p$ and using Equation (5) in (7) we obtain modified electron density as

$$n_e(z) = n_0(z)\left[1 + \frac{c^2}{\omega_p^2(z)}\nabla_\perp^2\gamma\right] \tag{8}$$

Equation (8) is valid in those regions of 'r' where the second term in the square bracket is greater then –1, otherwise $n_e = 0$. Thus there is complete electron evacuation in region where

$$\left[1 + \left(c^2/\omega_p^2(z)\right)\nabla_\perp^2\gamma\right] < 0 \tag{9}$$

Assuming that the pulse propagates without changing shape and employing a Gaussian constant shape ansatz for amplitude as

$$q^2 = \frac{q_0^2}{f^2}\exp\left(\frac{-r^2}{r_0^2 f^2}\right) \tag{10}$$

where, f is the dimensionless beamwidth parameter which is unity at $z = 0$, $r^2 = (x^2 + y^2)$ is the radial component in cylindrical coordinate system, r_0 is the initial beamwidth and

$q_0 \cong (e A_{00} / m\omega c)$ is the axial amplitude of the laser giving the modified electron density as

$$n_e(z) = n_0\left[1 - \frac{c^2}{\omega_p^2 r_0^2 f^2}\frac{q^2}{\gamma} \times \left(1 - \frac{r^2}{r_0^2 f^2}\frac{1+q^2/4}{\gamma^2}\right)\right] \tag{11}$$

Using Equation (11) in (4) and making Taylor expansion of dielectric function in the radial direction for arbitrary large nonlinearity under paraxial approximation can be written as

$$\varepsilon = \varepsilon_0^{'} - \frac{\varepsilon_1 r^2}{r_0^2} \tag{12}$$

where,

$$\varepsilon_0^{'} = 1 - \frac{\omega_p^2/\omega^2}{\gamma_0^{'}}\left(1 - \frac{c^2}{\omega_p^2 r_0^2 f^2}\frac{q_0^2/f^2}{\gamma_0^{'}}\right) \tag{13}$$

$$\varepsilon_1 = \frac{\omega_p^2}{4\omega^2}\frac{q_0^2/f^4}{\gamma_0^{'3}}\left(1 + \frac{c^2}{\omega_p^2 r_0^2 f^2}\frac{8 + q_0^2/f^2}{\gamma_0^{'}}\right) \tag{14}$$

and

$$\gamma_0^{'} = \left(1 + \frac{q_0^2}{2f^2}\right)^{1/2} \tag{15}$$

3. Characteristic Beam Propagation Equation

A laser beam has usually Gaussian intensity distribution along its wavefront. Its propagation characteristics are therefore most conveniently analyzed in a cylindrical coordinate system, with the axis of the cylinder coinciding with the direction of propagation, which may be assumed to be the z-axis. The vector **E** satisfies the wave

$$\nabla^2\mathbf{E} + \frac{\omega^2}{c^2}\varepsilon(r,z)\mathbf{E} = 0 \tag{16}$$

which comes directly through Maxwell's equation neglecting the term $\nabla(\nabla\cdot\mathbf{E})$. For a transverse field $\nabla\cdot\mathbf{E} = 0$, even if **E** has a longitudinal component the term $\nabla(\nabla\cdot\mathbf{E})$ can be neglected keeping in mind $(c^2/\omega^2)\ |(1/\varepsilon)\nabla^2 ln\ \varepsilon| << 1$. A condition found to be true in almost all the cases of interest in electromagnetic wave propagation through plasma. We express r and z dependence of **E** following earlier analysis as [25],

$$E(r,z) = A(r,z)\frac{\varepsilon_0^{'1/4}(0)}{\varepsilon_0^{'1/4}(z)}\exp\left[i(\omega t - \frac{\omega}{c}\int_0^z \varepsilon_0^{'1/2} dz)\right] \tag{17}$$

Introducing an eikonal

$$A = A_r(r,z)\exp\left[-i\frac{\omega}{c}\varepsilon_0^{'1/2} S(r,z)\right] \tag{18}$$

$S(r, z)$ is called the eikonal and related to the curvature of the wave front. Substituting for $E(r, z)$ and $A(r, z)$ from (17) and (18) in (16), neglecting $(\partial^2 A/\partial z^2)$ and derivatives of $\varepsilon_0(z)$, which is justifiable for slowly converging and diverging beams and separating real and imaginary parts of the resulting equation one gets.

$$2\left(\frac{\partial S}{\partial z}\right) + \left(\frac{\partial S}{\partial r}\right)^2 = \frac{c^2}{\omega^2 \varepsilon_0^{'} A_r}\left(\frac{\partial^2 A_r}{\partial r^2} + \frac{1}{r}\frac{\partial A_r}{\partial r}\right) - \frac{\varepsilon_1}{\varepsilon_0^{'}}\frac{r^2}{r_0^2} \tag{19}$$

and

$$\frac{\partial A_r^2}{\partial z^2} + \frac{\partial S}{\partial r}\frac{\partial A_r^2}{\partial r} + A_r^2\left(\frac{\partial^2 S}{\partial r^2} + \frac{1}{r}\frac{\partial S}{\partial r}\right) = 0 \tag{20}$$

The solution of Equations (19) and (20), for an initially Gaussian beam, can be written as [25, 26]

$$S = \frac{r^2}{2}\beta(z) + \varphi(z) \tag{21}$$

$$A_r^2 = \frac{A_{00}^2}{f^2}\exp\left(\frac{-r^2}{r_0^2 f^2}\right) \tag{22}$$

and

$$\beta = \frac{1}{f}\frac{df}{dz} \tag{23}$$

where 'β' represents the inverse of the radius of curvature of the wave front and $(r_0 f)$ is the width of the beam. In the geometrical optics approximation, $r = r_0 f(z)$ represents a ray in a plane containing the z-axis. We now substitute for S and A_r from Equations (21-23) in Equation (19) and making use of the paraxial ray approximation *i.e.*, $[(r/r_0 f)^4 << 1]$. Equating the coefficients of r^2 on both sides, we get resulting equation governing the beamwidth parameter as

$$\varepsilon_0^{'} \frac{d^2 f}{dz^2} = \left[\frac{c^2}{\omega^2 r_0^4 f^3} - \frac{\varepsilon_1}{r_0^2} f \right] \tag{24}$$

In Equation (24) the first term on the right-hand side is the diffraction term and the second term is the nonlinear term on account of relativistic nonlinearity. By choosing the beam and plasma parameters, one can find a condition such that the diffraction term [viz. the first term on the right-hand side of Equation (24)] and the nonlinear term [the second term on the right-hand side of Equation (24)] balance each other. In that situation $f_0 = 1$ and beam propagates without convergence/divergence. This is said to be propagating in "self-trapped mode". In the general case, the differential equation governing the beam width parameter of the pump laser beam [Equation (24)] can be solved numerically, following the boundary condition $f_0 = 1$ and $df_0 / dz = 0$. Transforming the coordinates z and the initial beam width r_0 to dimensionless forms:

$$\xi = (zc / r_0^2 \omega) \text{ and } \rho_0 = (r_0 \omega / c) \tag{25}$$

Equation (24) reduces to

$$\varepsilon_0^{'} \frac{d^2 f}{d\xi^2} = \left[1 + \rho_0^2 r_0^2 \; \varepsilon_1 f^4\right] \frac{1}{f^3} \tag{26}$$

Equation (26) is a second order differential equation amenable to mathematical manipulations. Knowing $\varepsilon_0^{'}(f)$ and $\varepsilon_1(r, f)$ one can integrate Equation (26) numerically to obtain 'f' as function of ξ.

For an initial plane wavefront of the beam the initial conditions on f are $f(\xi = 0) = 1$ and $(df/d\xi)_{\xi=0} = 0$. When the terms on right hand side cancel each other hence at $z = 0$, $(d^2f/d\xi^2)_{\xi=0} = 0$; and $(df/d\xi)$ is also zero and $f = 1$ at $z = 0$, $f = 1$ for all values of z; in other words beam can propagates without convergence or divergence in plasma channel. Therefore the condition for self-trapping is

$$\rho_0^2 = \frac{4\omega^2}{\omega_p^2} \frac{\left(1 + \frac{q_0^2}{2f^2}\right)^{3/2}}{q_0^2} \left[1 - \frac{c^2}{r_0^2 \omega_p^2} \frac{\left(8 + q_0^2\right)}{\left(1 + \frac{q_0^2}{2f^2}\right)^{1/2}} \right] \tag{27}$$

The dependence of ρ_0 versus q_0 according to Equation (27) gives the critical power curve. For an initial beamwidth (r_0) below a certain value say r_{0m} (minimum), no self-focusing can occur regardless for any initial power of the beam. When the initial beamwidth (r_0) is greater than r_{0m}, there are two values of critical beam power say q_{0cr1} and q_{0cr2} (such that $q_{0cr1} < q_{0cr2}$) for which the beam propagates in the uniform waveguide mode without any change in beamwidth. If the initial power q_0 of the beam lies between q_{0cr1} and q_{0cr2} (with $r_0 > r_{0m}$) the beam propagates in an oscillatory converging mode with the beamwidth varying between the original and a minimum, this is referred to as self-focusing. For $q_0 > q_{0cr2}$ (when $r_0 > r_{0m}$) and for some values of power (for $r_0 < r_{0m}$) the beam propagates in an oscillatory guided diverging mode with the beamwidth varying between the original value and a maximum.

4. Stimulated Raman Scattering

Consider the interaction of high power laser beam of frequency ω_0, wave number k_0, with the high frequency mode of the plasma (electron-plasma wave) of frequency ω, wave number k and resulting stimulated Raman scattered wave of frequency ω_s, and wave number k_s. The wave equations governing the pump laser beam and the scattered beam follows

$$\nabla^2\mathbf{E}_0 + \frac{\omega_0^2}{c^2}\left[1-\left(\frac{n}{n_0}\right)\left(\frac{\omega_{p0}^2}{\gamma\,\omega_0^2}\right)\right]\mathbf{E}_0 = \frac{-4\pi e i\omega_0}{2c^2}\left(n_{es}v_s\right) \tag{28}$$

and

$$\nabla^2\mathbf{E}_s + \frac{\omega_s^2}{c^2}\left[1-\left(\frac{n}{n_0}\right)\left(\frac{\omega_{p0}^2}{\gamma\,\omega_s^2}\right)\right]\mathbf{E}_s = \frac{-4\pi e i\omega_s}{2c^2}\left(n_{es}^*v_0\right) \tag{29}$$

where

$$\left(\frac{n_{es}}{n_0}\right) = i\left(\frac{n}{n_0}\right)\left[\frac{e^2k^2\left|E_sE_0^*\right|}{8\Gamma_e m_0^2\gamma^2\omega\omega_s\omega_0}\right] \tag{30}$$

The pump and scattered beams are obtained by solving the nonlinear energy differentials [1, 10] as

$$\nabla^2\mathbf{E}_0 + \frac{\omega_0^2}{c^2}\left[1-\left(\frac{n}{n_0}\right)\frac{\omega_{p0}^2}{\omega_0^2\gamma}\right]\mathbf{E}_0 + \left(\frac{n}{n_0}\right)\frac{ie^2\omega_{p0}^2k^2\left|E_sE_s^*\right|E_0}{16\Gamma_e m_0^2\gamma^3c^2\omega\omega_s^2} = 0 \tag{31}$$

and

$$\nabla^2 \mathbf{E}_s + \frac{\omega_s^2}{c^2}\left[1 - \left(\frac{n}{n_0}\right)\frac{\omega_{p0}^2}{\omega_s^2 \gamma}\right]\mathbf{E}_s - \left(\frac{n}{n_0}\right)\frac{ie^2 \omega_{p0}^2 k^2 \left|E_0 E_0^*\right| E_s}{16\Gamma_e m_0^2 \gamma^3 c^2 \omega \omega_0^2} = 0 \tag{32}$$

here, Γ_e is the damping factor. These coupled equations (31) and (32) are solved using the paraxial ray approximation. Expressing

$$E_0 = A_{00}(r,z)\exp\left[i\left(\omega_0 t - k_0\{z + S_0(r,z)\}\right)\right] \tag{33}$$

$$E_s = A_{s0}(r,z)\exp\left[i\left(\omega_s t - k_s\{z + S_s(r,z)\}\right)\right] \tag{34}$$

Substituting these E_0 and E_s in Equations (31) and (32), respectively and separating real and imaginary parts, we then obtained a of coupled equations. The coupled equation form yields:

Equation (31) gives

$$2\left(\frac{\partial S_0}{\partial z}\right) + \left(\frac{\partial S_0}{\partial r}\right)^2 = \frac{1}{\varepsilon_0}\left[1 - \left(\frac{n}{n_0}\right)\frac{\omega_{p0}^2}{\omega_0^2 \gamma}\right] + \frac{1}{k_0^2 A_{00}}\left(\frac{\partial^2 A_{00}}{\partial r^2} + \frac{1}{r}\frac{\partial A_{00}}{\partial r}\right) \tag{35}$$

and

$$\frac{\partial A_{00}^2}{\partial z} + \frac{\partial S_0}{\partial r}\frac{\partial A_{00}^2}{\partial r} + A_{00}^2\left(\frac{\partial^2 S_0}{\partial r^2} + \frac{1}{r}\frac{\partial S_0}{\partial r}\right) + \left(\frac{n}{n_0}\right)\frac{e^2 \omega_{p0}^2 k^2 \left|A_{s0}\right|^2 A_{00}^2}{16\Gamma_e m_0^2 \gamma^3 c^2 \omega \omega_s^2 k_0} = 0 \tag{36}$$

Similarly, Equation (32) yields

$$2\left(\frac{\partial S_s}{\partial z}\right) + \left(\frac{\partial S_s}{\partial r}\right)^2 = \frac{1}{\varepsilon_s}\left[1 - \left(\frac{n}{n_0}\right)\frac{\omega_{p0}^2}{\omega_s^2 \gamma}\right] + \frac{1}{k_s^2 A_{s0}}\left(\frac{\partial^2 A_{s0}}{\partial r^2} + \frac{1}{r}\frac{\partial A_{s0}}{\partial r}\right) \tag{37}$$

and

$$\frac{\partial A_{s0}^2}{\partial z} + \frac{\partial S_s}{\partial r}\frac{\partial A_{s0}^2}{\partial r} + A_{s0}^2\left(\frac{\partial^2 S_s}{\partial r^2} + \frac{1}{r}\frac{\partial S_s}{\partial r}\right) - \left(\frac{n}{n_0}\right)\frac{e^2 \omega_{p0}^2 k^2 \left|A_{00}\right|^2 A_{s0}^2}{16\Gamma_e m_0^2 \gamma^3 c^2 \omega \omega_0^2 k_s} = 0 \tag{38}$$

The solutions of Equations (35) and (36) can be written as

$$S_0 = \frac{r^2}{2}\left(\frac{1}{f_0(z)}\frac{df_0(z)}{dz}\right) + \varphi_0(z) \tag{39}$$

$$A_{00}^2 = \frac{E_{00}^2(z=0)}{f_0^2}\exp\left(-\frac{r^2}{r_{00}^2 f_0^2} - 2g_0 z\right) \tag{40}$$

with

$$g_0 = \left(\frac{n}{n_0}\right)\frac{e^2\omega_{p0}^2 k^2 |A_{s0}|^2}{32\Gamma_e m_0^2 \gamma^3 c^2 \omega\omega_s^2 k_0} \tag{41}$$

Similarly, for the backscattered wave, the system of Equations (37) and (38) gives

$$S_s = \frac{r^2}{2}\left(\frac{1}{f_s(z')}\frac{df_s(z')}{dz'}\right) + \varphi_s(z') \tag{42}$$

$$A_{s0}^2 = \frac{E_{s0}^2}{f_s^2(z')}\exp\left(-\frac{r^2}{r_{s0}^2 f_s^2(z')} - 2g_s z'\right) \tag{43}$$

with

$$g_s = -\left(\frac{n}{n_0}\right)\frac{e^2\omega_{p0}^2 k^2 |A_{00}|^2}{32\Gamma_e m_0^2 \gamma^3 c^2 \omega\omega_0^2 k_s} \tag{44}$$

or

$$g_s = -G_{SRS}\left(\frac{n}{n_0}\right)\frac{\alpha_0 E_{00}^2}{f_0^2 \gamma^3}\exp(-2g_0 z') \tag{45}$$

and

$$G_{SRS} = \frac{\omega_{p0}^2 k^2 q_0}{32\Gamma_e k_s \omega} \tag{46}$$

here, the relativistic factor γ is written as

$$\gamma = \left(1 + \frac{q_0^2}{f_0^2}\exp(-2g_0 z)\right)^{1/2} \tag{47}$$

$$\varepsilon_s = \left(1 - \frac{\omega_{p0}^2}{\omega_s^2}\right) \tag{48}$$

and $z' = L_z - z$, L_z is the interaction length and r_{00} and r_{s0} are the beamwidth of pump and scattered beam respectively. The differential equations governing the beamwidth parameter of pump beam f_0 and scattered beam f_s are given by

$$\frac{1}{f_0}\frac{d^2 f_0}{dz^2} = \left[\frac{c^2}{\omega_0^2 r_{00}^4 \varepsilon_{0p}' f_0^4} - \frac{\varepsilon_{1p}}{\varepsilon_{0p}' r_{00}^2}\right] \tag{49}$$

$$\frac{1}{f_s}\frac{d^2 f_s}{dz'^2} = \left[\frac{c^2}{\omega_s^2 r_{s0}^4 \varepsilon_{0s}' f_s^4} - \frac{\varepsilon_{1s}}{\varepsilon_{0s}' r_{s0}^2}\right] \tag{50}$$

where

$$\varepsilon_{0p}' = 1 - \frac{\omega_{p0}^2}{\omega_0^2}\left(1 + \frac{q_0^2 \exp(-2g_0 z)}{2f_0^2}\right)^{-1/2}$$

$$\times\left\{1 - \frac{c^2}{\omega_{p0}^2 r_{00}^2 f_0^2}\frac{q_0^2 \exp(-2g_0 z)}{f_0^2}\left(1 + \frac{q_0^2 \exp(-2g_0 z)}{2f_0^2}\right)^{-1/2}\right\} \tag{51}$$

$$\varepsilon_{1p} = \frac{\omega_{p0}^2}{4\omega_0^2}\frac{q_0^2 \exp(-2g_0 z)}{f_0^4}\left(1 + \frac{q_0^2 \exp(-2g_0 z)}{2f_0^2}\right)^{-3/2}$$

$$\times\left\{1 + \frac{c^2}{\omega_{p0}^2 r_{00}^2 f_0^2}\frac{8 + q_0^2 \exp(-2g_0 z)}{f_0^2}\left(1 + \frac{q_0^2 \exp(-2g_0 z)}{2f_0^2}\right)^{-1/2}\right\} \tag{52}$$

$$\varepsilon_{0s}' = 1 - \frac{\omega_{p0}^2}{\omega_s^2}\left(1 + \frac{q_s^2 \exp(-2g_s z')}{2f_s^2}\right)^{-1/2}$$

$$\times\left\{1 - \frac{c^2}{\omega_{p0}^2 r_{s0}^2 f_s^2}\frac{q_s^2 \exp(-2g_s z')}{f_s^2}\left(1 + \frac{q_s^2 \exp(-2g_s z')}{2f_s^2}\right)^{-1/2}\right\} \tag{53}$$

$$\varepsilon_{1s} = \frac{\omega_{p0}^2}{4\omega_s^2} \frac{q_s^2 \exp(-2g_s z')}{f_s^4} \left(1 + \frac{q_s^2 \exp(-2g_s z')}{2f_s^2}\right)^{-3/2}$$

$$\times \left\{1 + \frac{c^2}{\omega_{p0}^2 r_{s0}^2 f_s^2} \frac{8 + q_s^2 \exp(-2g_s z')}{f_s^2} \left(1 + \frac{q_s^2 \exp(-2g_s z')}{2f_s^2}\right)^{-1/2}\right\} \quad (54)$$

here the subscript 0p, 1p, 0s and 1s in ε denotes the corresponding expressions for pump beam and scattered beam respectively. The integrated reflectivity R given by

$$R = \frac{Scattered\ Power}{Input\ Power}$$

$$R = \frac{\int |E_{s0}(z'=0)|^2 dr}{\int |E_{00}(z=0)|^2 dr} \quad (55)$$

or

$$R = \frac{E_{s0}^2(z'=0)}{E_{00}^2(z=0)} \left(\frac{r_{s0}^2}{r_{00}^2}\right) \exp(-2g_s(z=0)L_z) \quad (56)$$

Equation (56) gives an expression for the SRS reflectivity by taking into account the finite beam size of the pump and scattered laser beam when both relativistic and ponderomotive nonlinearities are operative. The actual values of f_0 and f_s at a particular z and are governed by the numerical solutions of Equations (49) and (50) [11].

5. Stimulated Brillouin Scattering

We now consider the interaction of high power laser beam with the low-frequency mode of the plasma (ion acoustic wave) of frequency ω, wave number k and resulting stimulated Brillouin scattered wave of frequency ω_s and wave number k_s. The wave equations governing the pump laser beam and the scattered beam are given by

$$\nabla^2 \mathbf{E}_0 + \frac{\omega_0^2}{c^2}\left[1 - \left(\frac{n}{n_0}\right)\left(\frac{\omega_{p0}^2}{\gamma\omega_0^2}\right)\right]\mathbf{E}_0 = \frac{-4\pi e i\omega_0}{2c^2}(n_{es} v_s) \quad (57)$$

and

$$\nabla^2 \mathbf{E}_s + \frac{\omega_s^2}{c^2}\left[1-\left(\frac{n}{n_0}\right)\left(\frac{\omega_{p0}^2}{\gamma\omega_s^2}\right)\right]\mathbf{E}_s = \frac{-4\pi ei\omega_s}{2c^2}\left(n_{es}^* v_0\right) \tag{58}$$

On similar lines as in previous section the equations for the pump and scattered beam are obtained as,

$$\nabla^2 \mathbf{E}_0 + \frac{\omega_0^2}{c^2}\left[1-\left(\frac{n}{n_0}\right)\frac{\omega_{p0}^2}{\omega_0^2\gamma}\right]\mathbf{E}_0 + \left(\frac{n}{n_0}\right)\frac{ie^2\omega_{p0}^2 k\omega\left|E_s E_s^*\right|E_0}{8\Gamma_i m_0 M\gamma^2 c^2\omega_s^2 c_s^2} = 0 \tag{59}$$

and

$$\nabla^2 \mathbf{E}_s + \frac{\omega_s^2}{c^2}\left[1-\left(\frac{n}{n_0}\right)\frac{\omega_{p0}^2}{\omega_s^2\gamma}\right]\mathbf{E}_s - \left(\frac{n}{n_0}\right)\frac{ie^2\omega_{p0}^2\omega\left|E_0 E_0^*\right|E_s}{8\Gamma_i m_0 M\gamma^2 c^2\omega_0^2 c_s^2} = 0 \tag{60}$$

Here Γ_i is the damping factor, M is the ionic mass, $c_s = (k_B T_e / M)^{1/2}$ is the speed of the ion-acoustic wave and other symbols have their usual meanings. Similarly, the differential equations governing the beam width parameter of pump f_0 and back-scattered f_s beam for SBS phenomena follows:

$$\frac{1}{f_0}\frac{d^2 f_0}{dz^2} = \left[\frac{c^2}{\omega_0^2 r_{00}^4 \varepsilon_{0p}' f_0^4} - \frac{\varepsilon_{1p}}{\varepsilon_{0p}' r_{00}^2}\right] \tag{61}$$

$$\frac{1}{f_s}\frac{d^2 f_s}{dz'^2} = \left[\frac{c^2}{\omega_s^2 r_{s0}^4 \varepsilon_{0s}' f_s^4} - \frac{\varepsilon_{1s}}{\varepsilon_{0s}' r_{s0}^2}\right] \tag{62}$$

where

$$\varepsilon_{0p}' = 1 - \frac{\omega_{p0}^2}{\omega_0^2}\left(1+\frac{q_0^2\exp(-2g_0 z)}{2f_0^2}\right)^{-1/2}$$

$$\times\left\{1-\frac{c^2}{\omega_{p0}^2 r_{00}^2 f_0^2}\frac{q_0^2\exp(-2g_0 z)}{f_0^2}\left(1+\frac{q_0^2\exp(-2g_0 z)}{2f_0^2}\right)^{-1/2}\right\} \tag{63}$$

$$\varepsilon_{1p} = \frac{\omega_{p0}^2}{4\omega_0^2}\frac{q_0^2\exp(-2g_0 z)}{f_0^4}\left(1+\frac{q_0^2\exp(-2g_0 z)}{2f_0^2}\right)^{-3/2}$$

$$\times\left\{1+\frac{c^2}{\omega_{p0}^2 r_{00}^2 f_0^2}\frac{8+q_0^2\exp(-2g_0z)}{f_0^2}\left(1+\frac{q_0^2\exp(-2g_0z)}{2f_0^2}\right)^{-1/2}\right\} \tag{64}$$

$$\varepsilon_{0s}^{'}=1-\frac{\omega_{p0}^2}{\omega_s^2}\left(1+\frac{q_s^2\exp(-2g_sz')}{2f_s^2}\right)^{-1/2}$$

$$\times\left\{1-\frac{c^2}{\omega_{p0}^2 r_{s0}^2 f_s^2}\frac{q_s^2\exp(-2g_sz')}{f_s^2}\left(1+\frac{q_s^2\exp(-2g_sz')}{2f_s^2}\right)^{-1/2}\right\} \tag{65}$$

$$\varepsilon_{1s}=\frac{\omega_{p0}^2}{4\omega_s^2}\frac{q_s^2\exp(-2g_sz')}{f_s^4}\left(1+\frac{q_s^2\exp(-2g_sz')}{2f_s^2}\right)^{-3/2}$$

$$\times\left\{1+\frac{c^2}{\omega_{p0}^2 r_{s0}^2 f_s^2}\frac{8+q_s^2\exp(-2g_sz')}{f_s^2}\left(1+\frac{q_s^2\exp(-2g_sz')}{2f_s^2}\right)^{-1/2}\right\} \tag{66}$$

The intensity and the SBS gain of the pump beam follows

$$A_{00}^2=\frac{E_{00}^2(z=0)}{f_0^2(z)}\exp\left(-\frac{r^2}{r_{00}^2 f_0^2}-2g_0z\right) \tag{67}$$

and

$$g_0=\left(\frac{n}{n_0}\right)\frac{e^2\omega_{p0}^2\omega|A_{s0}|^2}{16\Gamma_i m_0 M\gamma^2 cc_s^2\omega_0\omega_s^2\varepsilon_{0p}^{1/2}} \tag{68}$$

Similarly, the intensity and SBS gain of the backscattered are written as

$$A_{s0}^2=\frac{E_{s0}^2(z'=0)}{f_s^2(z')}\exp\left(-\frac{r^2}{r_{s0}^2 f_s^2(z')}-2g_sz'\right) \tag{69}$$

$$g_s=-\left(\frac{n}{n_0}\right)\frac{e^2\omega_{p0}^2\omega|A_{00}|^2}{16\Gamma_i m_0 M\gamma^2 cc_s^2\omega_s\omega_0^2\varepsilon_{0s}^{1/2}} \tag{70}$$

or

$$g_s = -G_{SBS}\left(\frac{n}{n_0}\right)\frac{\alpha_0 E_{00}^2}{f_0^2\gamma^2}\exp(-2g_0 z') \quad (71)$$

and

$$G_{SBS} = \frac{\omega_{p0}^2 \omega\omega_s c}{16\Gamma_i M\omega_s^2 c_s^2 \varepsilon_{0s}^{1/2}} \quad (72)$$

By using the solution of pump and scattered field from above equations, one obtains the following expression for integrated (over the space only) reflectivity for SBS by taking into account the finite beam size of the pump and scattered laser beams (f_0 and f_s) as,

$$R = \frac{E_{s0}^2(z'=0)}{E_{00}^2(z=0)}\left(\frac{r_{s0}^2}{r_{00}^2}\right)\exp\left(-2g_s(z=0)L_z\right) \quad (73)$$

6. Results and Discussion

The analysis presented here characterize the focusing behavior and related stimulated scattering phenomena arising due to an intense Gaussian laser beam in a self-created plasma channel. To check the feasibility and validly computations are performed of Equation (26) to see combined effects of relativistic self-focusing and ponderomotive channeling corresponding to different values of self-trapped beam radius given by Equation (27). The numerical values are chosen for typical values of relativistic laser-plasma parameters applicable for underdense plasma, namely 1.06 μm Nd: glass laser of intensity **I** $\approx 1.21 \times 10^{18}$ W/cm^2, electron density $n_e \approx (10^{19} - 10^{20})$ cm^{-3}, $\omega = 1.7 \times 10^{14}$ sec^{-1}, initial beam radius $r_0 = 3$ μm and normalized critical parameter $(\omega_p / \omega)^2 \approx (0.1 - 0.5)$.

Figure 1 illustrates variation of dielectric function with relativistic factor γ at different concentration of electron densities. It is seen that for lower concentration of electron density, value of dielectric constant is more while for higher concentration of electron density it shifts downwards. It is evident from the figure, initially there is an increase in dielectric constant for all different concentration of electron densities and after a certain values it attains a constant value. Because of increase in relativistic factor due to very strong fields and the cumulative effects of ponderomotive force the nonlinearity saturates. As a result the electromagnetic wave drives all of the plasma electrons out of the regions of large field intensity and establishes an equilibrium with radiation inside a vacuum channel surrounded by plasma.

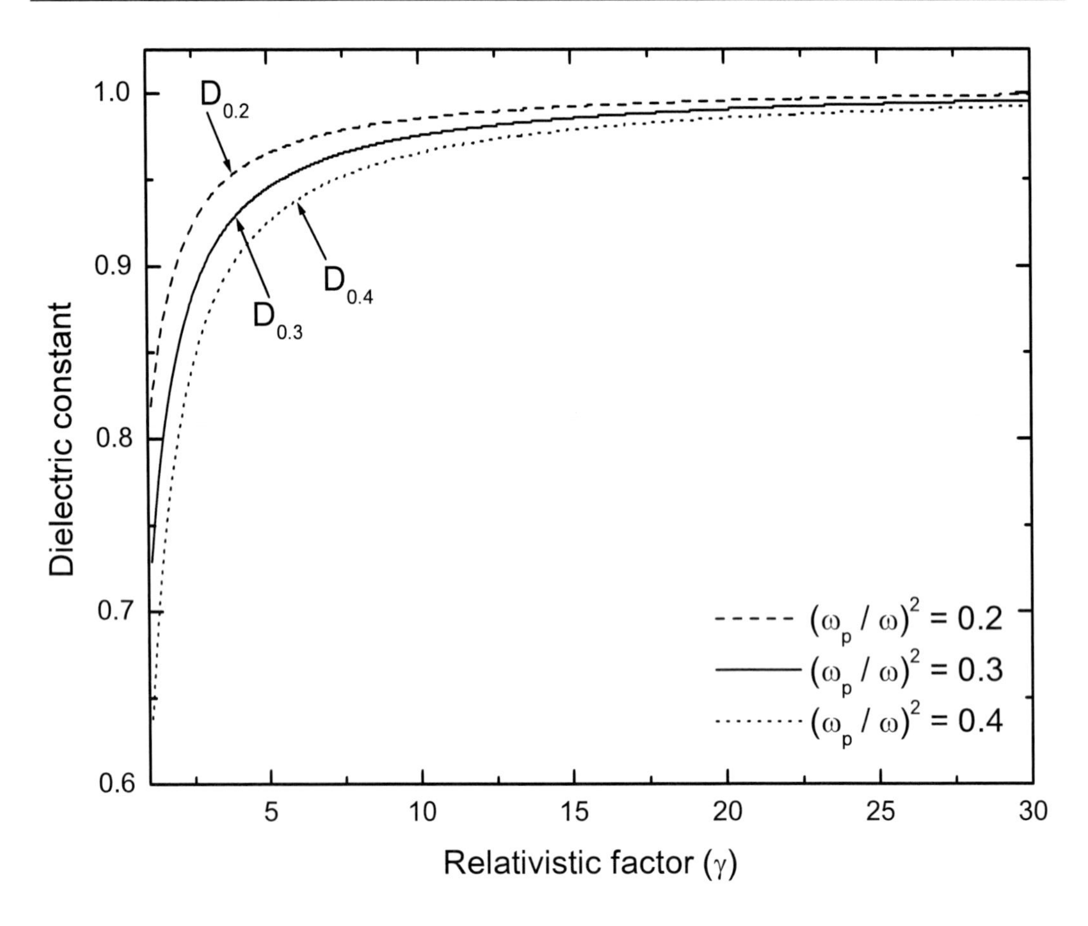

Figure 1. Variation of dielectric function ε as function of relativistic factor γ. Curves $D_{0.2}$, $D_{0.3}$ and $D_{0.4}$ corresponds to $(\omega_p / \omega)^2 = 0.2$, 0.3 and 0.4 respectively.

The critical curve Equation (27) is represented by Figure 2 for $\Omega_p^2 (= \omega_p / \omega)^2 = 0.1$ and 0.3 (*i.e.* a plot of the inverse of initial beam width $(1/\rho_0)$ against the axial beam irradiance (q_0). Referring to straight line drawn from the peak of both the curves to the right, parallel to q_0 axis, regions 1, 2 and 3 are characterized by self-focusing, oscillatory divergence and steady divergence of the beam respectively.

Considering different points in different regions the variation of beamwidth parameter with distance of propagation (f versus ξ) are illustrated in Figures 3. If these points exactly lie on critical curve then beam gets self-trapped and it propagates without convergence or divergence. As the initial value of q_0 and $(1/\rho_0)$ of a laser beam are such that the point $(q_0, 1/\rho_0)$ lies on the critical curve the value of $(d^2f / d\xi^2)$ will vanish at $\xi = 0$ $(z = 0)$ since the initial value of $(df / d\xi)$ (in case the wave front in plasma) is zero, the value of $(df / d\xi)$ continues to be zero as the beam propagates through the plasma. Hence, the initial value of 'f', which is unity (at z or $\xi = 0$), will remain unchanged. Thus, the beam propagates without any change in its beam width.

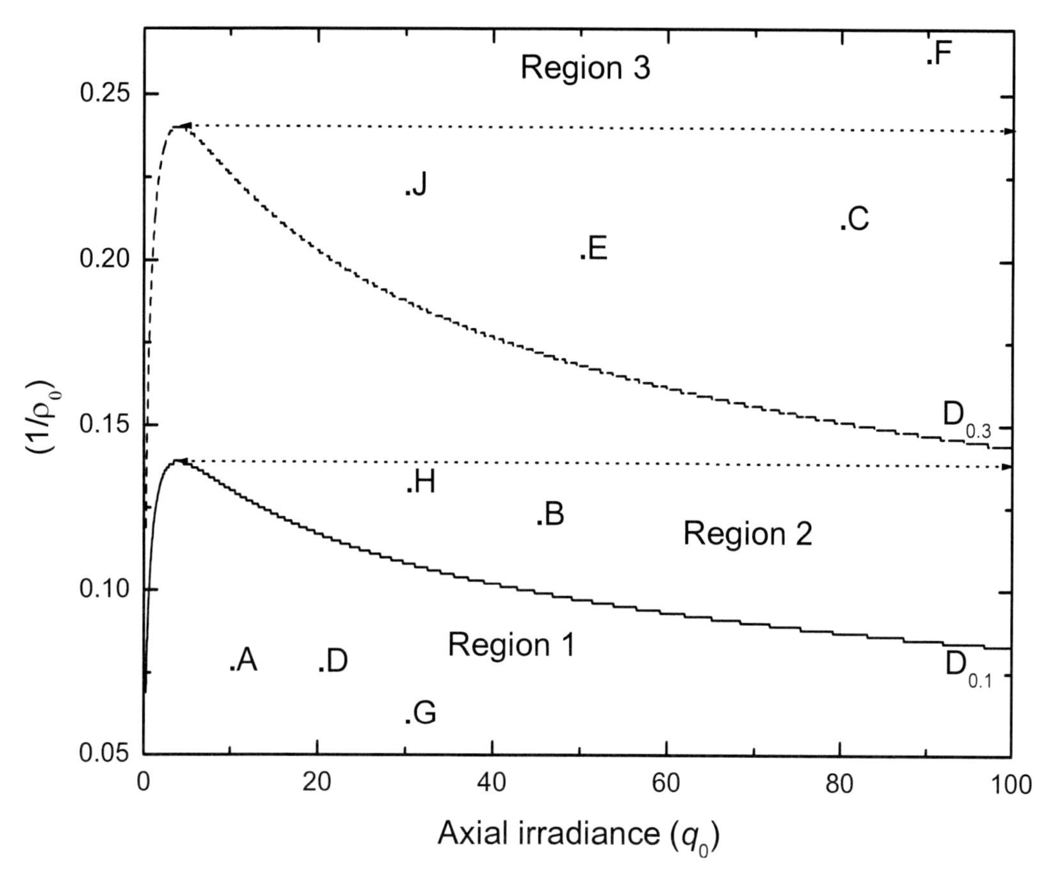

Figure 2. Relationship between axial irradiance q_0 and inverse of dimensionless beam radius $(1/\rho_0 = c/\omega r_0)$ indicating different regimes of propagation for varied density. Curves labelled $D_{0.1}$ (solid line) and $D_{0.3}$ (dash line) corresponding to $\Omega_p^2 = 0.1$ and 0.3 respectively. Region 1, Region 2 and Region 3 corresponding to self-focusing (SF), oscillatory divergence (OD) and steady divergence (SD) respectively.

The propagation of an intense short pulse laser in self-created plasma channel exhibits to stimulated scattering phenomena. The spot size of the laser inside the plasma depends on the intensity of the pump laser beam. The studies related to relativistic self-focusing of the pump laser beam and backscattered electromagnetic waves in the process of SRS / SBS are formulated through the coupled system of Equations [(49 - 50) and (61 - 62)] with corresponding expressions for integrated reflectivity Equations [56 and 73]. To have numerical appreciation of the results the coupled system of equations are solved numerically using the fourth order Runge – Kutta method for the initial plane wave fronts of the pump and scattered beams, following boundary conditions on the pump and the backscattered beams, $f_0 = f_s = 1$ and $df_0 / dz = df_s / dz = 0$.

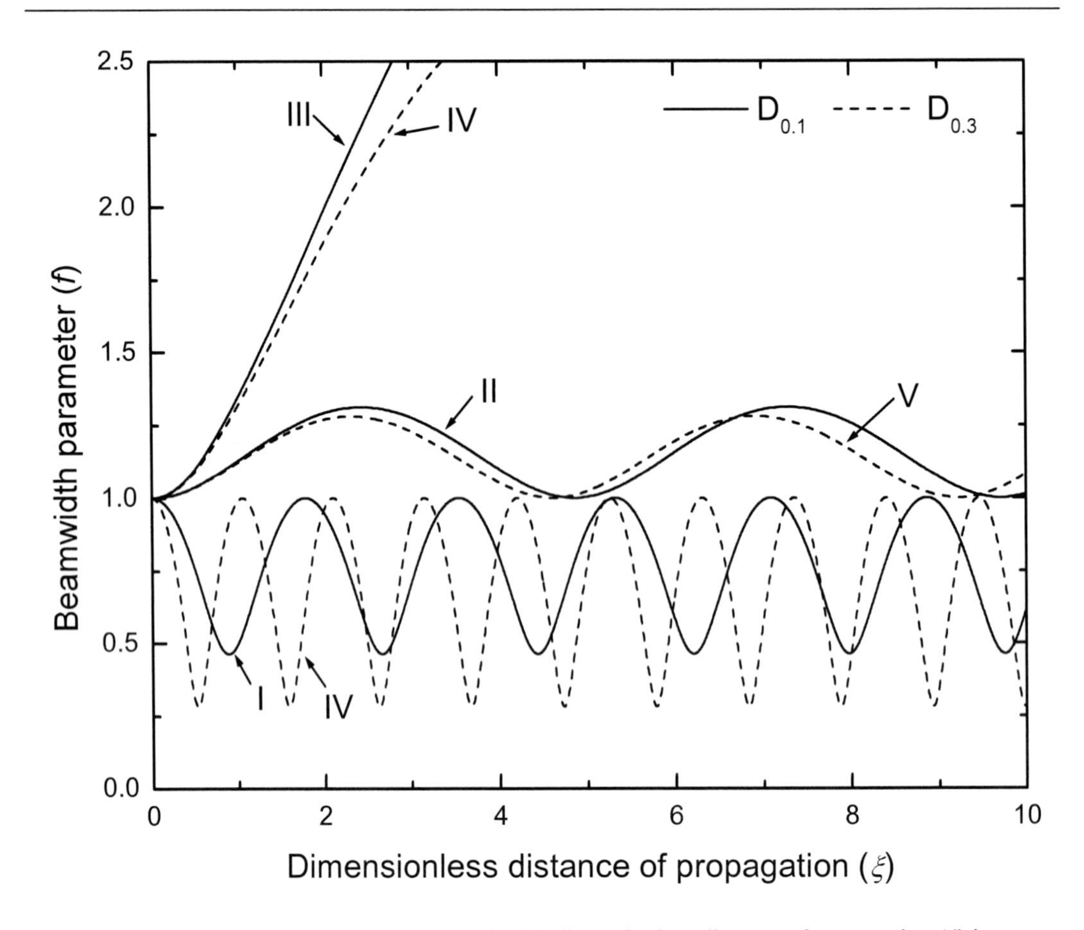

Figure 3. Variation of beamwidth parameter (f) with dimensionless distance of propagation (ξ) in different regimes for variable laser intensity and density. Curves labelled I, II and III corresponding to ($q_{0,}$ $1/\rho_0$) = A(10, 0.075), B(45, 0.12) and C(80, 0.21) for Ω_p^2 = 0.1 ($D_{0.1}$) and Curve IV, V and VI correspond to ($q_{0,}$ $1/\rho_0$) = D(20, 0.075), E(50, 0.2) and F(90, 0.26) for Ω_p^2 = 0.3 ($D_{0.3}$).

When a laser beam propagates through the plasma, the density of the plasma varies through the channel due to ponderomotive force. The ponderomotive force results from the lowering of the channel density, therefore the refractive index increases and laser gets focused in the plasma. Initially relativistic nonlinearity effects the focusing / defocusing of pump laser beam which further leads to cross-focusing of pump and scattered beams. The nonlinear effects and diffraction effects due to finite beam size of pump laser and scattered laser affects the dynamical evolution of each other. Figure 4 and 5 illustrate the variation of dimensionless beamwidth parameter (f_0) of the pump wave and back scattered wave (f_s) against the normalized distance of propagation for SRS. For the present computation the chosen value of pump irradiance are q_0 = 1.2 and 1.7 for $q_s = 1 \times 10^{-5}$. It is seen that the Raman back-scattered wave focused faster for higher value of q_0 and pump wave focused faster for lower value of q_0. A graphical plot of gain versus back reflectivity for SRS is sketched in Figure 6.

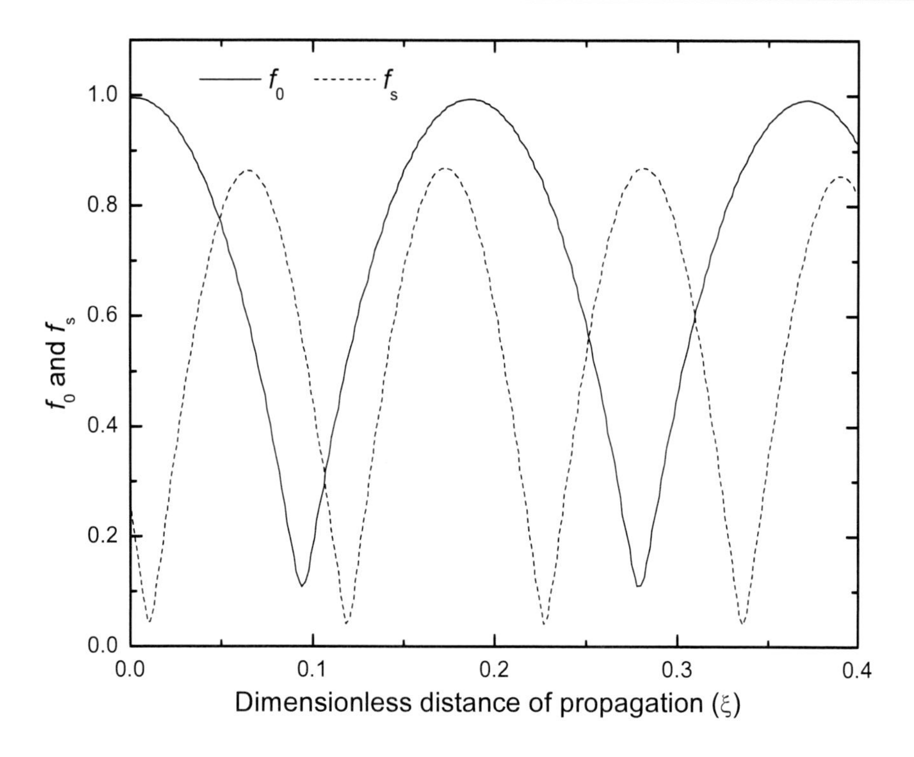

Figure 4. Variation of beamwidth parameter of pump wave f_0 and scattered wave (SRS) f_s with dimensionless distance of propagation ξ for $q_0 = 1.7$ and $q_s = 1 \times 10^{-5}$.

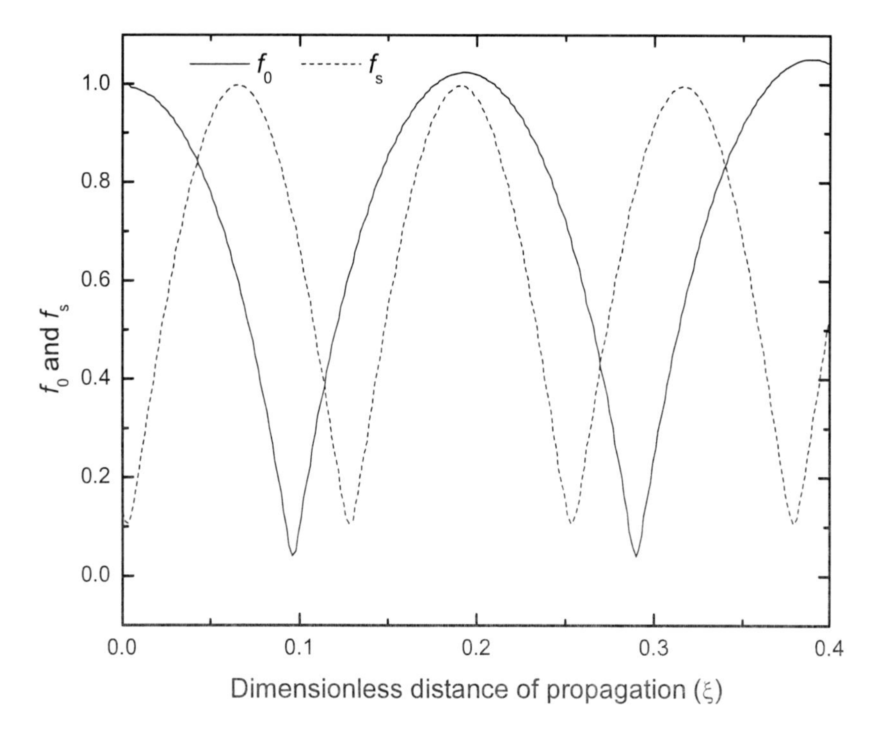

Figure 5. Variation of beamwidth parameter of pump wave f_0 and scattered wave (SRS) f_s with dimensionless distance of propagation ξ for $q_0 = 1.2$ and $q_s = 1 \times 10^{-5}$.

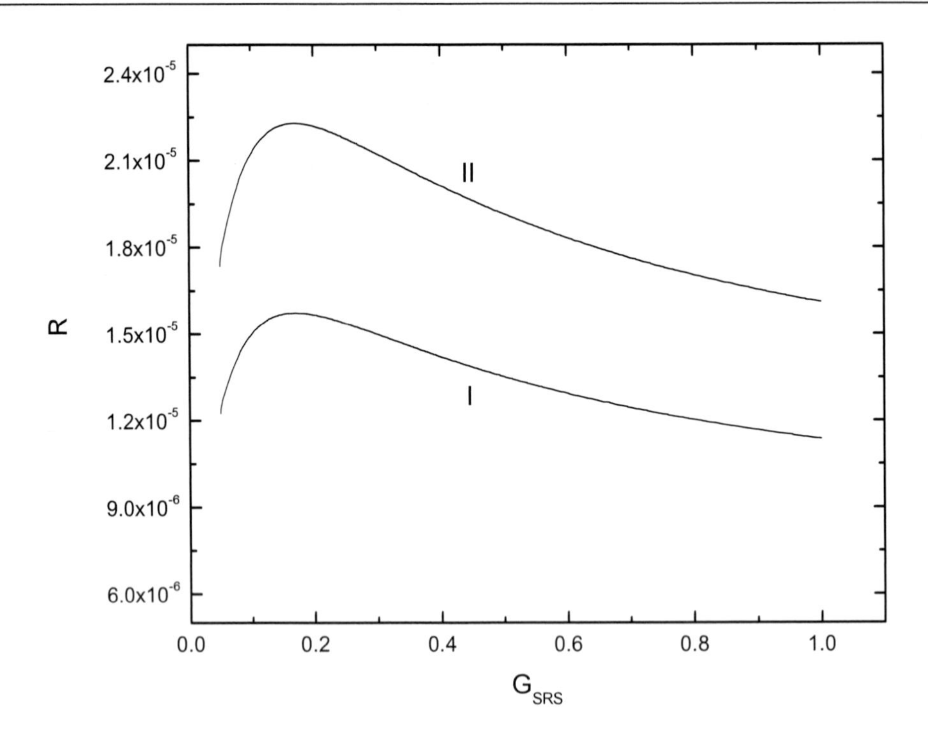

Figure 6. Variation in SRS back-reflectivity with G_{SRS} at $z = 0$ for $q_0 = 1.7$ (curve I) and 1.2 (curve II) for $q_s = 1 \times 10^{-5}$.

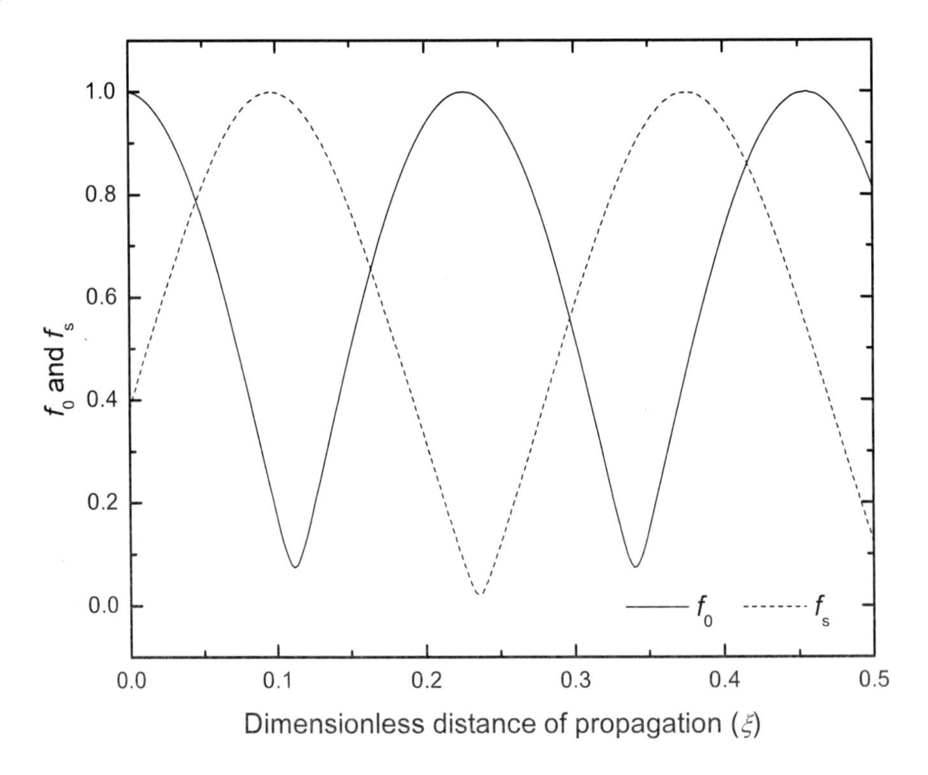

Figure 7. Variation of beamwidth parameter of pump wave f_0 and scattered wave (SBS) f_s with dimensionless distance of propagation ξ for $q_0 = 3$ and $q_s = 2 \times 10^{-5}$.

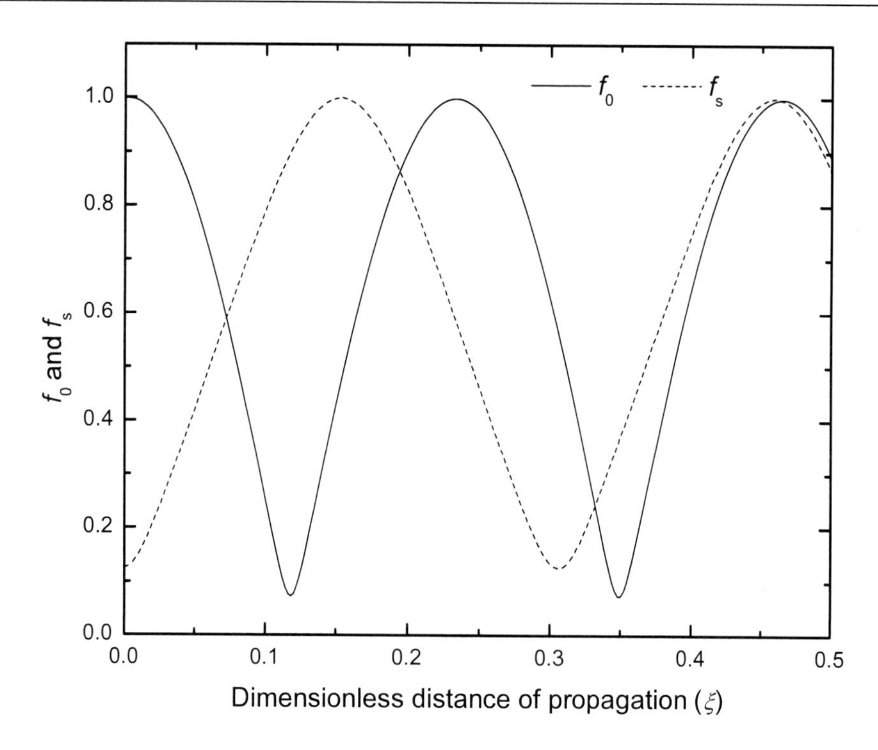

Figure 8. Variation of beamwidth parameter of pump wave f_0 and scattered wave (SBS) f_s with dimensionless distance of propagation ξ for $q_0 = 2$ and $q_s = 2 \times 10^{-5}$.

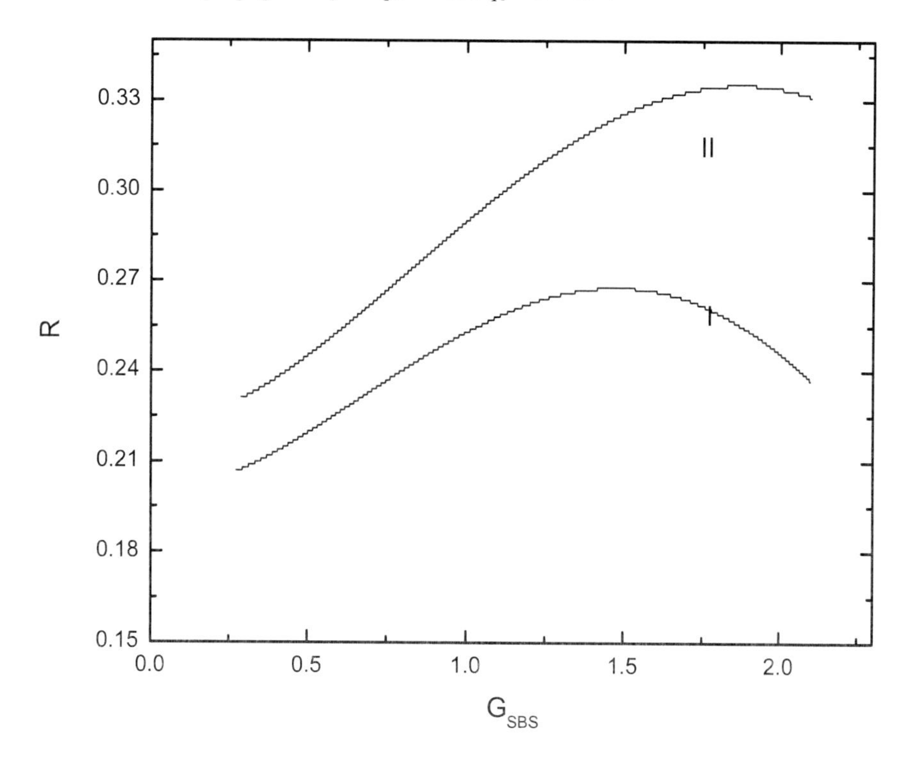

Figure 9. Variation in SBS back-reflectivity with G_{SBS} at $z = 0$ for $q_0 = 3$ (curve I) and 2 (curve II) for $q_s = 2 \times 10^{-5}$.

Due to ponderomotive force, which is responsible for change in plasma density it is inferred that reflectivity initially increases, attains a maximum and then falls off. The decrease in reflectivity at high intensities appears due to the depletion of electrons from the central region. As evident from Equation (44), the gain factor (g_s) depends upon the beamwidth parameter of pump wave (f_0), relativistic factor 'γ', gain of pump wave (g_0) and intensity. Due to relativistic self-focusing, the intensity of laser beam increases which increases the electron quiver velocity and hence relativistic factor is increased. Therefore at a particular value of G_{SRS} (for SRS) the product of f_0 and γ, results in a maximum value. The gain of the pump and scattered beams are coupled through Equation (41) and (44). Thus, as the gain of a scattered beam attains maximum value, the pump beam gain will be increased. Consequently, the maximum value of the pump gain leads to a decrease in the scattered gain due to the exponential term in Equation (45), hereby back-reflectivity decreases.

Based on the equations derived for stimulated Brillouin back scattering [Equations 61 – 73] similar plots are sketched for another set of parameters: q_0 = 2 and 3, and $q_s = 2 \times 10^{-5}$. The qualitative nature of focusing and reflectivity resembles to that of SRBS. The plasma channel formed by the laser pump guides only the fundamental mode of the sideband and the higher order modes leak out from the channel as they have a large transverse wave vector. Hence, the scattered modes are mostly in backward direction. The angular spread of the back-scattered laser intensity may provide an estimate of the radius of the plasma channel. Laser propagation in channels needs further understanding and insight, as plasma channels are increasingly used for diffraction guiding of intense laser pulses over extended distances.

The physical mechanisms responsible for the occurrence of relativistic self-focusing, SRS and SBS in the plasma channel is self-sufficient to be understood under WKB and the simple semi-analytical paraxial approximation for a wide range of normalized parameters used here. The instabilities as SRS and SBS that arise due to propagation of intense laser pulse propagation through plasma in view of the significant loses of incident laser beam are substantial. The experimental situation where intense lasers are focused in a plasma requires a relativistic description of the central part of the beam, where the main contribution to self-focusing occurs. This region is well described by the paraxial approximation used here. The analytical model is very instructive in understanding channel dynamics near the axis for wide range of normalized parameters applicable for underdense plasma.

The analytical and numerical calculations suggest that the relativistic increase of the electron mass and electron expulsion by the laser ponderomotive force governs the initial phase of plasma self-channeling. Both these effects increase the refractive index of a plasma on axis, focusing the laser pulse and trapping a substantial part of the incident power propagating in a self-trapped mode. In the focused regime beyond the focus, the nonlinear refraction starts weakening, and the spot size of the laser increases, showing focusing / defocusing behavior with distance of propagation. Further showing stimulated scattering phenomena. The results obtained based on realistic physical parameters are relevant for laser plasma interaction experiments, where self-guiding is necessary and laser beam can be used for plasma heating.

ACKNOWLEDGMENTS

We have been benefited from extensive and fruitful discussions with Dr. A. Giulietti, Prof. D. Giulietti, Dr. L. A. Gizzi, Prof. T. Desai, Prof. M. S. Sodha, Prof. A. K. Ghatak, Prof. V. K. Tripathi, Prof. R. K. Singh, and Dr. V. K. Sanecha at various stages. Valuable and useful interactions with group researchers K. A. Qureshi, B. Rathore and S. Asthana are also gratefully acknowledged.

APOLOGY

With the advent of high power lasers, a fantastic situation arose which is unique in the history of laser-plasma physics, materials processing as well as of non-linear physics. Several scientists working on wide-angle topics have accepted the sense of excitement and extensive studies on the properties have already been made. In this situation, we cannot claim that this is an unbiased review. More or less, we have concentrate and discussed our own contributions at great length and share bulk of the phenomena's. We sincerely apologize to many of the researchers whose work, while relevant, has not been cited.

REFERENCES

[1] Kruer, W. L. *The Physics of Laser Plasma Interactions*; Addison-Wesley: Redwood City, CA, 1988.
[2] Umstadter, D. *J. Phys. D. Appl. Phys.* 2003, *36*, R151.
[3] Gibbon, P.; Forster, E. *Plasma Phys. Control Fusion* 1996, *38*, 769.
[4] Labaune, C.; Fuchs, J.; Depierreux, S.; Baldis, H. A.; Pesme, D.; Myatt, J.; Huller, S.; Tikhonchuk, V. T.; Lavalc, G. *C. R. Acad. Sci.* 2000, *1*, 727.
[5] Lindl, W. L. *Phys. Plasmas* 1995, *2*, 3933.
[6] Adam, J. C.; Heron, A.; Laval, G.; Mora, P. *Phys. Plasmas* 2001, *8*, 1664.
[7] Thomson, J. J. *Phys. Fluids* 1978, *21*, 2082.
[8] Karttunen, S. J. *Phys. Rev. A* 1980, *23*, 2006.
[9] Antonsen, T. M.; More, P. *Phys. Rev. Lett.* 1992, *69*, 2204.
[10] Kolber, T.; Rozmus, W.; Tikhonchuk, V. T.; *Phys. Fluids* 1993, *B5*, 138.
[11] Fernandez, J. C.; Cobble, J. A.; Failor, B. H.; DuBois, D. F.; Montgomery, D. S.; Rose, H. A. *Phys. Rev. Lett.* 1996, *77*, 2702.
[12] Divol, L.; Mounaix, P. *Phys. Rev. E* 1998, *58*, 2461.
[13] Tzeng, K. C.; Mori, W. B.; *Phys. Plasmas* 1999, *6*, 2105.
[14] Berger, R. L.; Lefebvre, E.; Langdon, A. B.; Rothenberg, J. E.; Still, C. H.; Williams, E. A. *Phys. Plasmas* 1999, *6*, 1043.
[15] Baker, K. L.; Drake, R. P.; Estabrook, K. G.; Sleaford, B.; Prasad, M. K.; LaFontaine, B.; Villeneuve, D. M. *Phys. Plasmas* 1999, *6*, 4284.
[16] Modena, A. *IEEE Trans. Plasma Sci.* 1996, *24*, 289.
[17] Rousseaux, C.; Rabec le Gloahec, M.; Baton, S. D.; Amiranoff, F.; Fuches, J.; Gremillet, L.; Adam, J. C.; Heron, A.; Mora, P. *Phys. Plasmas* 2002, *9*, 4261.

[18] Miyakoshi, T.; Jovanovic, M. S.; Kitagawa, Y.; Kodama, R.; Mima, K.; Offenberger, A. A.; Tanaka, K. A.; Yamanaka, T. *Phys. Plasmas* 2002, *9*, 3552.

[19] Guerin, S.; Laval, G.; Mora, P.; Adam, J. C.; Heron, A. *Phys. Plasmas* 1995, *2*, 2807.

[20] Powers, L. V.; Turner, R. E.; Kauffman, R. L.; Berger, R. L.; Amendt, P.; Back, C. A.; Bernat, T. P.; Dixit, S. N.; Eimerl, D.; Harte, J. A.; Henesian, M. A.; Kalantar, D. H.; Lasinski, B. F.; MacGowan, B. J.; Montgomery, D. S.; Munro, D. H.; Pennington, D. M.; Shepard, T. D.; Stone, G. F.; Suter, L. J.; Williams, E. A. *Phys. Rev. Lett.* 1995, *74*, 2957.

[21] Riconda, C.; Weber, S.; Tikhonchuk, V. T.; Adam, J. C.; Heron, A.; *Phys. Plasmas* 2006, *13*, 083103.

[22] Weber, S.; Riconda, C.; Tikhonch, V. T.; *Phys. Plasmas* 2005, 94, 055005.

[23] Tsytovich, V. N.; Stenflo, L.; Wilhelmsson, H.; Gustavsson H. G.; Ostberg, K.; *Physica Scripta* 1973, *7*, 241.

[24] Rose, H. A.; DuBois, D. F. *Phys. Rev. Lett.* 1994, *72*, 2883.

[25] Varshney Asthana, M., Varshney D., and M. S. Sodha, *Laser and Particle Beams* 2000, 18, 101; Varshney Asthana M., Qureshi K. A., and Varshney D., *J. Plasma Physics* 2006, 72, 195; Varshney Asthana M., Rathore B., and Varshney D., *Journal of Modern Optics* 2009, 56, 1613; Varshney Asthana M., Rathore B., Sen S, and Varshney D., 2009, *Journal of Modern Optics* 2009, 56, 2368; Varshney Asthana M., Sen S., Rathore B., and Varshney D., *J. Phys.: Conf. Ser.* 2010, 208 012088 (pp1 –6); Varshney Asthana M., Sen S., Rathore B., and Varshney D., *J. Phys.: Conf. Ser.* 2010, 208 Pp. 012088 (pp1 –9); Varshney Asthana M., Rathore B., Sen S, and Varshney D., *Optik Int. J. Light Electron Optics* 2011, 122, 395; Varshney Asthana M., Rathore B., and Varshney D., *Optik Int. J. Light Electron Optics* 2011, 122, 1

[26] Sodha M S, Ghatak A K and Tripathi V K 1974 *Self-Focusing of Laser Beams in Dielectrics, Semiconductors and Plasmas*, Delhi: Tata McGraw-Hill

In: Laser-Induced Plasmas
Editor: Ethan J. Hemsworth, pp. 231-292
ISBN 978-1-61324-851-5

Chapter 9

ADVANCED CO_2 LASER-INDUCED PLASMA PROCESSING AND ITS APPLICATION

Chen-Kuei Chung
National Cheng Kung University, Taiwan

ABSTRACT

The CO_2 laser of 10.6 μm in wavelength is an inexpensive, rapid and flexible one for the soft polymer and hard glass and ceramic related materials processing. It has been widely applied to the fabrication of microchannel ablation, cutting, microhole drilling, material annealing and modification in the categories of MEMS, bio-chip, optical/ optoelectronic devices, displays and laser dentistry. The basic CO_2 laser physics is photo-thermal mechanism for material removal therefore some defects of debris, bulges, cracks and scorches around ablated microstructure are formed during laser processing in air which degrades the device yield and quality for bonding. In this article, some advanced laser processing methods have been proposed for improving the microstructure quality of fabrication including Liquid Assisted Laser Processing (LALP), cover-layer protection processing and Glass Assisted CO_2 LAser Processing (GACLAP) for eliminating the cracks and scorches defects, diminishing bulges height and reducing feature size, even making the transparent-in-nature silicon material to be etched, drilled and cut. LALP can effectively reduce the temperature, heat-affected zone, thermal gradient and stress via water for hindering the crack and scorch formation together with the laser heating induced stronger natural convection in water for carrying debris away to reduce bulge height. The feature size can be reduced from 400–500 μm via traditional processing in air to 150–200 μm even smaller via LALP. Combing LALP and low-temperature bonding techniques have been used for the fabrication of capillary-driven glass-based microfluidic chip for the application of low-to-high viscosity fluid actuating and biomedical blood coagulation testing. GACLAP can change light absorption behavior of Si and make Si be etched from the top surface toward the interface whose new mechanism is discussed in viewpoint of the variation of electronic band structure, surface oxidation and light absorption of Si at high temperature. A simple thermal model and ANSYS software are adopted for the analysis of thermal and stress distribution on specimen during the laser irradiation in air and water ambient. Also, a simple CO_2 laser annealing process for titanium dioxide treatment has also been developed instead of conventional expensive short wavelength laser annealing and non-selective high-temperature furnace annealing.

Both crystalline rutile and anatase titanium dioxide transformation from amorphous titanium oxide can be controlled by the sol-gel composition and laser annealing parameters for material property adjustment which is potentially used for the photocatalyst and optoelectronic application. The relationship between process, microstructure and phase transformation of titanium oxide is discussed and established.

1. Introduction and Motivation

Brittle materials of Si, glass and ceramics and soft polymer material have been widely used for the micro (/nano) electromechanical systems (MEMS/NEMS), microelectronics, optoelectronics, optics, communication, microfluid and biochip. The micromachining technologies are used to pattern these materials depending on the material crystallography and properties. Semiconductor processes of photolithography, etching and deposition are applied to Si based materials while LIGA-like, excimer laser or μ-EDM is performed for the non-Si based materials. Regarding glass material, the etching of glass, especially deep etching of Pyrex glass, was studied by many researchers using hydrofluoric acid (HF) based wet solution [1]-[3] or dry plasma etching [4]-[6] for the silicon-glass devices and microsystems application. But the traditional glass biochips fabricated by the photolithography for dry/wet etching processing were at low etching rate together with high cost especially for deep through-wafer etching or large area microstructure. In the fluidic chips, many researchers have paid much attention to fluid pumping and control for the lab-on-a-chip application these decades [7]. High-hydrophilic surface of channels is a crucial issue for capillary-driven microfluidic systems to effectively actuate liquid without any additional pump. The intrinsic hydrophobic behavior of polydimethylsiloxane (PDMS) or SU8 polymer fluidic channels was frequently modified by surface treatments to obtain the hydrophilic surface for a better desired capillary performance [8]-[12]. However, the modified PDMS surface exhibits the time-limited hydrophilic behavior using traditional surface treatment methods of exposure to energy [8] including oxygen plasma [9], ultra-violet [10] and corona discharges [11], and chemical solution treatment [12]. Therefore, besides PDMS and SU8 materials, glass is another popular material for fluidic chip fabrication but encountered more complicated processing procedure [13],[14]. Complex glass chip fabrication as partly mentioned above generally includes cleaning, deposition of etching mask material, photolithography patterning, HF etching, hole drilling and high-temperature (650 °C) glass-glass fusion bonding. It is involved in high temperature process and makes fabrication time-consuming and high cost as well as a challenge for commercialization of glass-to-glass microchips. It is noted that all the dry plasma or wet solution etching processes need photolithography to define the pattern geometry and select the suitable mask material of resist or hard mask to meet the different requirements of chemical resistance and the desired etching depth and patterns. Their etching rates are generally in the range of 0.5-3 μm/min, dependent on the chemistry, process conditions and sample geometry. Such etching rates are still low, especially for a through Si or glass wafer etching of 525 μm, and it will take several hours. The equipment cost of dry plasma etching is high and the etched geometry for wet etching is limited to crystallography and directionality.

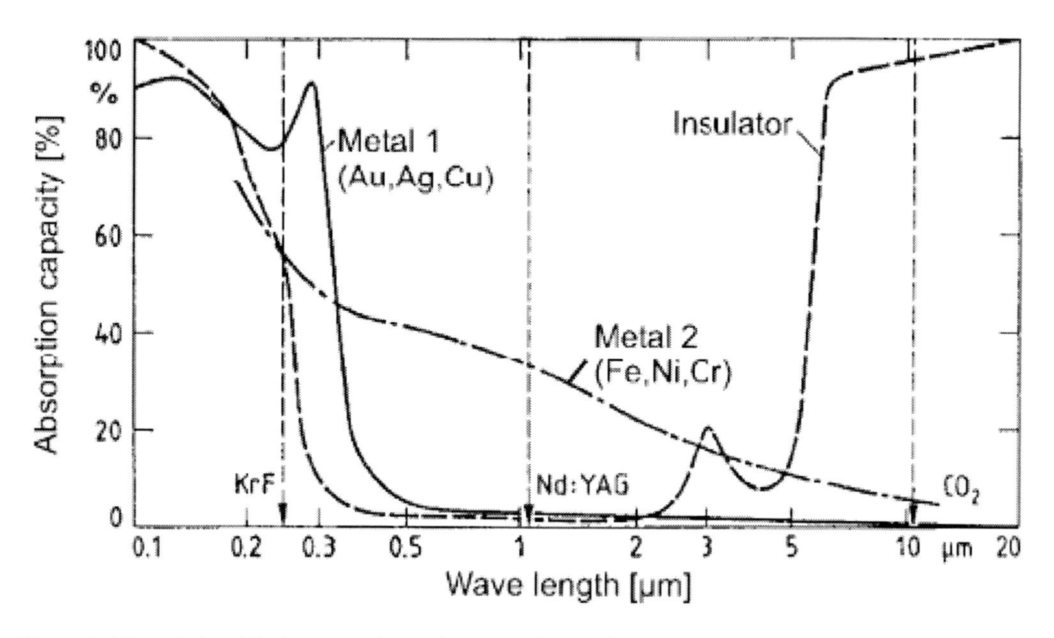

Figure 1. The relationship between absorption capacity and laser source wavelength [15].

Compared to the dry plasma or wet solution etching, laser machining has advantages of simple, maskless, fast and direct-write for different geometrical shapes. It includes various modes for the microfabrication of different materials. The wavelength of laser is related to the absorption rate of metal or insulator materials. The relationship between absorption capacity and laser source wavelength is shown in Figure 1. The nanosecond and microsecond pulses of different wavelengths are related to laser machining system [15]. The material is removed during laser processing, which heats it up to the vaporization temperature and melts material at the same time. Figure 2 shows the diagram of power intensity effect during laser processing. The higher laser intensity output or longer pulse duration will induce bigger plasma formation [15],[16]. It affects the etching rate and product quality during laser machining. Many methods were presented to obtain improved experiment results for industry application. Laser functioning can be discriminated between photoexcitation and absorption mechanism due to pulse duration difference. One is the energy transfer from electron to phonon during photoexcitation. The thermal energy increases as the pulse duration becomes longer than the timescale of electron-phonon coupling (<1 ps). So the ultra-shot laser system has advantage of less plasma effect to get a good product quality. It is noted that the conventional Si etching by laser can be performed by various wavelength laser ablation including Nd:YAG [17]-[20], excimer [21],[22] or femtosecond lasers [23]-[24] with wavelengths between 1.06 μm and 193 nm. No CO_2 laser of a 10.6 μm wavelength was directly used for Si etching. Moreover, the femtosecond laser or excimer laser for etching the brittle material had low etching rate together with high cost especially for deep through-wafer etching or large area microstructure. The other characteristic of laser ablation is related to the distribution of thermal energy due to difference of dominant light absorption mechanism. One is multiphoton absorption by valence electrons, and another is linear absorption by photoexcited electrons [31],[32]. The laser manufacture technique has been applied to micromachining for various microelectronics, optoelectronic and display markets as shown in Figure 3 [33],[34]. Pulse excimer laser was used to etch glass in silica and glass but the etching rate was only 6-10 nm per pulse [35]. Femtosecond laser is another technology for etching silica glass [36],[37]. The ultra-short laser has the advantage of small heat affected zone (HAZ) on the surrounding material due to the minimized heat conduction during a short pulse. Several publications used the assisted method to etch transparent silica glass at low repetition rate or higher output laser energy [38]-[40]. But the expensive femtosecond laser or excimer laser [41] used for etching glass had low etching rates together with high cost especially for deep through-wafer etching. For example, Cheng et al. [42] reported the acid solution assisted UV-laser wet etching glass substrate for microfluidic application but the etching rate was still low together with occurrence of traditional high-bulge defects. The basic CO_2 laser physics is photo-thermal mechanism for material removal therefore some defects of debris, bulges, cracks and scorches around ablated microstructure are formed during laser processing in air which degrades the device yield and bonding quality. The CO_2 laser micromachining is related to the photothermal melting and evaporation due to the long wavelength of 10.6 μm. It results in the HAZ in solid, molten zone in liquid and evaporated gas with high pressure in the laser working area. The formation of bulge is attributed to molten polymer resolidified by atmospheric air cooling on the rim of channel after splashing from high-pressure gas or surface tension driven flow. Because of the high absorption in polymer and glass, low cost and easy operation, the CO_2 laser processing has widely been used in many kinds of industry application, and fabrication of microchannel ablation, cutting, microhole drilling, material annealing and modification in the categories of MEMS, bio-chip, optical/optoelectronic devices, displays and laser dentistry. Continuous wave CO_2 laser has been used for thermal treatment before wet etching of silica glass [43], but rarely in directly linearly etch glass. Although the CO_2 laser was widely used for transparent material ablation [44], it was rarely used to directly etch through amorphous Pyrex glass at room temperature in air due to the thermal stress induced crack problem [45]. Ogura and Yoshida [46] reported crystalline and amorphous glass hole drilling using CO_2 laser and found that crystalline synthetic quartz had the best drilling quality together with through-hole drilling while the amorphous Pyrex glass had the worst result together with the highest bulge around the hole as well as no through-hole cross section being demonstrated. Yen et al. [47] reported the substrate heating method to improve the defects of CO_2 laser ablation on glass for microfluidic application. But no through-hole etching was shown and such method would affect the cutting performance and increase instrument complexity. The liquid-assisted laser processing (LALP) which was applied to the welding, surface cleaning and cutting [[48],[49]] can reduce debris, re-deposition and HAZ. Chung et al. [50] reported the linearly through-wafer etching of Pyrex glass using liquid-assisted CO_2 laser process but no through-hole etching. Barnes et al. [51] propose the water assisted CO_2 laser cutting alumina specimen compared with laser processing in air, then a finite element approach was used to analyze the temperature and thermal stress distributions during both cutting. Chung et al. [52] used the combination of water-assisted CO_2 laser processing for ablating glass slides and the modified thermal bonding for capillary-driven bio-fluidic application. The feasibility of pure water-assisted laser processing in linear through-wafer deep etching of Pyrex glass without mask materials to obtain a crackless surface has been demonstrated [50]. The deep discussion of mechanism of water cooling and shear-force enhanced bulge reduction is discussed. The optical spectrum of pure water indicates that the transmission in 10.6 μm of CO_2 laser wavelength is very low [49]. Therefore, the interaction of CO_2 laser radiation with water caused water vaporization [53]. The energy loss due to water absorption can result in the decrease of etching rate during ablation. In addition to water, other liquids were reported for enhancing etching rate. The etching rate of single crystal Si in dimethyl sulfoxide (DMSO, $(CH3)_2SO$) [54] and dimethyl formamide (DMFA, $HCON(CH3)_2$) [55] was higher than in water. The aqueous solution of surfactant in other category has been used for formation of gold or silver nanoparticles by laser ablation [56],[57], rather than glass etching process.

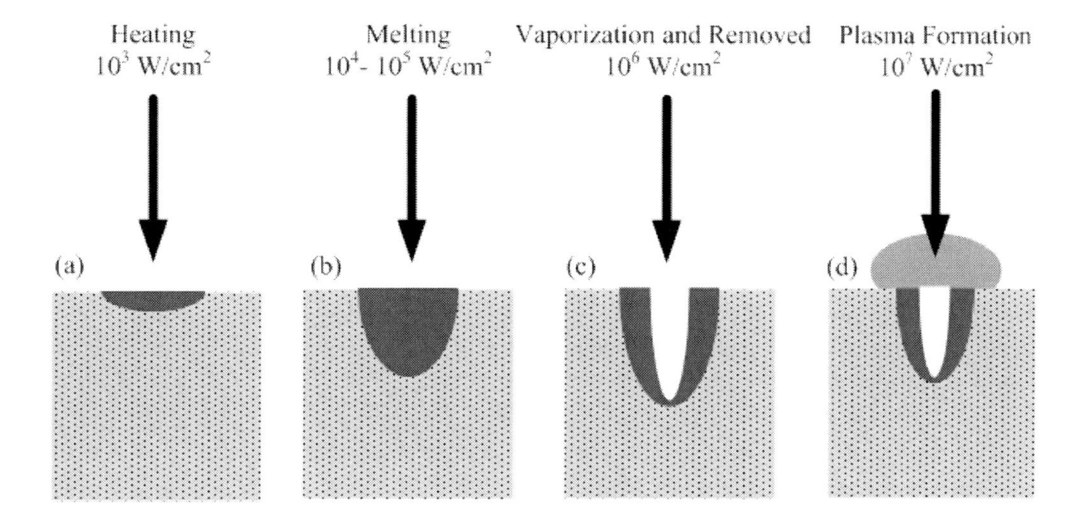

Figure 2. The diagram of intensity effect during laser processing [15],[16].

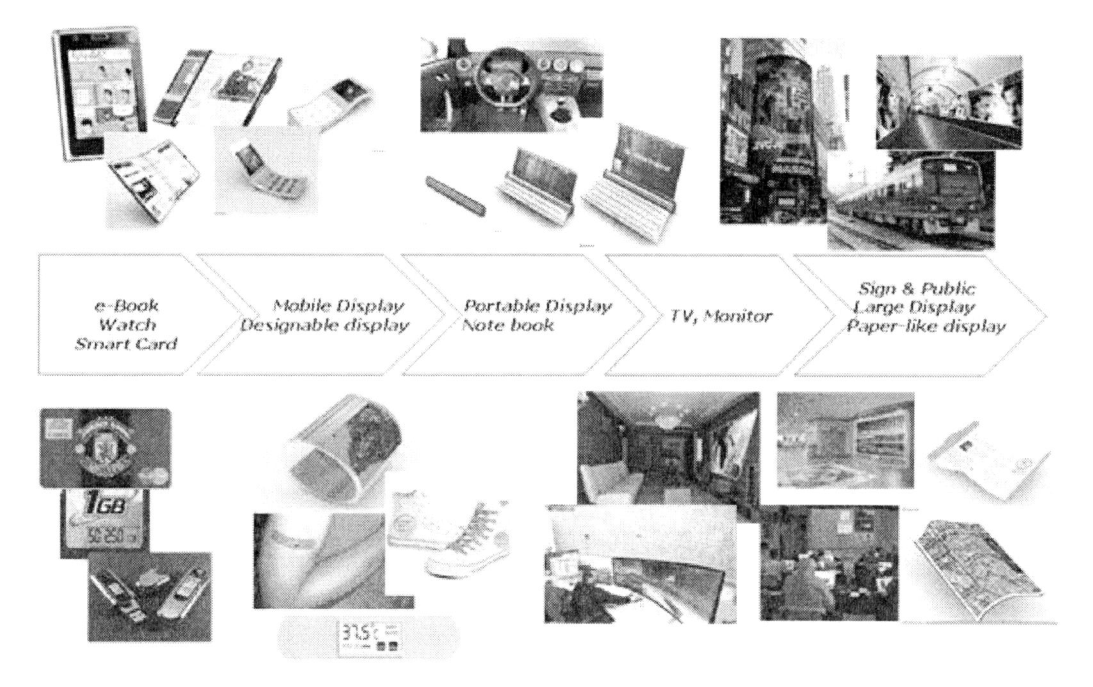

Figure 3. The application market using laser manufacture technique [33].

In this article, some modified or advanced laser processing methods have been proposed for improving the microstructure quality of fabrication including Liquid Assisted Laser

Processing (LALP), cover-layer protection processing and Glass Assisted CO_2 LAser Processing (GACLAP) for eliminating the cracks and scorches defects, diminishing bulges height and reducing feature size, even making the transparent-in-nature silicon material to be etched, drilled and cut. Also, the LALP is also used to fabricate a capillary-driven bio-fluidic chip. The laser treatment of sol-gel processing combined with CO_2 laser annealing is also applied for enhancing titanium dioxide microstructure and properties.

2. LIQUID-ASSISTED CO_2 LASER PROCESSING

Liquid waster assisted laser processing has been presented for the commercial application of laser cutting, shock processing and surface cleaning since 1970s, and there are many publications studied with LALP technique. Figure 4 show the pure water transmission spectrum of varied wavelength [49], that CO_2 laser of 10.6 μm wavelength has well absorption efficiency and it can provide a protection effect during laser processing in above preview function. Table 1 show the comparison of water-assisted laser processing with processing in air.

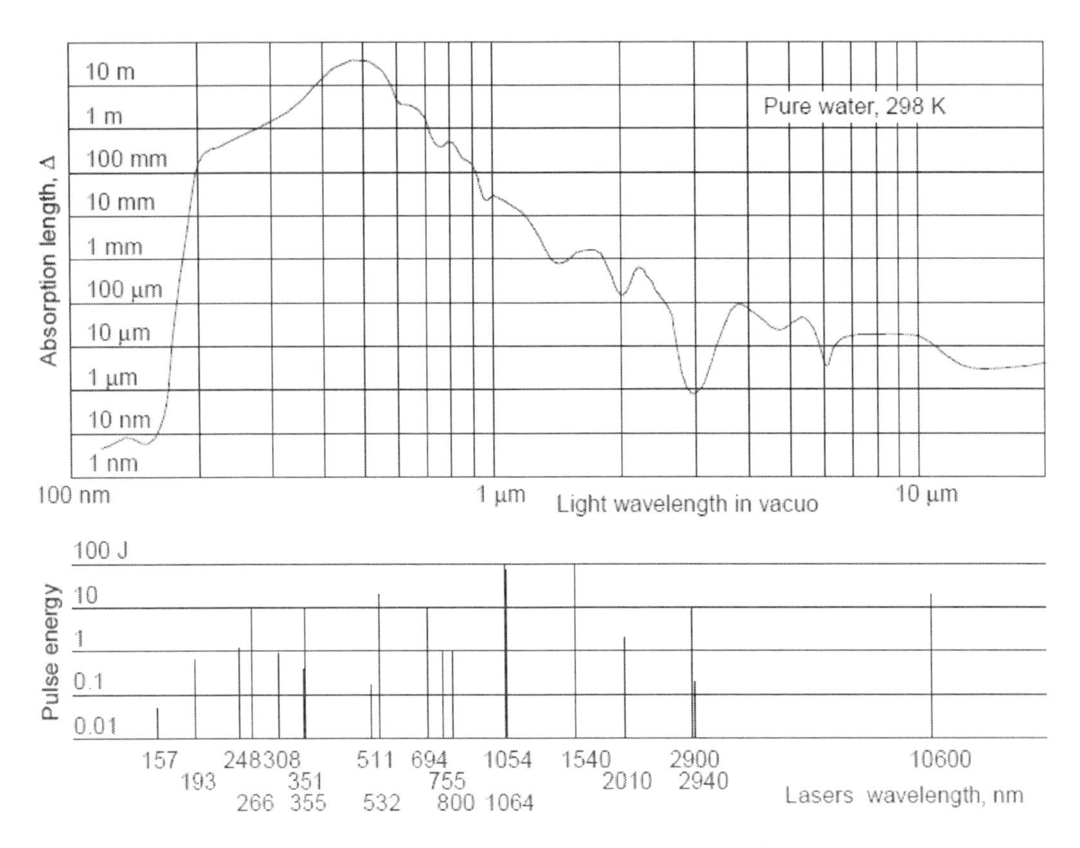

Figure 4. The pure water transmission spectrum of varied wavelength. [49].

Table 1. Comparison of water-assisted laser processing with processing in air [49]

Advantages of water-assisted processing	Disadvantages of water-assisted processing
• Light transport is possible along a water jet. • Higher plasma pressure due to confinement. • Water convection/explosive evaporation carries debris away. • More effective cooling of workpiece and ejected material. • Useful chemical reactions with water. • Reduced atmospheric pollution by waste gases and aerosols. • Higher optical breakdown threshold tan in air. • Smaller focal spot size.	• Light is absorbed by water (which may also be useful) and by the debris produced during processing. • Light may be scattered by the water surface, suspensions and bubbles. • Power loss due to water cooling. • Harmful chemical reactions with water. • Water photolysis → explosion hazard. • Slower processing due to the long relaxation time of the vaporized state of water. • Water and vapor are hazardous to electronics. • Possible corrosion of materials.

2.1. Liquid-Assisted CO_2 Laser Cutting

In this section, we have demonstrated the effective method of linear/through-hole etching of amorphous Pyrex glass using the low-cost CO_2 laser based on LALP technology. The liquid of water was used to obtain good quality of linear/through-hole glass etching for the sake of cooling and cleaning. In order to understand the effect of surfactant addition on glass ablation in water-assisted CO_2 laser processing, different weight percent of surfactant solution was used in assistance liquid. We also discuss adding the surfactant technology to water-assisted CO_2 laser processing for promoting glass ablation. The difference of etching performance between the pure water and the surfactant solution was compared and discussed. In order to further investigate the temperature distribution, maximum temperature and thermal stress during the drilling machining process, a simple thermal model is adopted for the analysis of thermal behavior during machining. The ANSYS software was used to solve this transient problem. The water cooling effect and debris removal mechanism were also discussed for drilling quality improvement. Figure 5 shows the schematic diagram of LALP setup for Pyrex glass etching. It includes the a computer, a continuous wave (CW) CO_2 laser source, a reflective mirror, a focal lens and an experimental sample immersed into water. The test sample is Pyrex 7740 glass with an average thickness of 500 μm. The commercial available air-cooled CO_2 laser equipment (VL-200, Universal Laser system Inc., U.S.A.) was used with a maximum laser power of 30 W. The CO_2 laser has a wavelength of 10.6 μm and a TEM_{00} output of beam. Also, the CO_2 laser uses a sealed-off, RF excited, slab design and a multi-pass, free space resonator for a good quality beam with a M2 value of 1.4±0.2. The maximum scanning rate of laser was 1140 mm/s and the largest working area was 409 x 304 mm^2. The focal length of the lens is about 38.1 mm and the smallest beam spot size of the commercial product after standard verification could reach 76 μm which is defined as the double distance from the center of Gaussian intensity distribution to that with $1/e^2$ maximum intensity. The laser spot was focused on the top surface of Pyrex glass during LALP. The glass was immersed into water in a container at room temperature, and then laser beam was performed to scan on the wafer to get the desired trench at proper parameters. A computer aided design program of CorelDraw software was available to set the experimental parameters of laser power, scanning rate and number of beam passes for auto-controlled processing. The 20 mm linear length was tested for through-wafer etching in the experiments. Optical microscope (OLYMPUS BX51M, Japan) was used to examine the morphology after laser processing. The profile and morphology of laser-machined glass channels were examined by the alpha-step (α-step) profiler (TENCOR, MA-1450, USA) and 3D confocal microscope (NanoFocus®, μSurf®, USA), respectively.

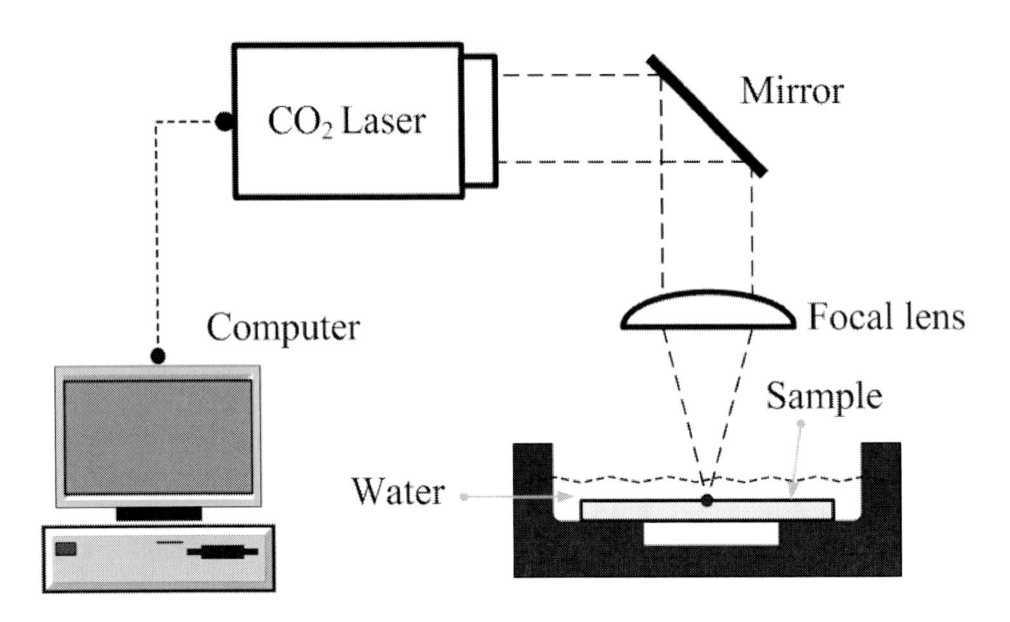

Figure 5. The schematic diagram of liquid-assisted laser processing for etching of Pyrex glass. The glass is immersed into water in a container at room temperature.

2.1.1. Pure-Water Assisted CO_2 Laser Cutting

At the pure-waster assisted laser cutting processing, the laser power was controlled at about 24 W for the scanning rate between 114 and 342 mm/s, and ablation passes from 1 to 100. Also, the water depth was controlled between 0.3 and 1.0 mm by total volume method of which depth is proportional to the volume at a constant area of holder. In order to keep water always wetting on the glass surface during ablation, 5 ml volume of hydrophilic surfactant agent of Lauramidopropyl Betaine was added to avoid the water layer broken into drops. Traditional laser processing in air was performed as a reference sample. Figures 6 (a)-(b) show the top-view optical micrographs of Pyrex glass etched by CO_2 laser in air at a fixed 24 W power with a 114 mm/s scanning rate for 2 passes and with a 228 mm/s rate for 5 passes, respectively. The typical crack mode of failure was observed at the initial stage at low passes (Figure 6(a)) and then the scorch mode of failure appeared at higher passes even with increased scanning rate (Figure 6(b)). Both crack and scorch failures sometimes may occur simultaneously as etching of glass at high passes. The reasons for crack formation is primarily attributed to the thermal stress induced crack [[45]] and the scorch formation are mainly from great accumulated heating energy for burning. Therefore, it is nearly impossible to etch though glass of 500 μm thick by CO_2 laser in air at room temperature. In order to reduce the effect of thermal stress, stress gradient and burning on glass etching using CO_2 laser processing, water cooling is adopted to assist laser ablation of glass and has potential for the though-wafer etching.

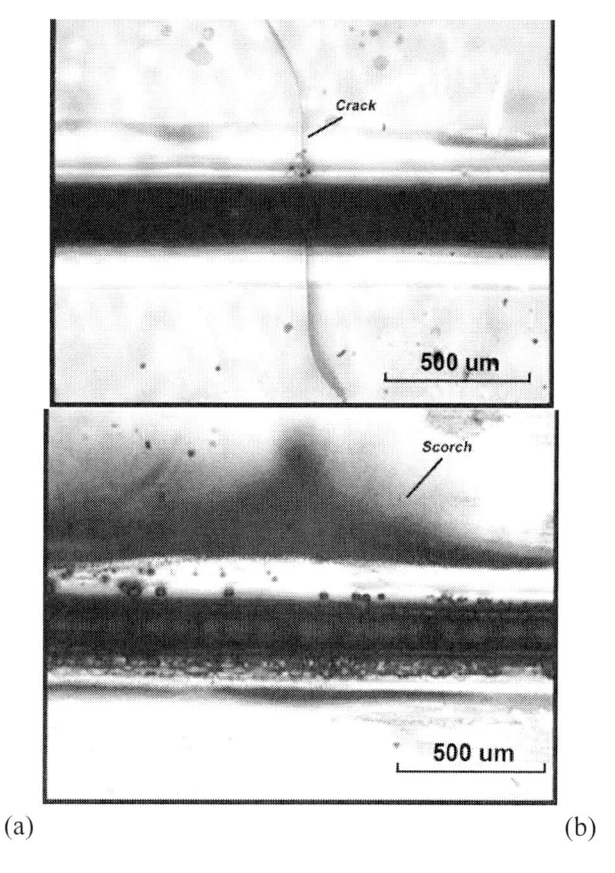

(a) (b)

Figure 6. The optical micrographs of Pyrex glass etched by CO2 laser in air at a 24 W power for: (a) 2 passes at a 114 mm/s rate to form crack failure mode and (b) 5 passes at a 228 mm/s rate to form scorch failure mode.

Figures 7(a) and (b) show the optical micrographs of the planar view and cross section of 500 μm thick Pyrex glass etched under water depth 0.5 mm at laser parameters of 24 W power, 114 mm/s scanning rate for 80 passes, respectively. The through-wafer etching of Pyrex glass has been obtained with crackless surface under the assistance of water cooling. From the end of cutting line towards the center, the existence of a clear grey white line at nearly center of width corresponds to through-wafer etching of glass. The black color around the grey white line is due to the V-like tapered profile (Figures 7(b)), which is concerned with Gaussian distribution of laser beam energy. Since the focal length of the lens is about 38.1 mm and the laser beam intensity is a Gaussian distribution, the etch rate is higher at center than that at sides of the trench. It may lead to the V-like tapered profile i.e. the shape of the trench is much wider at the top than at the bottom of the trench. Moreover, the absorption of the laser beam at the sides of the trench near the surface may prevent absorption of the laser energy deeper inside the sample. This could also result in the enlargement of the V-like tapered profile of trench near the surface. In addition, a HAZ region is found in fig. 7(b) due to the existence temperature gradient. The higher temperature gradient leads to the larger HAZ region even cracks. The LALP method can reduce the temperature gradient, HAZ region and crack formation. The aspect ratio is up to 3 for the through-wafer etching, which is higher than previous reports [1]-[2],[5]A,[6].

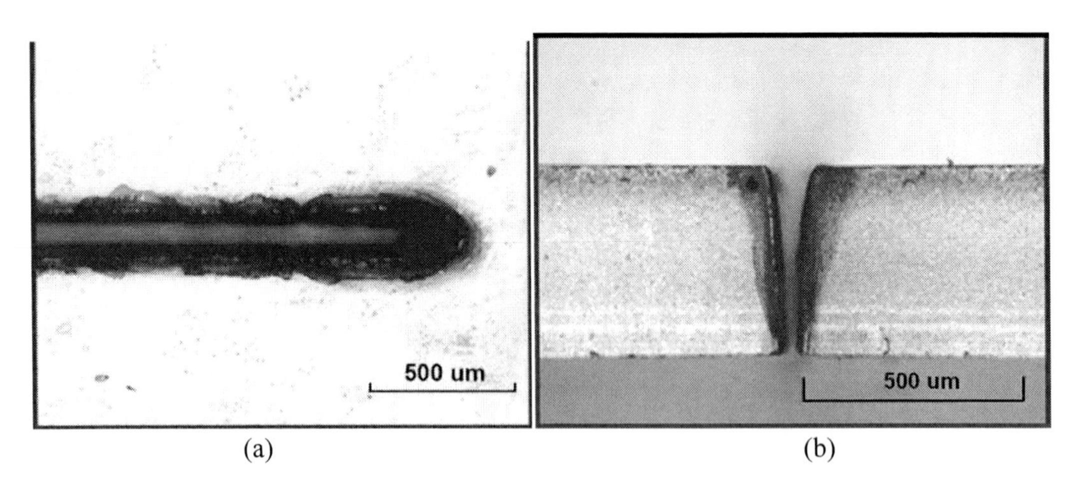

Figure 7. The top-view optical micrographs of: (a) the planar view and (b) cross section of 500 μm thick Pyrex glass etched under water depth 0.5 mm at laser parameters of 24 W power, 114 mm/s scanning rate for 80 passes.

The etching result using LALP is influenced by the laser scanning rate and water depth. Figure 8 shows the effect of laser scanning rate on the etching depth of glass at a fixed 24 W laser power, 0.5 mm water depth for 10-50 passes. The depth of etch through is defined at the wafer thickness of 500 μm drawn by a black dash line. The scanning rate varies from 114 to 342 mm/s. The etching depth increases with the cut passes. As the depth is divided by the passes for the depth rate or etching rate (μm/pass), it indicates that the etching rate is non-linear and gradually decreases with increasing passes. The lower scanning speed, the higher etching rate is. At 114 mm/s, nearly through etching of glass occurs at around 30 passes. The increased scanning rate results in increase of the pass of through etching. For example, around 50 passes are needed for 228 mm/s to etch through glass. It is because the absorption of photo-thermal energy per second at unit length increases with decreasing scanning rate. That is, if we convert the three scanning speeds into linear laser energy in order, they will be 0.210, 0.105 and 0.070 J/mm, respectively. The higher linear laser energy leads to the higher etching rate, too. The decrease of etching rate with passes may be attributed to the defocus effect of beam intensity. Since the water can absorb part of CO_2 laser energy and cool the glass temperature, the water depth may affect the passes of through-glass etching. Figure 9 shows the effect of water depth on the etching depth of glass at a fixed 24 W laser power, 228 mm/s scanning rate for 10-100 passes. The water depth varies from 0.3 to 1.0 mm. The etching depth increases with passes at the fixed water depth. And the lower the water depth, the smaller the pass of through etching is. The passes of through etching of glass at water depth of 0.3, 0.5 and 0.7 mm are around 30, 50 and 100, respectively. The etching depth of glass at 1.0 mm water depth for 100 passes is around 390 μm. Although the low water depth increases the etching rate of glass, too shallow water may lead to crack formation at higher passes due to the insufficient cooling as shown in Figure 6. The crack of glass is observed for the etching of glass into water of 0.3 mm depth at 24 W power and 342 mm/s rate for 30 passes.

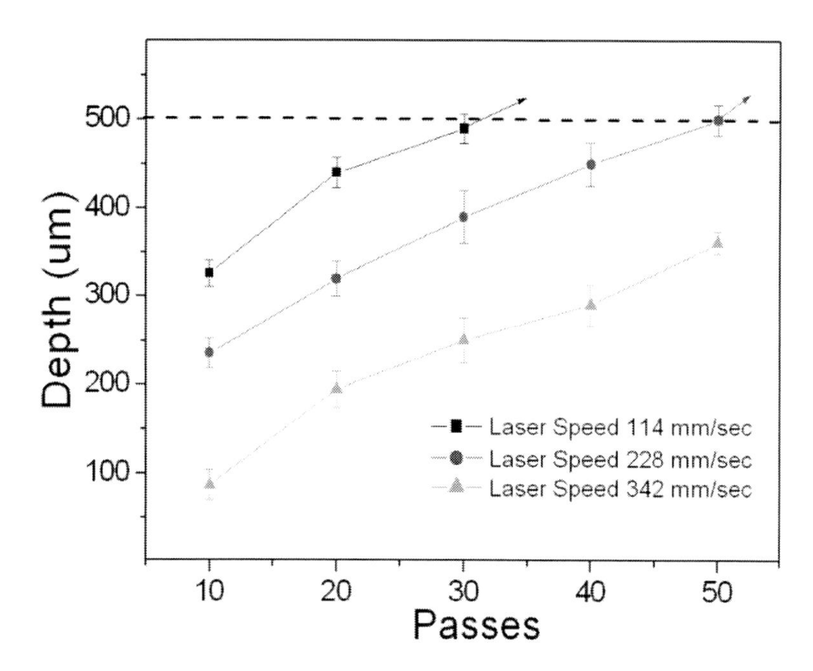

Figure 8. The effect of laser scanning rate on the etching depth of glass at a fixed 24 W laser power, 0.5 mm water depth for 10-50 passes.

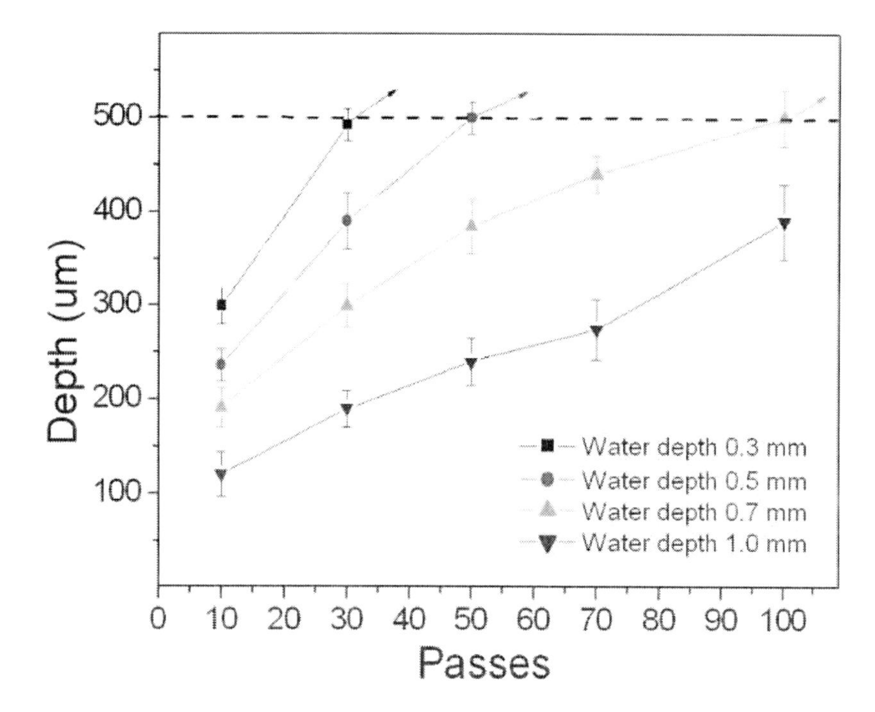

Figure 9. The effect of water depth on the etching depth of glass at a fixed 24 W laser power, 228 mm/s scanning rate for 10-100 passes.

The formation of crack on the laser-heated glass is most likely caused by the in-plane residual surface stress during cooling. The residual thermal stress (σ_{th}) is tensile and can be expressed in equation (1) [45].

$$\sigma_{th}(=\sigma_{xx}=\sigma_{yy})=\frac{E\alpha}{1-\nu}(\Delta T) \quad (1)$$

Where E, α and v are the Young's modulus, coefficient of temperature expansion (CTE) and Poisson ratio of Pyrex 7740 glass. ΔT is the increased temperature at one position during laser ablation. For Pyrex 7740 glass, E is around 63 GPa, α is around 3.2 x 10^{-6} K^{-1} and v is around 0.2. ΔT is dependent on the laser energy absorbed for glass etching. The laser energy is concerned with the laser power, scanning rate, absorption of material. Because the CO_2 laser ablation is a photothermal process, the surface temperature glass is above the softening for significant material removal. The softening and working points of Pyrex 7740 glass in air are around 820 °C and 1252 °C, respectively. So, the estimated σ_{th} at softening point is around 200 MPa (63GPa x 3.2 x 10^{-6} x (820-25)/(1-0.2)) and that at working points is around 308 MPa. The maximum σ_{th} increases with temperature gradients, ΔT, and is affected by the air or water environment. The temperature gradient in water is smaller than that in air because of the cooling effect of water to enhance the heat loss and reduce the overheating and thermal stress. Therefore, the thermal stress can be diminished during glass etching by LALP. The Biot number (*Bi*) is the dimensionless number that can be used to characterize the magnitude of the established thermal gradient [58] and defined as the ratio of the average surface heat transfer coefficient (*h*) for convection from the entire surface, to the conductivity (*k*) of the solid, over a characteristic dimension *L* in equation (2)

$$Bi=\frac{hL}{k} \quad (2)$$

Since *h* of water (500 to 10000 W/m^2K) is 2-3 orders larger than air (10 to 100 W/m^2K) [59], the great increase of *Bi* in water with pronounced heat loss leads to the decrease of both temperature itself and its gradient. Therefore, temperature gradient in water is expected to decrease much to reduce the thermal stress and to improve the crackless cutting quality. Although the 10.6 μm CO_2 laser can be absorbed in water with a static absorption coefficient (α) about 500 cm^{-1} [60], we cannot use the Beer's law ($I_x=I_0\exp(-\alpha x)$) to quantitatively estimate the amount of transmitted CO_2 light intensity (I_x) on glass after the original power intensity (I_0) penetrates water thickness (x) because LALP is a dynamic process accompanied by the complicated mechanism of plasma formation, expansion and quenching together with the possible shock wave and bubble involved in the laser ablation. However, it is right for the quality discussion on the laser energy variation with water thickness to explain the experimental data aforementioned. Moreover, the higher water depth can reduce more thermal stress but it may reduce the etching rate (Figure 9). In contrast, the cutting rate increases at lower water depth with higher surface temperature. Moreover, deficiency in water depth can lead to the crack at deep etching due to insufficient cooling (Figure 10). Therefore, it needs a compromise between the water depth and crackless etching rate of glass. In our experiments, 0.5 mm thick water is the candidate for the crackless glass etching at high etching rate.

Figure 10. The optical micrographs of glass with crack occurs at the etching with a 0.3 mm water depth, 24 W power, 342 mm/s rate for 30 passes.

Besides the cooling effect, water can be beneficial for the debris removal during LALP. Figures 11(a) and (b) show the alpha-step profiles of Pyrex glass linearly ablated at 24 W power and 228 mm/s scanning rate in air for 5 passes and into 0.5 mm water for 10 passes, respectively. The bulges at both rims of cutting line in air are around 8.7 and 13.3 μm while those in water are around 1.9 and 2.2 μm. The bulges on the rims of a trench are mainly from resolidification of evaporated debris. Much reduced bulge height can be obtained using LALP because the laser heating induces stronger natural convection in water to help carry debris away. Assume laser moving is along x-direction and water depth is for the z-direction. The shear stress (τ) from natural convection along x-direction in cross-sectional x-z plane is defined in equation (3)

$$\tau = \mu(\frac{\partial u_x}{\partial z} + \frac{\partial u_z}{\partial x}) \tag{3}$$

Where μ is water viscosity, u_x and u_z are the water velocities along the x-direction and z-direction, respectively. In water environment, temperature rises as the heat source passes. The temperature gradient in water occurs during laser ablation because of heat accumulation on the surface of glass which arises from the heat loss through thermal diffusion less than the heat input of laser source. The highest temperature is located at the moving laser spot on surface and decreases with the distance away from heat source. As laser moves, it induces the natural convection with shear stress which is proportional to water viscosity (μ) and the gradient of water velocity to distance ($\frac{\partial u_x}{\partial z} + \frac{\partial u_z}{\partial x}$). The lower laser moving rate or larger heat input may lead to the higher shear stress. The maximum shear stress occurs at the surface of laser spot with the highest water x-velocity. Therefore, it helps to carry away the debris in water to reduce the bulge height (Figure 11(b)). Comparing laser ablation in air and in water, the natural convection of the former is much smaller than the later. Therefore, no effective shear force in air moves the resolidified debris away to result in the large budge (Figure 11(a)). The vaporization for the fast etching could occur in the local laser induced high-temperature plasma environment. And the glass vapor will rapidly condense due to water cooling effect. Then much of condensed glass debris can be carried away by the natural convection of water during LALP as aforementioned. Note that the spot of maximum shear stress in water moves with the heat source and is expected to enhance debris removal for improvement of surface quality.

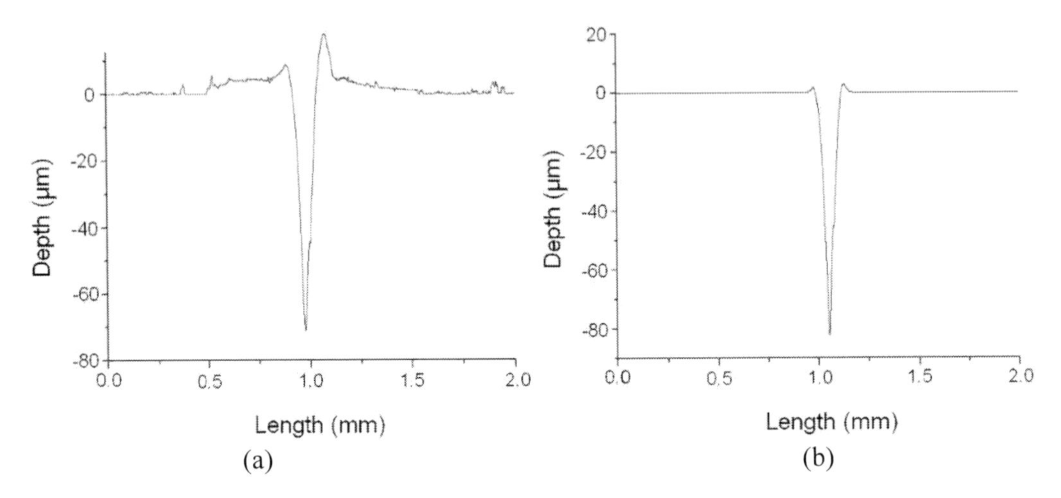

Figure 11. The alpha-step profiles of Pyrex glass linearly ablated at 24 W power and 228 mm/s scanning rate: (a) in air for 5 passes and (b) into 0.5 mm water for 10 passes.

Crackless through-wafer deep etching of Pyrex 7740 glass has been demonstrated using liquid-assisted laser processing. It improves the drawbacks of low etching rate, masking, surface imperfections and high cost in conventional HF/BOE wet etching, plasma dry etching, femtosecond or excimer laser ablation. It also solves the problems of cracks, scorches and bulges formed in traditional laser processing in air. The CO_2 laser ablated glass in air results in crack or scorches due to the sharp temperature gradient and high temperature while the bulges on the rims of etching line are from resolidification of evaporated debris. LALP can cool down samples during ablation for less temperature gradient to eliminate the crack and carry away the debris to much reduce the bulge height. Proper water depth of 0.5 mm is beneficial for high etching rate of crackless through-glass etching with an etching rate of a 20 mm long trench more than 25 μm/s and the aspect ratio up to 3.

2.1.2. Surfactant and Pure Water Mixing Liquid Assisted CO_2 Laser Cutting

In order to understand the effect of surfactant addition on glass ablation in water-assisted CO_2 laser processing, different weight percent of surfactant solution was used in assistance liquid. The difference of etching performance between the pure water and the surfactant solution was compared and discussed. The laser power was controlled at 18 to 27 W for the scanning rate between 57 and 114 mm/s, and ablation passes from 2 to 80. Because the cations and anions in ionic surfactant may react with the ablated glass, it causes precipitates in solution and affects following laser process. Therefore, the nonionic surfactant whose major composition is Lauramidopropyl Betaine ($RCONH(CH_2)_3N^+(CH_3)_2CH_2COO$-) was chosen the addition of solution to water. The varied concentration ranged at 0% (pure water), 7.5%, 15%, and 30% weight percent. The assisted liquid depth was controlled at 1.0 mm by total volume method of which depth is proportional to the volume at a constant area of holder. Conventional laser processing in air was performed as a reference sample. Figure 12 shows the effect of surfactant concentration on the etching depth of Pyrex glass at a fixed 24 W laser power, 114 mm/s scanning rate and 1.0 mm liquid depth for 10–80 passes. The depth of etch through is defined at the wafer thickness of 500 μm drawn by a black dash line. The surfactant addition varies from 0% (pure water) to 15% weight percent. The etching depth increases with the surfactant concentration at the same passes until etch through Pyrex glass wafer. At around 80 passes, there occurs nearly through etching of glass in the 7.5% and 15% surfactant solutions except for in pure water. The etching depth increased with laser passes but it is a non-linear behavior. The etching depth can be converted into etching rate by divided by the passes. Figure 13 shows the etching rate of glass as a function of pass at surfactant concentrations of 0, 7.5% and 15% at a fixed 24 W laser power, 114 mm/s scanning rate and 1.0 mm liquid depth. The etching rate gradually decreases at higher passes. In the experiment, the maximum etching rate is 25 μm/pass at 10 passes in 15% surfactant solution. Compared with the pure water (13.6 μm/pass), the surfactant enhances around 1.8 times etching rate. In addition, the etching rate at 80 passes in 15% surfactant solution is 1.33 times that in pure water. It reveals that the effect of surfactant addition decreases with the increasing passes or the increasing etching depth.

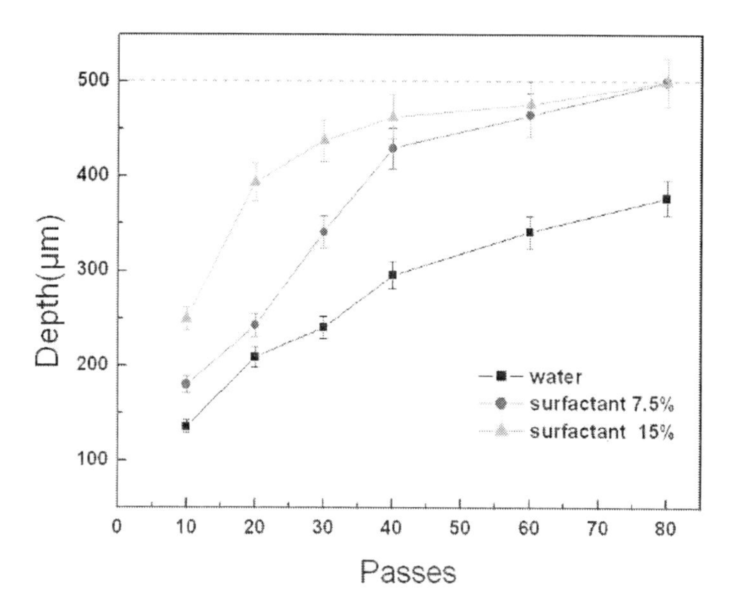

Figure 12. Effect of surfactant concentration on the etching depth of glass at a fixed 24 W laser power, 114 mm/s scanning rate, 1.0 mm liquid depth for 10–80 passes.

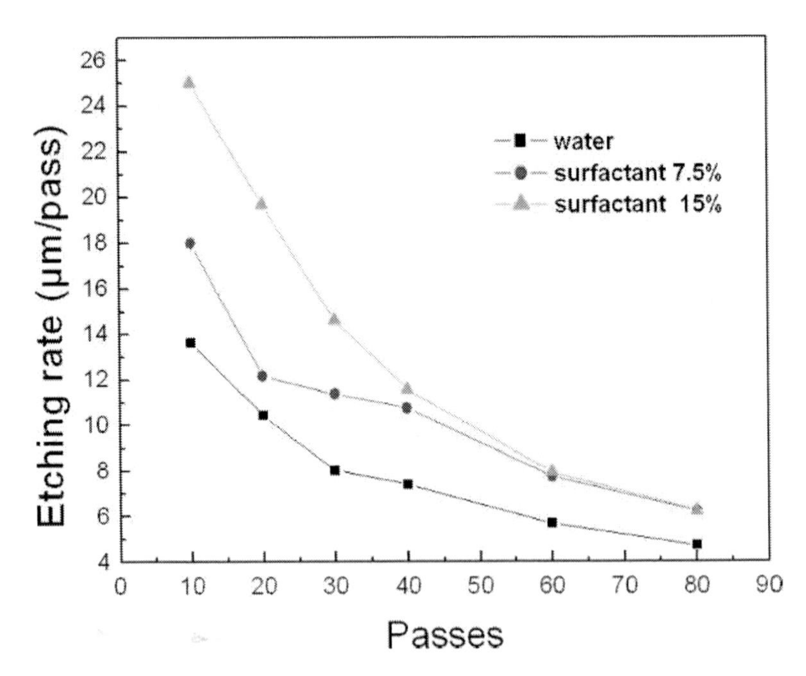

Figure 13. Etching rate of glass as a function of pass at surfactant concentrations of 0, 7.5% and 15% at a fixed 24 W laser power, 114 mm/s scanning rate and 1.0 mm liquid depth.

The increase of etching rate of glass due to surfactant addition in LALP has been demonstrated. In order to understand more of the effect of surfactant addition, we have changed the linear laser energy density during etching process. Figure 14 shows the effect of linear laser energy density on the etching depth of glass in 15% surfactant solution, 1.0 mm liquid depth for 10–80 passes. The linear laser energy density can be defined as the laser power divided by the scanning rate [61] and the value varies from 1.84 J/cm to 4.21 J/cm. The etching depth increases with linear laser energy density. Especially, the amount of increased depth is nonlinear. For example, at the 10 passes of laser etching glass, the increase of 1.7 times energy density (1.84 to 3.16 J/cm) results in the increase of 2.7 times etching depth (161 to 447 μm). Furthermore, the increase of energy density from 2.11 to 4.21 J/cm results in the decrease of the through-etching number of pass from 80 to 10. In other words, the etching rate increases about 8 times at through wafer etching. The nonlinear variation of etching rate is due to the absorption of laser light. The increased linear energy density seems beneficial for more enhanced light absorption to result in faster etching. It implied that the chemical or optical changes caused by surfactant addition are not major reason for the increase of etching rate. However, too high light absorption may do negative effect on surface quality of glass etching. Figures 15(a) and (b) show the optical micrographs of the planar view and cross section of the 500 μm thick Pyrex glass etched at the 1.0 mm depth of 15% surfactant solution at the 3.16 J/cm linear laser energy density (18 W power, 57 mm/s scanning rate) for 20 passes, respectively. The through-wafer etching of Pyrex glass in surfactant solution has been obtained with a crackless surface like in water. Moreover, the increase of linear laser energy density from 3.16 to 4.21 J/cm (power from 18 to 24 W, see Figure 14) results in the decrease of the pass number of through etching from 20 to 10, but there occur cracks in the glass surface as shown in Figure 16(a). The cracks can be explained that the cooling rate of assisted-liquid is not high enough to carry unnecessary heat away. The similar situation occurs when the concentration of surfactant solution increases to 30%. Even there happen many scorches on the glass surface due to heat accumulation [50], [62] as shown in Figure 16(b). It indicates that the suitable concentration of surfactant addition can enhance etching rate of glass with crackless surface in LALP, but too high concentration will cause the defects of cracks and scorches.

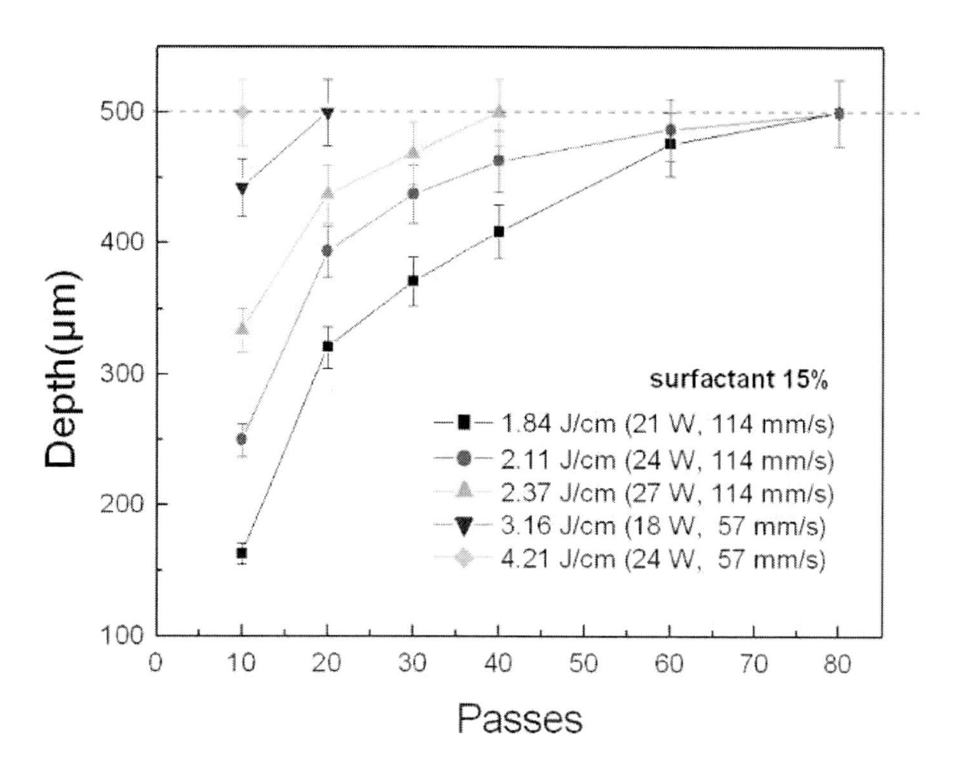

Figure 14. Effect of linear laser energy density on the etching depth of glass in 15% surfactant solution, 1.0 mm liquid depth for 10–80 passes.

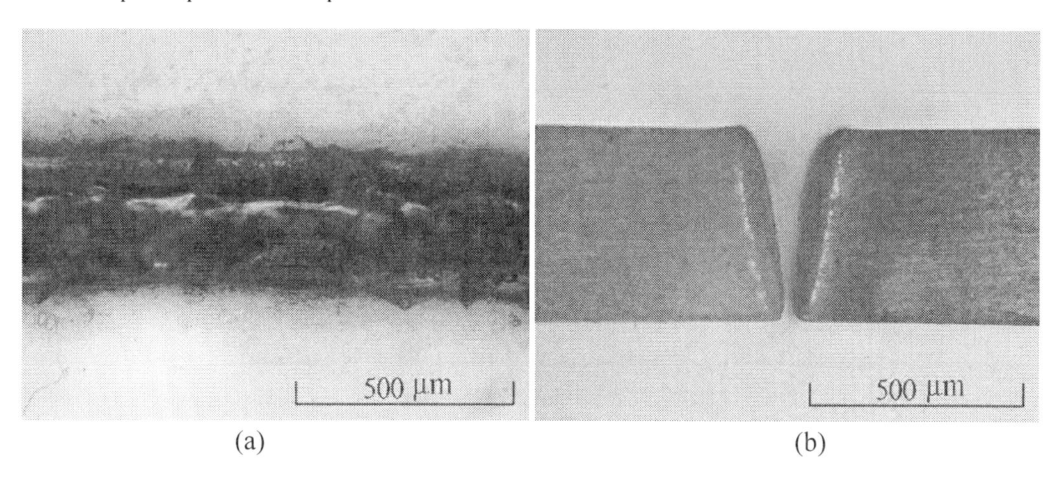

Figure 15. Optical micrographs of the: (a) planar view and (b) cross section of the 500 μm thick Pyrex glass etched at the 1.0 mm depth of 15% surfactant solution at the 3.16 J/cm linear laser energy density (18 W power, 57 mm/s scanning rate) for 20 passes.

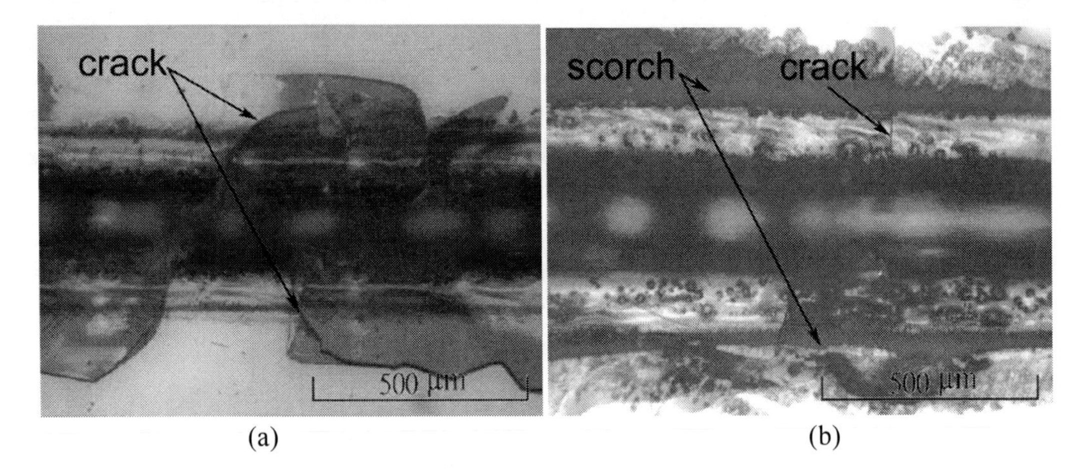

Figure 16. Optical micrographs of the planar view of pyrex glass etched by CO2 laser in: (a) 15% surfactant solution at the 4.21 J/cm linear laser energy density (24 W laser power, 57 mm/s scanning rate) for 10 passes and (b) 30% surfactant solution, the 2.11 J/cm linear laser energy density (24 W laser power, 114 mm/s scanning rate) for 50 passes.

The formation of crack on the laser-ablated glass is most likely caused by the thermal induced in-plane residual surface stress during ablation. The thermal stress (σ_{th}) is tensile and can be expressed in equation (1). The ΔT is the increased temperature at one position during laser ablation. The equation shows that the residual thermal stress is proportional to ΔT. The temperature gradient in water is smaller than that in air because of the cooling effect of water to enhance the heat loss and reduce the overheating and thermal stress. A simple thermal model is adopted for the analysis of thermal distribution during the straight-line machining process. The laser beam focusing on the plane surface maintains a constant TEM_{00} mode and travels in the one direction at a constant velocity v. The density of laser power can be described by Gaussian distribution in the following equation (4), where p_0 and r are the power and the radius of the CO_2 laser beam, respectively.

$$I(x,y,z,t) = \frac{p_0}{\pi r^2} \exp\left(-\frac{(x - vt)^2 + y^2}{r^2} \right) \tag{4}$$

The thermal conductivity coefficient, absorption coefficient and specific heat of pyrex 7740 properties are constant. Therefore, the temperature distribution and maximum temperature can be solved by ANSYS software. Figures 17 (a) and (b) show the difference of maximum temperature and thermal stress between presence and absence of 1.0 mm depth water, respectively. When linear laser energy density is fixed at 2.37 J/cm, the maximum temperature significantly decreases from 8773 °C in air to 4655 °C with water assisted machining. This variation of temperature results in the thermal stress greatly decreases from 1046 to 474 MPa. Therefore, the crack can be eliminated in water-assisted CO_2 laser processing.

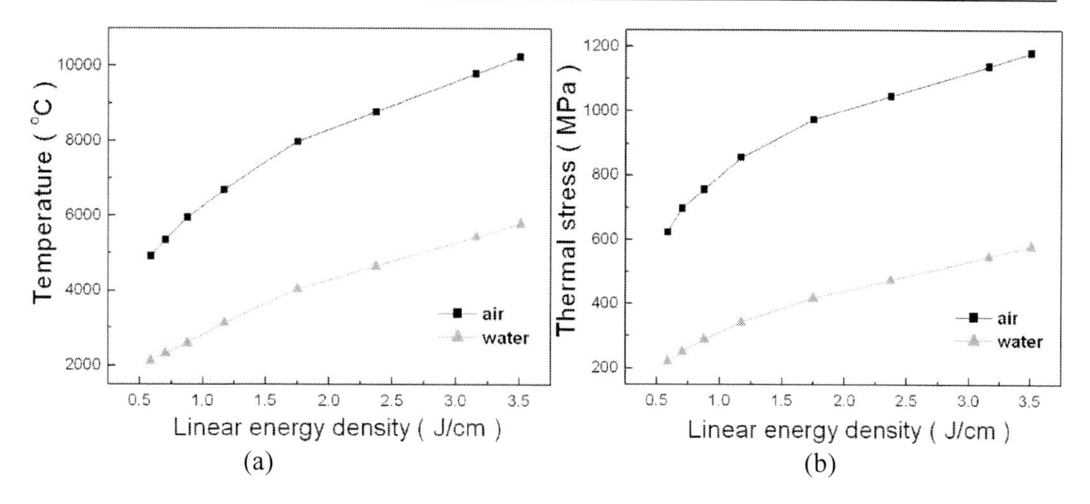

Figure 17. Relationship between the linear laser energy density and: (a) maximum temperature and (b) thermal stress in the presence and absence of 1.0 mm depth water.

Figure 18(a) shows the schematic diagram of transient liquid state when laser energy reaches the surface of glass in LALP in order to understand the mechanism of laser etching glass under assistance liquid. According to the optical spectrum of water [[49]], the absorption of CO_2 laser is over 99%. Therefore, laser energy will first evaporate the water over the glass during etching process. The way of laser becomes air cone, and then the glass is etched. Tending to reach the equilibrium state of liquid level, water moves to the vacant position continuously, so the laser energy is unceasingly loss by the evaporation process. The larger linear laser energy density causes the more evaporation volume of water and larger contact area with air as well as higher heat accumulation. Too much heat accumulation without conduction via liquid will lead to higher thermal stress for crack formation. The cracks increase with laser power or linear energy density. Figure 18(b) shows schematic movement of liquid toward vacant position during WACLAP. The Newton's law of viscosity for one dimension is defined in equation (5) as following, where τ is the shear stress for glass to liquid, u is liquid velocity, and μ is the viscosity of liquid.

$$\tau_{yx}=\mu\frac{\partial u}{\partial y} \tag{5}$$

When the surfactant is added into water, the fluid viscosity was changed much. The viscosity of Lauramidopropyl Betaine which was used in the experiment is 3000-4000 cps [[63], and water is 1 cps. The more the surfactant percentage, the higher the liquid viscosity is. If the liquid has higher viscosity, it will has larger shear stress τ_{yx} to cause the lower flow rate of liquid to vacant position and then reduce the energy loss of glass in laser etching. If the concentration of surfactant is too high, there's not enough time to let liquid recover equilibrium state of level. It causes a long time that the glass surface is in touch with air. Therefore, some typical defects of cracks and scorches in laser processing in air appear on condition of high-concentration surfactant. Figures 19(a) and (b) show the alpha-step profiles of Pyrex glass linearly ablated at 3.16 J/cm linear laser energy density (18 W power, 57 mm/s scanning rate) in 1.0 mm depth of water and 15% surfactant solution for 10 passes, respectively. The glass etching in water and surfactant show the similar bulges. The height of bulge at both rims of cutting line in water are around 1.6 and 2.1 μm and in surfactant are around 3.2 and 1.5 μm. Compared to typical bulge height more than 10 μm formed in laser ablation in air [[50]], the influence of surfactant addition to water in LALP can not only effectively reduce the bulge height but also increase much the etching rate (Figures 12-13).

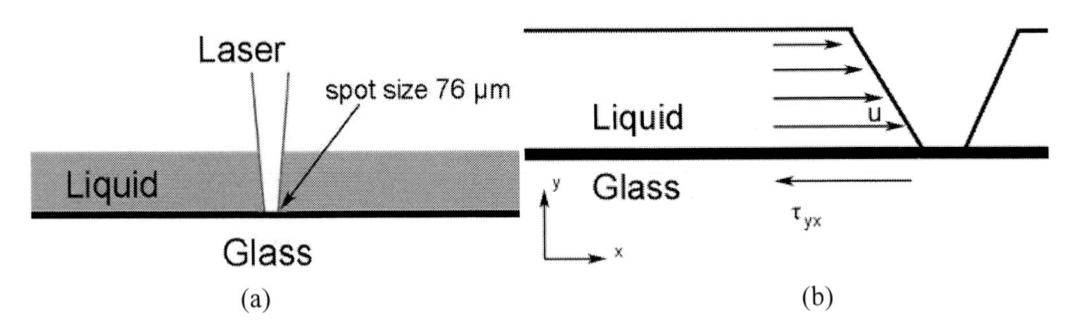

Figure 18. Schematic diagram of: (a) transient liquid state when laser energy reaches the surface of glass in LALP to form air cone, and (b) the motivation of liquid toward vacant position..

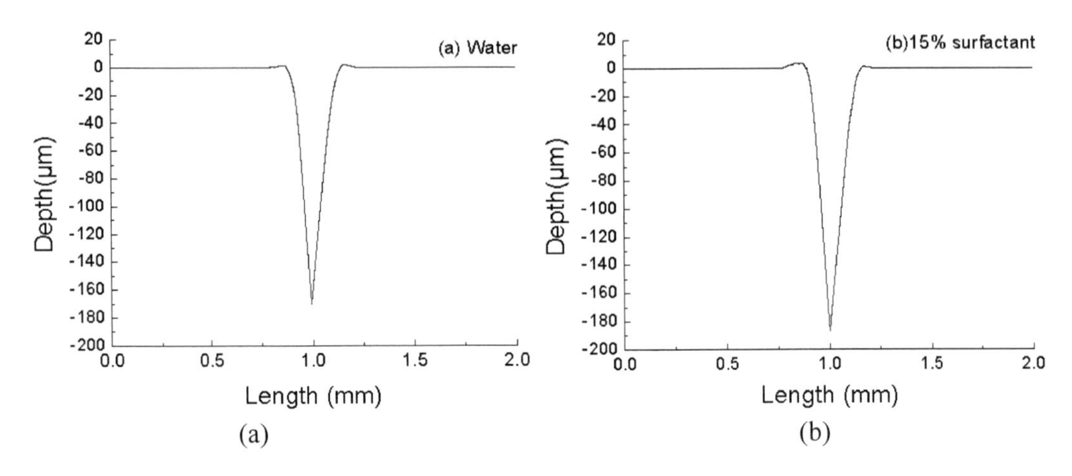

Figure 19. Alpha-step profiles of Pyrex glass linearly ablated at 3.16 J/cm linear laser energy density (18 W power, 57 mm/s scanning rate) in the 1.0 mm depth: (a) water (b) 15% surfactant for 10 passes.

The effect of nonionic surfactant addition to water in LALP on glass ablation has been investigated. Different weight percent of surfactant solution from 0% (pure water) to 30% was used as assistance liquid for LALP. The laser power was controlled at 18 to 27 W for the scanning rate between 57 and 114 mm/s, and ablation passes from 2 to 80. The etching depth increases with the surfactant concentration at the same passes until etch through Pyrex glass wafer. The etching depth also increases with linear laser energy density from 1.84 J/cm to 4.21 J/cm. But too high linear laser energy density may lead to cracks and scorches on surface of glass as that in conventional laser ablation in air. A schematic mechanism of laser etching glass under surfactant assistance liquid has been proposed in viewpoint of Newton's law of viscosity. The more surfactant percentage of solution has the higher liquid viscosity. It will have larger shear stress to impede liquid flow to vacant position so as for more heat accumulation to enhance etching rate. But if the concentration of surfactant is too high, too much heat accumulation will lead to the formation of cracks and scorches. The proper surfactant addition can not only effectively reduce the bulge height but also increase etching rate.

2.2. Pure-Water Assisted CO_2 Laser Drilling

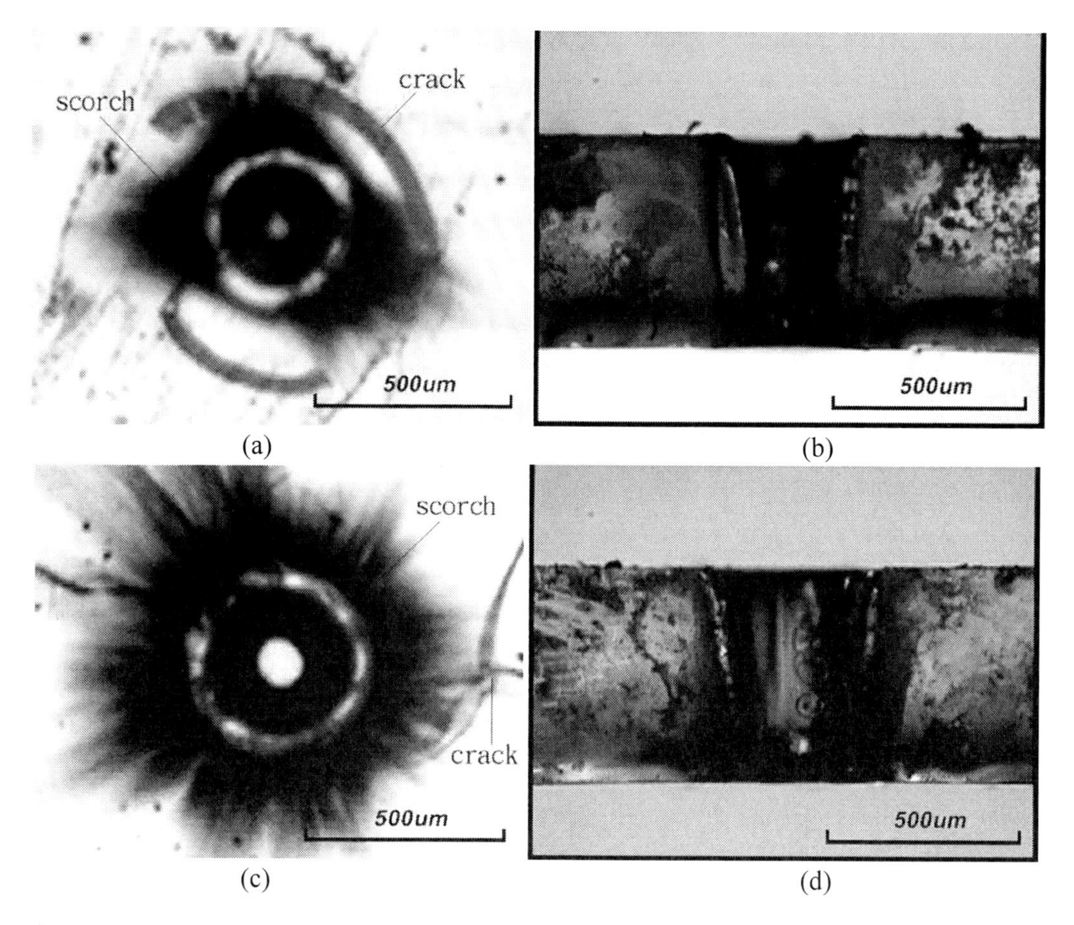

Figure 20. The top-view and cross-sectional optical micrographs of CO2 laser etched glass holes in air at a power of 6 W, a speed of 11.4 mm/s and 10 passes with drawn diameters of: (a, b) 100 μm, and (c, d) 200 μm. Cracks and scorches frequently appeared in this process.

A computer aided design program of CorelDraw software was available to set the experimental parameters for auto-controlled processing. Laser drilling processing, a constant laser power of 6 W and 11.4 mm/s scanning speed at different scanning numbers from 1 to 10 passes were applied to the glass drilling. Both kinds of water depth were controlled at about 0.5 mm and 1.0 mm above glass surface by total volume method of which depth is proportional to the volume at a constant area of holder. In order to keep water always wetting on the glass surface during ablation, 5 ml volume of hydrophilic surfactant agent of Lauramidopropyl Betaine ($RCONH(CH_2)_3N^+(CH_3)_2CH_2COO$-) was added to avoid the water layer broken into drops. Traditional laser processing in air was performed as a reference sample. The ANSYS software was used to investigate the temperature distribution and thermal stress field in air and water ambient during glass hole machining. The laser beam focusing on the plane surface maintains a constant TEM_{00} mode, that laser intensity distribution is a Gaussian function model and the heating phenomena due to phase changes are neglected. The beam of CO_2 laser is regarded as a surface heating source but the surface of Pyrex 7740 glass without laser heating and the superficial heat irradiation is negligible. Figures 20(a)-(d) show the top-view and cross-sectional optical micrographs of CO_2 laser etched glass holes in air with drawn diameters of 100 μm (Figures 20(a)-(b)) and 200 μm (Figures 20(c)-(d)), respectively, at a power of 6 W, a speed of 11.4 mm/s and 10 passes. Cracks and scorches frequently appeared on the glass surface during laser processing in air. The cracks occur prior to through-hole glass etching because of the high thermal stress [45] while the scorch formation is mainly from great accumulated heating energy for burning, especially more in circular etching than in linear cutting [61]. Since the focal length of the lens is about 38.1 mm and the laser beam intensity is a Gaussian distribution, the etch rate is higher at center than that at the rim of hole. In addition, a clear HAZ area can be found in Figures 20(b) and (d) due to the existence of temperature gradient. The higher temperature gradient leads to the larger HAZ area even cracks because the drawn diameter of 100 μm has larger thermal accumulation energy than that of 200 μm.

The CO_2 laser micromachined glass hole using LALP on a drawn diameter of 100 μm is shown in Figures 21(a)-(b) in terms of top view and cross section. The liquid is water and about 0.5 mm above the surface of glass. The laser operation parameters are the same as those in air in Figures 20(a)-(b). No cracks and scorches appear. And the HAZ area by LALP is smaller than that in air. The central white spot indicates the through-hole etching and it has a smaller diameter than the total circle due to the tapered profile (Figure 21(b)) formed by laser ablation. Similar results can be obtained by etching of a hole in larger diameter. Figures 22(a)-(b) show the top-view and cross-sectional optical micrographs of the CO_2 laser etched Pyrex glass with a drawn diameter of 200 μm using LALP, respectively, at a power of 6 W, a speed of 11.4 mm/s and 10 passes. The water depth is the same 0.5 mm above the surface of glass. Also no crack or scorch is formed under LALP. The tapered profile is more diffusive than that in the small diameter hole. The water depth will affect the efficient laser power and etching rate. Figures 23(a) and (b) show the etching depths of the holes in drawn diameters of 100 μm and 200 μm, respectively, as a function of etching passes of 1 to 10 using LALP at the 6 W laser power and 11.4 mm/s scanning rate under water depths of 0.5 mm and 1.0 mm. The dash line at 500 μm depth indicates the through-hole etching. The etching depth increases with passes at a fixed water depth. The average etching rate can be defined by the etching depth divided by the total passes (μm/pass) and obtained from the slope of Figure 23. It reveals that the etching rate normally decreases with increasing passes to be the non-linear behavior. The through etching of glass in both diameters of holes at the water depth of 0.5 mm occurs at around 5 passes. As the pass increases to 10, all the through holes etching are finished even at the water depth of 1.0 mm. That is, the passes of through-hole etching of glass at water depth of 0.5 and 1.0 mm are around 5 and 10, respectively. The effect of water depth on the passes of through-hole glass etching is concerned with the absorption of CO_2 laser energy through water and the glass temperature cooled by water. The higher the water depth, the lower the etching depth, the more the pass of through etching is. The drawn hole diameter will also influence the etching rate due to the variation of thermal accumulation energy. The small drawn diameter of 100 μm has lager thermal accumulation energy than that of 200 μm for higher etching depth on the same process condition. For instance, the etching depth of the hole in a drawn diameter of 100 μm at 5 scanning passes and 1.0 mm thick water is around 398 μm while that in 200 μm diameter is reduced to around 320 μm. The benefit of water adopted to assist laser through-hole etching of Pyrex glass is the cooling effect for reducing both temperature and its gradient. Therefore, LALP method can reduce the HAZ area and eliminate formation of cracks and scorches. Fig. 24 shows the schematic procedure of CO_2 laser etched through-hole glass during LALP in order to interpret its mechanism. Because water can absorb a little CO_2 laser, it is locally evaporated as laser energy goes through water layer to the surface of glass during LALP (Figure 24(a)-(b)). Therefore, the way of laser becomes air cone, and then the glass is etched through photothermal mechanism of melting and evaporation until through hole (Figure 24(c)-(d)). Tending to reach the equilibrium state of liquid level, water moves to the vacant position continuously, so the laser energy is unceasingly loss by the evaporation process for reducing the etching rate. Also, the water covering around the micro-hole may reduce the temperature and its gradient for diminishing surface defect.

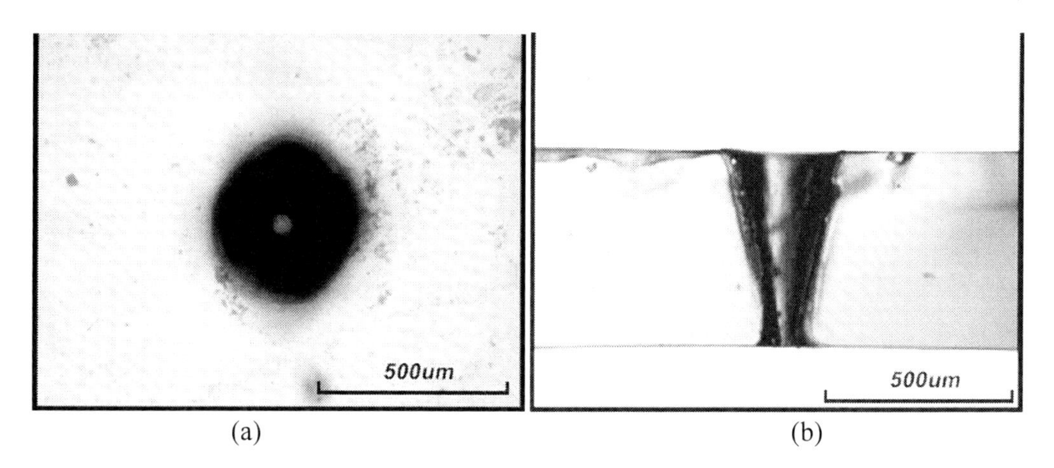

Figure 21. (a) Top-view and (b) cross-sectional optical micrographs of the CO2 laser etched Pyrex glass material using LALP with a drawn diameter of 100 μm at a power of 6 W, a speed of 11.4 mm/s and 10 passes. The water thickness is about 0.5 mm..

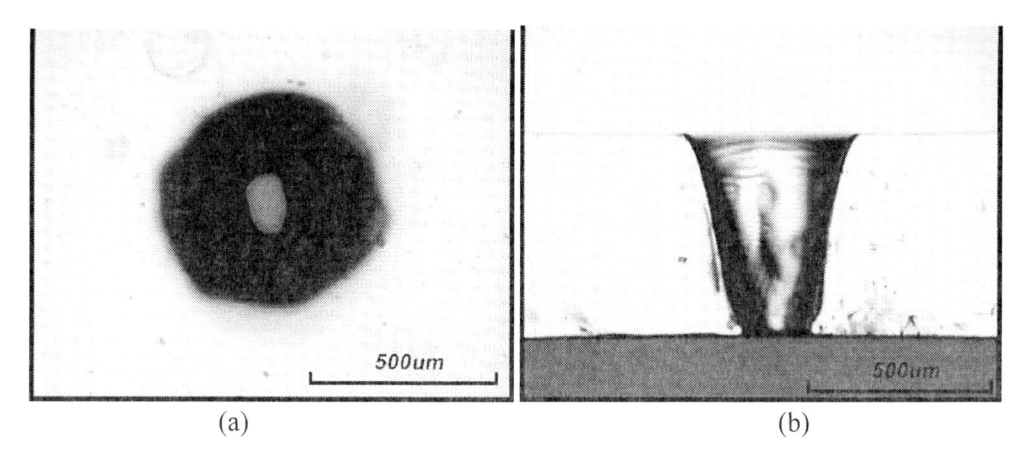

Figure 22. (a) Top-view and (b) cross-sectional optical micrographs of the CO2 laser etched Pyrex glass material using LALP with a drawn diameter of 200 μm at a power of 6 W, a speed of 11.4 mm/s and 10 passes. The water thickness is about 0.5 mm..

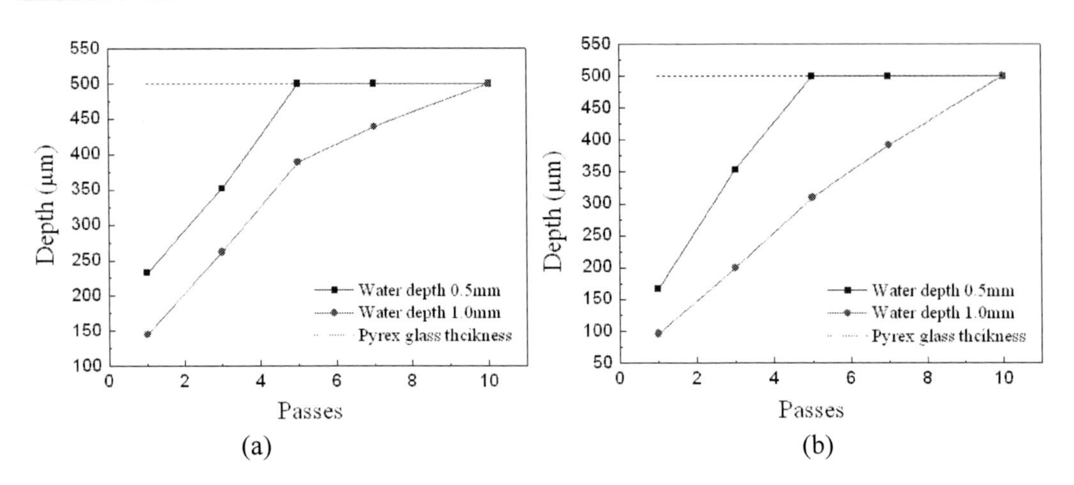

Figure 23. The etching depth of glass as a function of scanning pass at a fixed 6 W laser power, 0.5 and 1.0 mm water depth for 1-10 passes with a drawn diameter of: (a) 100 μm and (b) 200 μm.

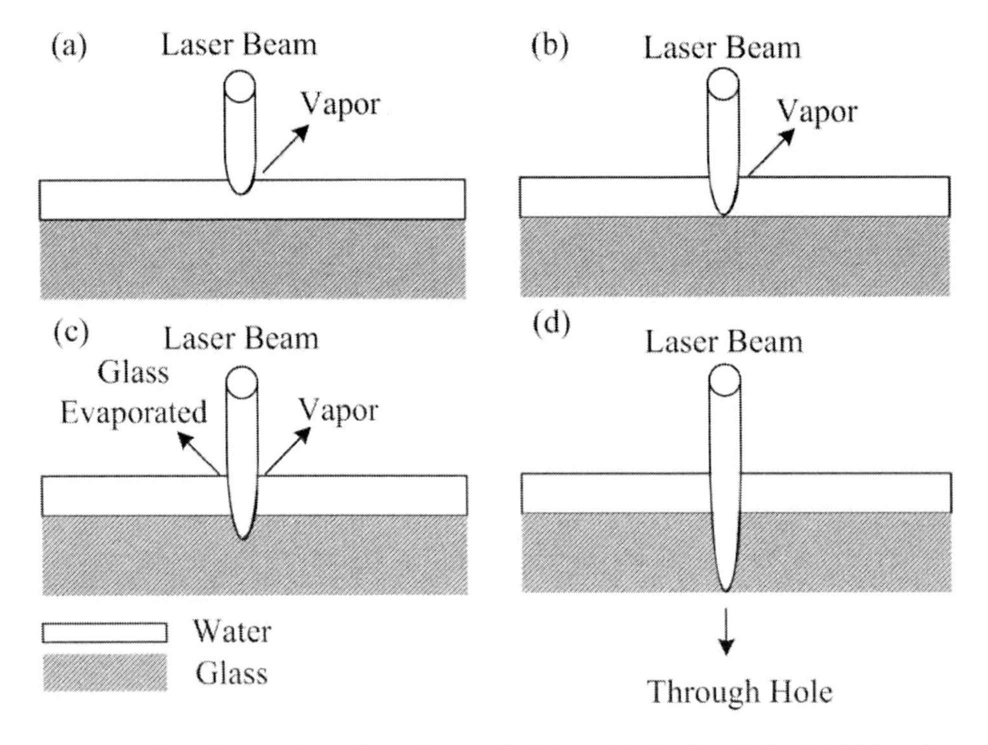

Figure 24. The schematic procedure of CO2 laser etched through-hole glass during LALP in order to understand its mechanism.

With regard to the temperature distribution and maximum temperature during glass hole machining, a simple thermal model is adopted for the analysis of thermal distribution in air and water ambient by means of the ANSYS software for solving this transient problem. It is assumed that the laser intensity distribution is a Gaussian function model and the heating phenomena due to phase changes are neglected. This simulation was performed for a laser beam power of 6 W and a scanning speed 11.4 mm/s at an initial reference temperature of 25 °C. The laser beam focusing on the plane surface maintains a constant TEM_{00} mode, that density of laser power can be described by a Gaussian distribution. The surface heat flux distribution $I_{(x,y)}$ can be calculated according to equation (4), where P_0 is the laser beam power and r is the laser beam radius, respectively. The CO_2 laser wavelength of 10.6 μm can be absorbed in water with a static absorption coefficient of about 500 cm^{-1} [52],[60]. However, we cannot employ the Beer's law to calculate the quantitative values of transmitted laser intensity on glass after the initial laser intensity penetrates water layer because the liquid-assisted laser processing is a dynamic issue including the complex photo-thermal interaction of water on glass related to the plasma shape, expansion and cooling along with others problems of undulation and bubble formation during the laser ablation. Therefore, we introduce a correction factor to modify the absorption coefficient of CO_2 laser in varied water depth. The correction factor is calculated from the ratio of experiment data in water to that in air.

$$I_{(x,y)} = \frac{p_0}{\pi r^2} \exp\left(-\frac{x^2 + y^2}{r^2} \right) \tag{6}$$

Figure 25 (a) shows the maximum temperature as a function of distance from beam center at a constant 6 W laser power and 11.4 mm/s scanning speed in air, in 0.5 mm and 1.0 mm thick water. The temperature decreases with increasing distance. The maximum surface temperature of glass with a drawn diameter of 100 μm can be reduced from 4009 °C in air to 2502 °C in 1.0 mm thick water for glass machining. When the diameter increases from 100 to 200 μm, the surface temperature of glass machining in air is reduced from 4009 °C to 2012 °C due to less thermal accumulation. The equation (1) is also used to analyze the heat transfer problem for temperature distribution of glass machining in air and water ambient by ANSYS software. Table 2 lists the material property of Pyrex 7740 glass for simulation. The maximum temperature is significantly reduced in water compared to in air. The thicker the water layer, the lower the temperature is. It indicates that LALP is an effective method to avoid thermal crack of glass through water cooling. It is noted that the distance over 300 μm was not suitable for drilling fabrication at above process conditions by LALP. It is because the thermal accumulation energy is not high enough to melt Pyrex glass at the center of the hole. In our larger drawn diameter test, it only etched some depth of glass but not through the wafer. The simulation result has a good agreement with our experiments. With regard to temperature gradient induced thermal stress, Fig. 25 (b) shows the residual thermal stress as a function of distance from beam center at a constant 6 W laser power and 11.4 mm/s scanning speed in air, in 0.5 mm and 1.0 mm thick water. The evolution of thermal stress is the same as the varied maximum temperature. The formation and propagation of crack on the laser-heated glass primarily results from the in-plane residual surface stress during ablation and cooling. The residual thermal stress (σ_{th}) is tensile and can be expressed in equation (1) [[45]]. ΔT is dependent on the laser energy absorbed for glass etching. The maximum σ_{th} increases with temperature gradients, ΔT, and is affected by the air or water surroundings. The temperature gradient in water is smaller than that in air because of the water cooling to enhance the heat loss and reduce the overheating and thermal stress. Therefore, the thermal stress can greatly be reduced during glass etching by LALP. For example in the simulation result in Figure 25 (b), the LALP method can reduce residual stress of glass with a drawn diameter of 100 μm from 359 MPa in air to 223 MPa in 1.0 mm thick water. Also, the Biot number (*Bi*) is the dimensionless number that can be helpful to characterize the relative magnitude of the established thermal gradient [[58]] and defined as the ratio of the average surface heat transfer coefficient (*h*) for convection from the entire surface, to the conductivity (*k*) of the solid, over a characteristic dimension *L* in equation (2). Since *h* of water (500 to 10000 W/m^2K) is 2-3 orders higher than air (10 to 100 W/m^2K) [[59]], the great increase of *Bi* in water with pronounced heat loss leads to the decrease of both temperature itself and its gradient. Therefore, temperature gradient in water is expected to decrease much to reduce the thermal stress and HAZ areas as well to improve the crackless through-hole etching quality as shown in Figures 21 and 22. It evidences that LALP is a very effective method for the through-hole etching of glass and is potential for the application of microfluidic inlets and outlets as well as the interconnection of metal deposition for feedthrough in package. The repeatability study of through hole etching is performed in the drawn circular arrays at the efficient water depth of 0.5 mm. Figures 26 (a)-(b) show the optical micrographs of CO_2 laser micromachined glass holes at the water depth of 0.5 mm at 10 x10 holes array in both drawn diameters of 100 μm and 200 μm, respectively. The insets of pictures are the magnification of holes in arrays. Good quality of uniform through holes of Pyrex glass without cracks and scorches are also obtained using LALP technology in this easy and fast approach.

Table 2. The material property of Pyrex 7740 glass

Property	Value	Property	Value
Density (kg/m^3)	2230	Specific heat (J/kg.K)	753.12
Thermal conduction (W/mK)	1.12968	Thermal expansion (1/°C)	32.5×10^{-7}
Young's modulus (Pa)	6.272×10^{10}	Poisson's ratio	0.23

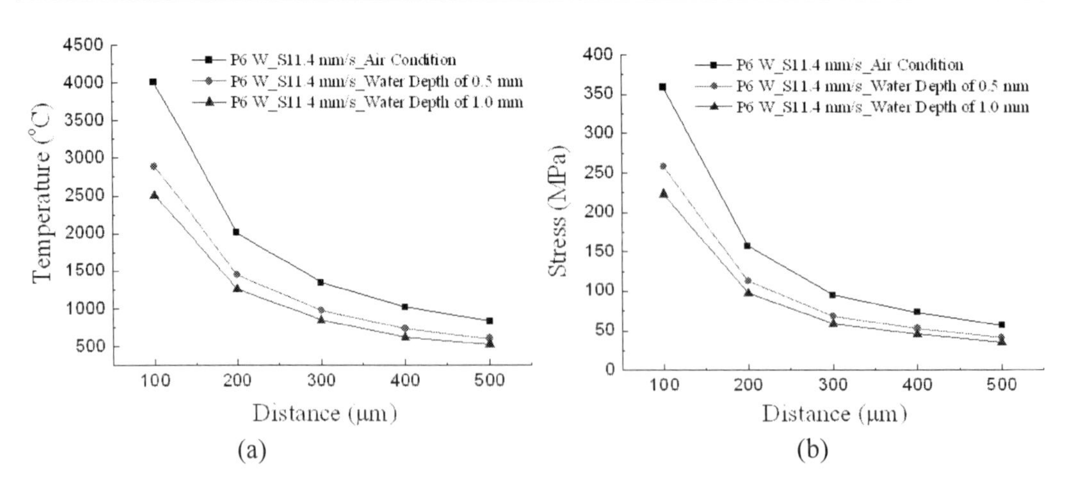

Figure 25. (a) The maximum temperature and (b) the thermal stress as a function of distance from beam center at a constant 6 W laser power and 11.4 mm/s scanning speed in air, in 0.5 mm and 1.0 mm thick water.

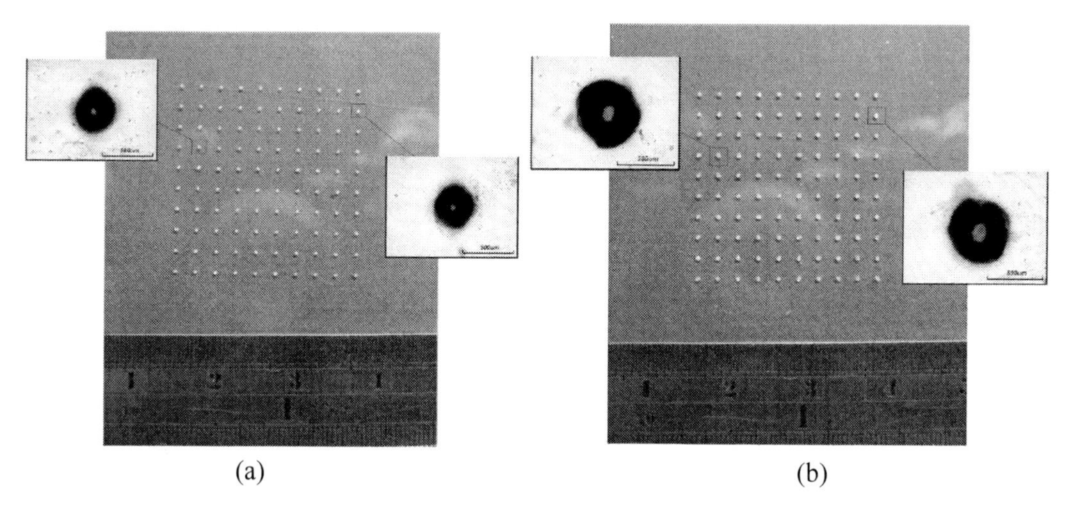

(a) (b)

Figure 26. The CO2 laser etched glass hole arrays using LALP at 0.5 mm thick water: (a) 10 x10 arrays in 100 μm, and (b) 10 x10 arrays in 200 μm.

Besides the cooling effect, water can be beneficial for the debris removal during LALP. Figures 27 (a)-(c) show the alpha-step profiles of through-hole glass etching with a drawn diameter of 100 μm at the 6 W power and 11.4 mm/s scanning rate for 10 passes in air, under 0.5 mm and 1.0 mm thick water, respectively. The bulges at both rims of etched hole in air are around 3.75 μm and 7.58 μm while those by LALP at 0.5 mm thick water are around 1.15 and 1.57 μm; and at 1.0 mm thick water for 0 and 1.58 μm, respectively. The formation of bulge on the rims is mainly from resolidification of evaporated debris. Much reduced bulge height can be obtained using LALP because the laser heating induces stronger natural convection in water to help carry debris away. Assume laser moving is along x-direction and water depth is for the z-direction. The shear stress (τ) from natural convection along x-direction in cross-sectional x-z plane is defined in equation (3). In water environment, temperature rises as the heat source passes. The temperature gradient of glass during laser ablation under water occurs because of heat accumulation on the surface of glass which arises from the heat loss through thermal diffusion less than the heat input of laser source. The highest temperature is located at the moving laser spot on surface and decreases with the distance away from heat source. As the laser moves, it induces the natural convection with shear stress which is proportional to water viscosity (μ) and the gradient of water velocity to distance ($\frac{\partial u_x}{\partial z} + \frac{\partial u_z}{\partial x}$). The lower laser moving rate or larger heat input may lead to the higher shear stress. The maximum shear stress occurs at the surface of laser spot with the highest gradient of water velocity to distance. Therefore, it helps to carry away the debris under water to reduce the bulge height (Figures 27 (b)-(c)). Comparing laser ablation in air and under water, the natural convection of the former is much smaller than the later. Therefore, no effective shear force in air moves the resolidified debris away to result in the large budge (Figure 27 (a)). The vaporization for the fast etching could occur in the local laser induced high-temperature plasma environment. The glass vapor will rapidly condense due to water cooling effect. Then much of condensed glass debris can be carried away by the natural convection of water during LALP. The spot of maximum shear stress of glass under water moves with the heat source and is expected to enhance debris removal for improving surface quality. The bulge improvement result is also verified in larger diameter hole etching. Figures 28 (a)-(c) show the alpha-step profiles of through-hole glass etching with a drawn diameter of 200 μm at the 6 W power and 11.4 mm/s scanning rate for 10 passes in air under 0.5 mm and 1.0 mm thick water, respectively. The bulges at both rims of the etched hole in air are around 15.95 and 16.65 μm while those in 0.5 and 1.0 mm water are around 1.59-2.51 μm and 0.05-2.04 μm, respectively. The evolution of bulge height of glass hole with a drawn diameter of 200 μm is similar to that in smaller diameter of 100 μm. But the magnitude is increased due to more debris production during ablation. The thicker water depth is beneficial for debris reduction but declines the etching rate. There is a compromising between the water depth and the etching rate for LALP performance at specific laser parameters. Overall, liquid-assisted laser processing for through-hole etching of Pyrex glass is a fast, simple and clean method at low cost.

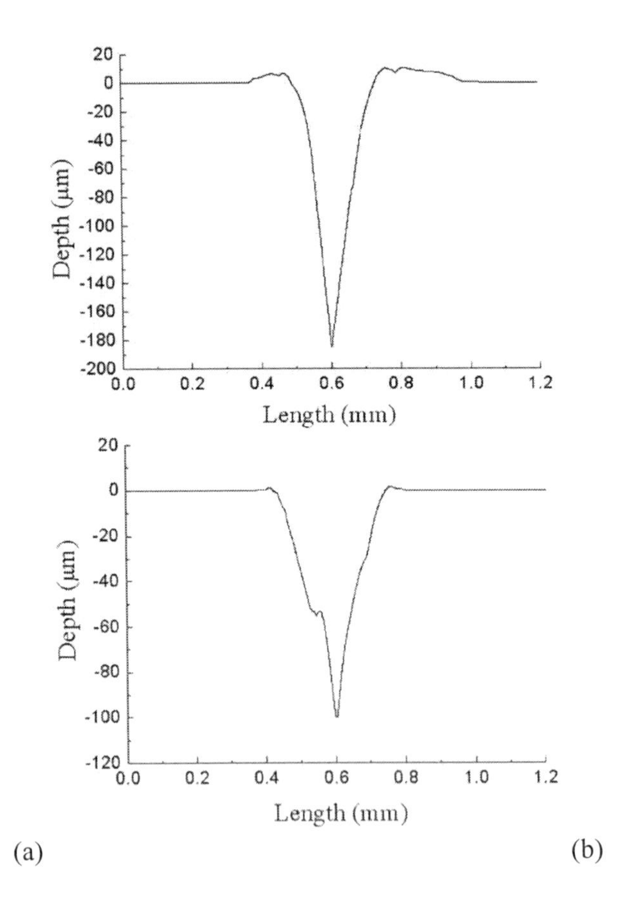

(a) (b)

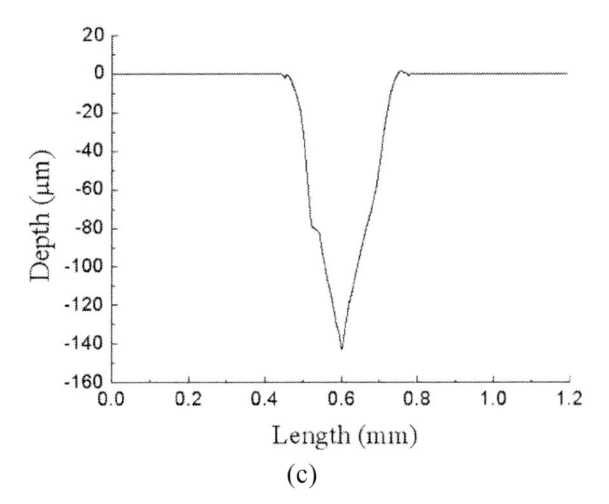

Figure 27. The alpha-step profiles of through-hole glass etching with a drawn diameter of 100 μm at the 6 W power and 11.4 mm/s scanning rate for 10 passes: (a) in air, (b) under 0.5 mm thick water, and (c) under 1.0 mm thick water.

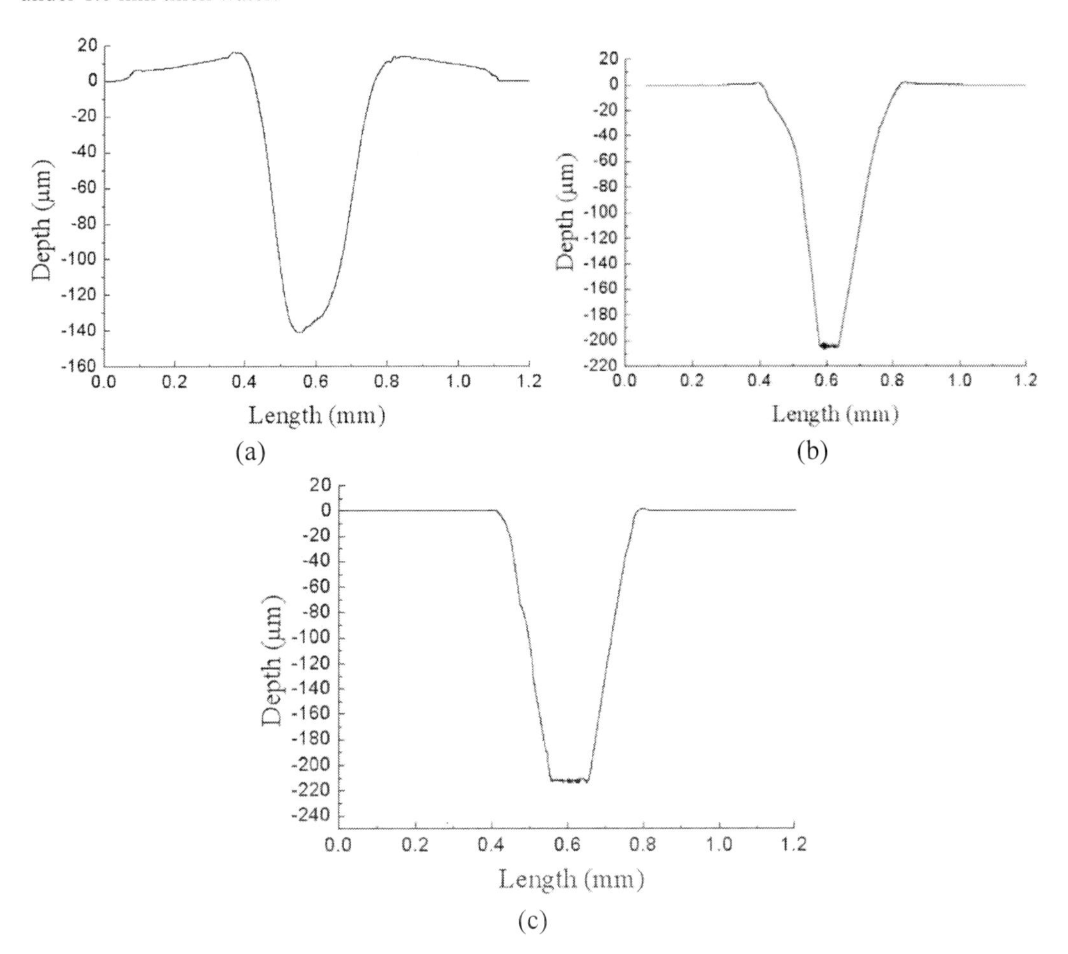

Figure 28. The alpha-step profiles of through-hole glass etching with a drawn diameter of 200 μm at the 6 W power and 11.4 mm/s scanning rate for 10 passes: (a) in air, (b) under 0.5 mm thick water, and (c) under 1.0 mm thick water.

A crucial issue in Pyrex glass etching is how to etch through holes of amorphous Pyrex glass for the inlet/outlet connection or package. Here, good quality of through hole etching of Pyrex 7740 glass without cracks and scorches has been demonstrated using liquid-assisted laser processing. The cracks or scorches are formed using CO_2 laser etched glass in air due to the high temperature induced thermal stress. The water is used to reduce the temperature and its gradient to avoid the formation of cracks or scorches. A through hole of Pyrex glass is easy to be obtained using LALP to etch a circular shape for 5-10 passes at a 6 W power and 11.4 mm/s scanning speed via the drawn circular diameter of 100-200 μm in several seconds. The hole arrays are also obtained to prove the repeatability of through hole etching of Pyrex glass. LALP technology has advantages of high etching rate, easy fabrication, and low cost with no crack and scorches. The ANSYS software was employed to analyze the temperature distribution and thermal stress field. The higher temperature gradient in air induced higher stress for crack formation while the smaller temperature gradient in water had the less heat-affected zone range and eliminated the crack during processing. The simulation result is in a good agreement with the experiments.

2.3. Liquid-Assisted CO_2 Laser Micromachining for Capillary-Driven Bio-Fluidic Application

At the bio-chip application, we present an easy method to fabricate the long-term high-hydrophilic microchannel for self-driven fluidic glass chips using low-power CO_2 laser ablation and low-temperature interlayer glass bonding. The medical-grade glass slide and LCP film were used because of the low contact angle and intrinsic hydrophilic behavior without using any modification treatment. Flow behavior of various viscosity fluids will be tested in our fluidic chips at the beginning and after 2 months. It can be used for self-driving the high-viscosity fluid of whole blood which is generally driven using external syringe pumps [64],[65]. The laser power was controlled at 6-24 W for the scanning speed between 114-342 mm/sec. Also, the water depth was controlled between 0.3 and 1.0 mm by total volume method in which depth is proportional to the volume at a constant area of a holder. Traditional laser processing in air was performed as a reference for comparison with LALP. Optical microscope and SEM were used to examine the morphology after laser processing. In the fluidic glass chip, the hydrophilic LCP material was used as the channel cover and adhesion layer. The main advantage of LCP film is that it is more hydrophilic than other polymer materials without any modification treatment and with close hydrophilicity to glass. Glass-LCP-glass bonding was finished using C-shaped clamps at 300 °C for 20 minutes. No water-leak phenomena were found during performing self-driven flow tests. The three-way glass chip was fabricated for the self-driven flow tests of low-viscosity deionized water added with 0.1 wt% Rhodamine B dye and high-viscosity whole blood bio-fluid. The Y-channel chip was also fabricated for the coagulation testing of the whole blood and 1 M $CaCl_2$ solution. Figures 29(a)-(b) show the optical micrographs of laser ablated 500 μm thick Pyrex glass using laser parameters of 24 W power and 228 mm/s speed in air ambient for linear and circular patterns. Cracks are easily formed at the first 5 passes in ablation of both patterns. In contrast, crack-free through-wafer ablation is obtained from glass under water 0.5 mm by LALP on both linear and circular patterns as shown in Figures 29(c)-(d). The occurrence of crack failure is primarily attributed to the thermal stress induced crack [45]. It leads to the difficulty of etch through glass of 500 μm thick by CO_2 laser in air at room temperature. Therefore, water cooling is adopted to assist laser ablation of glass in order to reduce the effect of thermal stress and stress gradient on glass ablation. It also has potential for the through-wafer etching. The through-wafer etching of Pyrex glass has been obtained with crackless surface using LALP. Besides the optical images, SEM images of Pyrex glass also show that no crack is observed on the surface of glass at high magnification up to 20000 X. The existence of a clear grey white line or spot at nearly center of patterns (Figures 29(c)-(d)) corresponds to through etching of glass. The black color around the grey white one is owing to the tapered profile, which is related to Gaussian distribution of laser beam energy. Moreover, the absorption of the laser beam at the sides of the trench near the surface may prevent the laser energy absorption deeper inside the sample. This could also result in the V-like tapered profile of trench near the surface. In order to further investigate the temperature distribution and maximum temperature during the straight-line machining process, a simple thermal model is adopted for the analysis of thermal distribution during machining. ANSYS software was used to solve this transient problem. Figure 30(a) shows the temperature distribution of Pyrex glass at a power of 15 W and a scanning speed of 57 mm/s in the absence of water. Surface of the glass shows the highest temperature of 7680 °C. Figure 30(b) shows the maximum temperature as a function of scanning speed at different powers in the presence and absence of water. Lower power and higher scanning speed will lead to the lower temperature. The maximum temperature decreases in the water assisted machining. Therefore, water cooling is an effective method to avoid thermal crack of glass. From the equation of thermal stress [45], residual stress is proportional to value of the temperature difference. Effect cooling can reduce the residual tensile stress.

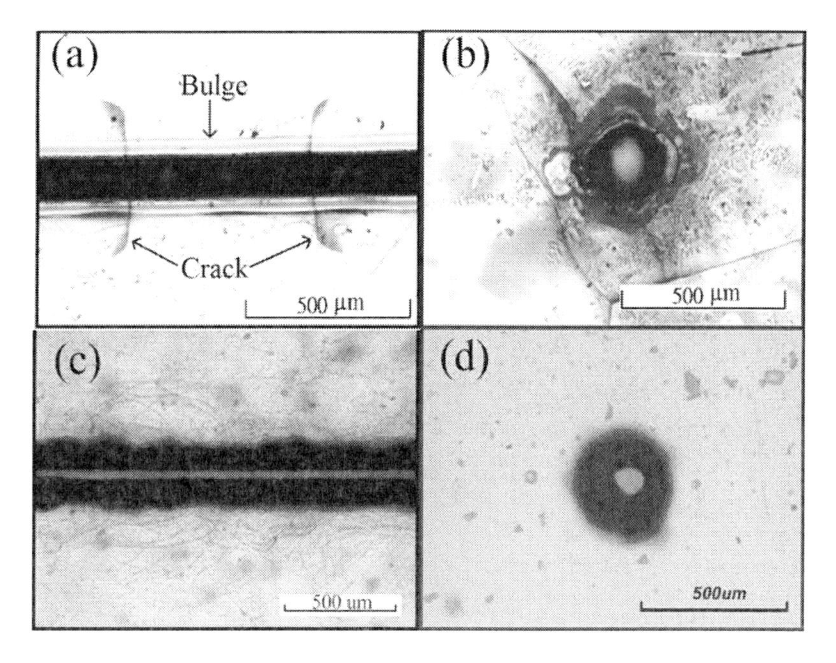

Figure 29. Optical micrographs of laser ablated 500 μm thick Pyrex glass using laser at 24 W power and 228 mm/s speed in air ambient ((a) for line, (b) for circle), and under water 0.5 mm through etching ((c) for line, (d) for circle).

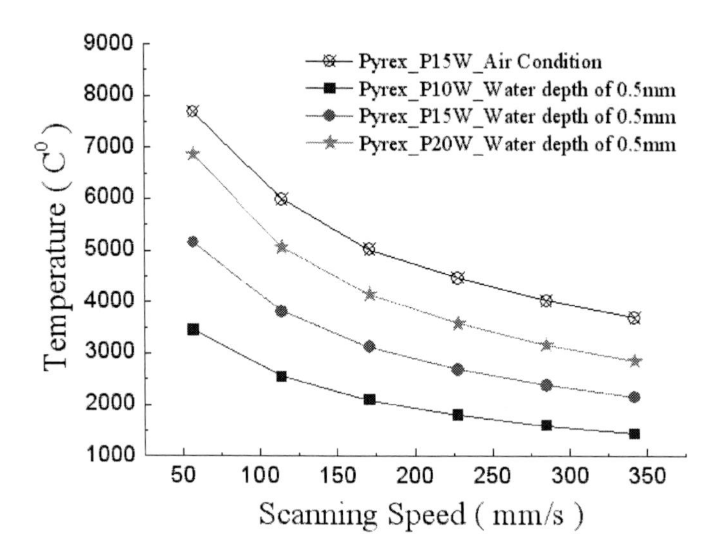

Figure 30. Relationship between maximum temperature and scanning speed at different laser powers in the presence and absence of water.

The formation of crack on the laser-ablated glass is most likely caused by the thermal induced in-plane residual surface stress during ablation. The residual thermal stress (σ_{th}) is tensile and can be expressed in equation (1). The laser energy is concerned with the laser power, scanning speed, absorption of material. Because the CO_2 laser ablation is a photothermal process, the surface temperature of the glass is above the softening temperature for significant material removal. The softening and working points of Pyrex 7740 glass in air are around 820 °C and 1252 °C, respectively. So, the estimated σ_{th} at softening point is around 200 MPa (63GPa $\times$ 3.2 $\times$ 10^{-6} $\times$ (820-25)/(1-0.2)) and that at working points is around 308 MPa. The maximum σ_{th} increases with temperature gradients, ΔT, and is affected by the air or water environment. The temperature gradient in water is smaller than that in air because of the cooling effect of water to enhance the heat loss and reduce the overheating and thermal stress. Therefore, the thermal stress can be diminished during glass etching by LALP. Although the 10.6 μm CO_2 laser can be absorbed in water with a static absorption coefficient (α) about 500 cm^{-1} [60], we cannot use the Beer's law (I_x=I_0 exp(-αx)) to quantitatively estimate the amount of transmitted CO_2 light intensity (I_x) on glass after the original power intensity (I_0) penetrates water thickness (x) because WALA is a dynamic process accompanied by the complicated mechanism of plasma formation, expansion and quenching together with the possible shock wave and bubble involved in the laser ablation. However, it is right for the qualitative discussion on the laser energy variation with water depth to explain the experimental data aforementioned. The higher water depth can reduce more thermal stress but it may reduce the etching rate. Generally, it needs a compromise between the water depth, glass thickness and etching rate of glass. Regarding the modified thermal bonding of glass-LCP-glass, the chemical nature and glass transition temperature (T_g) of LCP is two critical issues. The chemical nature of LCP is thermo-plastic property and belongs to liquid crystal polyester class. From the Material Safety Data Sheet (MSDS), it is essentially odorless, insoluble in water, no health hazard with normal handling and no hazardous polymerization. Also, the LCP was a popular polymer and widely used in the electronic package because of its excellent thermal stability, chemical resistance, low moisture absorption (0.04% at 25 °C for 25 hrs), low coefficient of thermal expansion, and high hermeticity. The glass transition temperature (T_g) of CT-25N LCP is around 240-280 °C. So, it can be a good material to bond both glass slides at a working temperature of around 300 °C for fluidic application due to the thermal plastic property. Too low or high temperatures are not good for bonding due to poor adhesion or broken polymer. Because of the intrinsic hydrophilic property of LCP, the composite glass-LCP channel is suitable for the capillary-driven fluidic chip without additional surface treatment. In order to understand the pressure related bonding strength of the glass-LCP channel, a syringe pump and a syringe were used to supply DI water into the glass-LCP channel. At an average flow velocity of 1 m/s which is much higher than conventional LOC operation condition, no any leakage is found because of good bonding between flat glass and LCP. Our chip can endure the high-flow-rate operation and suitable for most microlfuidic system.

Figures 31(a)-(c) show the optical micrographs of the static contact angle of high-viscosity whole blood about 0.004-0.005 Pa·s [67] in healthy human on the surfaces of the ablated glass, LCP and PDMS without using additional surface modification, respectively, for intrinsic hydrophilic examination. The whole blood was extracted from a healthy human and 5 μl volume of each droplet was controlled by a micropipette for contact angle measurement. In order to prevent the whole blood from coagulation, whole blood was mixed with sodium citrate anticoagulant. The largest static contact angle about 90° was observed in the PDMS materials due to intrinsic hydrophobicity. In contrast, both the LCP film and ablated glass exhibit much low contact angle owing to their native hydrophilic behavior. Driving force (ΔP) of capillary-driven fluid is related to the degree of hydrophilicity or contact angle. The driving force at the meniscus interface from the Laplace-Young equation is given by equation (6), where ΔP the pressure drop across the interface, σ the surface tension of the fluid, θ the contact angle and h the height of a microchannel.

$$\Delta P = \frac{2\sigma \cos\theta}{h} \tag{6}$$

If the contact angle is low than 90°, pressure drop across the interface is positive and liquid will be capillary-driven. The lower the contact angle, the higher the driving force is. So, low contact angles of the ablated glass and LCP material correspond to high driving force and are favorable for capillary-driven fluidic chips. Contact angles of DI water at different materials had also been explored and show same trend for these measurements. Also, the machined surface is not characteristic of a specific texture. Therefore, the texture effect can be negligible in this capillary driven chip. Figure 32 (a)-(d) show the optical micrographs of capillary-driven low-to-high viscosity fluid flow as a function of time in the three-fork glass chip: (a)-(c) for low-viscosity DI water with 0.1 wt% Rhodamine B dye and (d) for high-viscosity whole blood. For the dye-added DI water, the meniscus interface flows from inlet to rapidly pass half single channel at 0.9 sec, to about one-third distance of three-fork channels at 4 sec and to the ends at 7.8 sec. Moreover, the whole blood flows to all the ends of three-way channels at about 81 sec. Figure 33 shows the relationship between flow time and meniscus moving distance of the whole blood and dye-added DI water. Fast capillary-driven velocity has been observed compared to conventional chip in need of the external pump [[64]]. DI water also has much higher moving velocity of meniscus interface than whole blood because of its low viscosity of 0.001 Pa·s. The meniscus velocity is non-linear. The average moving velocities of DI water and whole blood in single channel from inlet to the junction about 18 mm are 9.52 mm/s and 1.89 mm/s, respectively. And the velocity decreases at the three-fork channels. The driving speed at different channel width had been further investigated. Following the mathematical model by Bouaidat et al in 2005 [67], relationship between time (t) and position (x) of the meniscus interface can be further described as $x = \sqrt{\frac{HW\sigma(\cos\theta) + H^2\sigma(\cos\theta)}{12W\eta}t}$. Where W is the channel width, H the channel depth, σ the surface tension, θ the contact angle and η the viscosity. As H/W is reduced to 0.1 or lower, the moving speed of the meniscus interface is insensitive to the channel width.

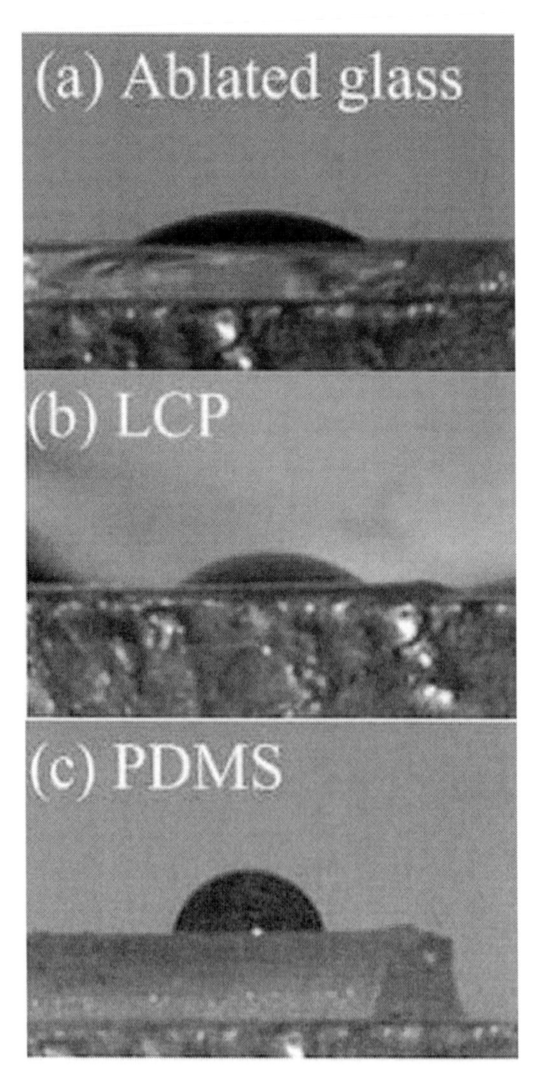

Figure 31. The static contact angle of whole blood on the surface of: (a) ablated glass, (b) LCP and (c) PDMS.

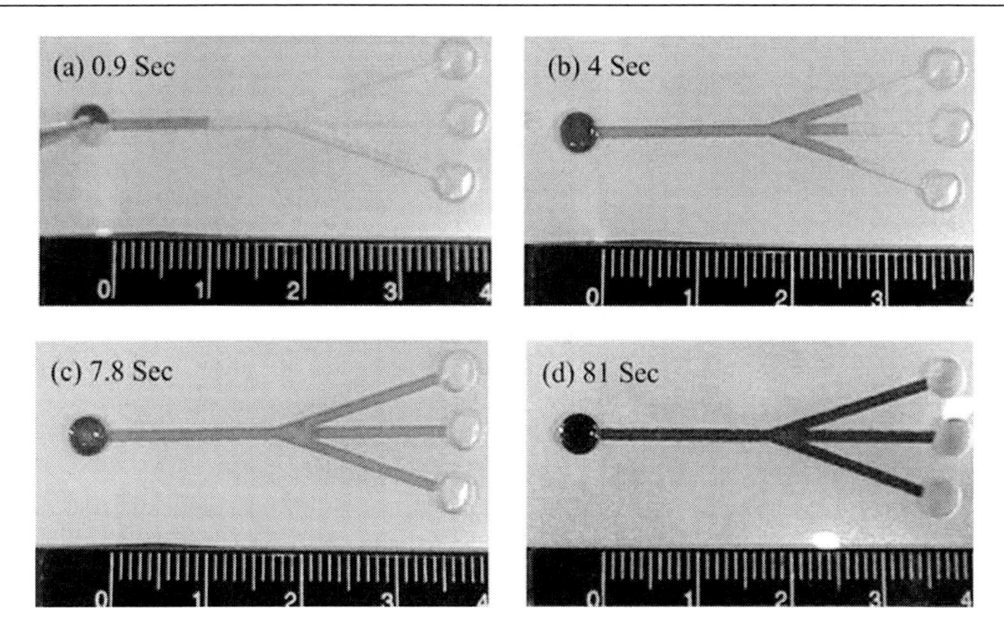

Figure 32. Optical micrographs of fluid flow as a function of time in the three-fork glass chip: (a)-(c) DI water with 0.1 wt% Rhodamine B dye, (d) whole blood.

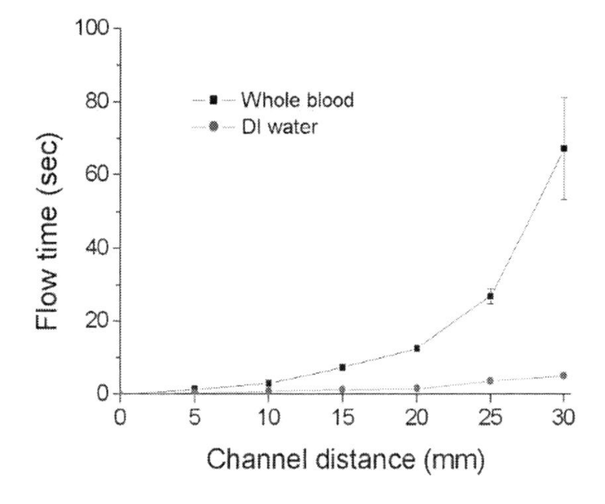

Figure 33. The flow time as a function of meniscus interface moving distance the three-fork glass chip for DI water and whole blood.

The Y-channel glass-LCP chip is also used for the whole blood flow test as shown in Figures 34(a)-(b). The whole blood takes less than 9 sec from one inlet hole through straight channel and turns into the main channel. Then it takes 40 sec to near the outlet end. The meniscus of whole blood in the straight channel is faster than that in the three-way channels. It is much easier fabrication and faster driving than some conventional external pumped blood chip [64] because of the high hydrophilic surfaces of glass-LCP chip using the LALP and modified glass-LCP-glass bonding technologies. Because whole blood contains sodium citrate anticoagulant, coagulation of whole blood is difficult and blood can arrive at the end of the channel. If a droplet is put on a glass, clotting time is about 5 minutes. In order to further test the clotting behavior, whole blood coagulation testing had been carried out using the glass-LCP-glass chip with coating $CaCl_2$ solution in the channel surface of the main straight channel. Figure 35 shows the optical micrograph of the initial coagulation testing using whole blood and $CaCl_2$ solution. 1 Mole $CaCl_2$ solution was first coated by capillary-flow filling. In order to dry the $CaCl_2$ solution in the main channel, baking at 90 °C had been performed using the hot plate. Before the clotting test, whole blood coating and baking were repeated for two times. Whole blood is supplied to the inlet reservoir using a micropipette. Whole blood flow into inlet channel rapidly (time = 1 sec), as shown in Figure 35(a). Then, moving speed of the whole blood decrease as whole blood flow into main channel because of the coating $CaCl_2$. At the same time, coagulation reaction is activated by clotting factors. The clotting stop occurs after 70 sec at about 22 mm position of the ruler. It indicates the mixing and coagulation reaction of the whole blood and $CaCl_2$ effectively occurs in the channel of glass-LCP chip. The capillary-driven glass chip offers an opportunity of increasing demand of the blood coagulation analyzer for point-of-care (POC) testing in future [68]. The whole blood test evidences that our highly hydrophilic fluidic chip is feasible for the application of the capillary driven bio-fluidic system in future without any additional surface treatment or external pumps. High-hydrophilicity chips can actuate high-viscosity whole blood which is generally driven by external syringe pumps [64], [65] or micropumps [69].

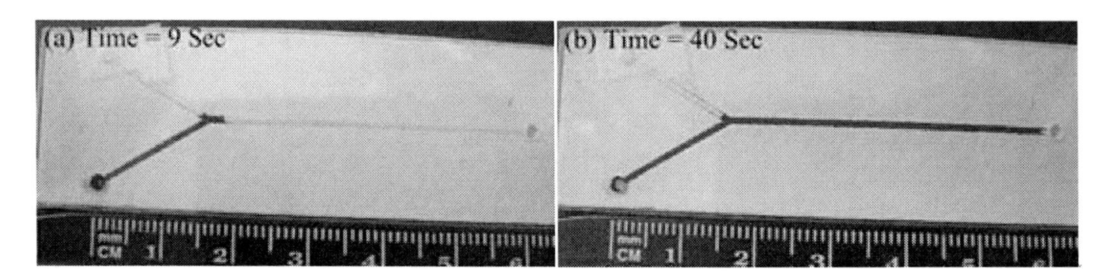

Figure 34. The whole blood flow test using the Y-channel glass-LCP chip.

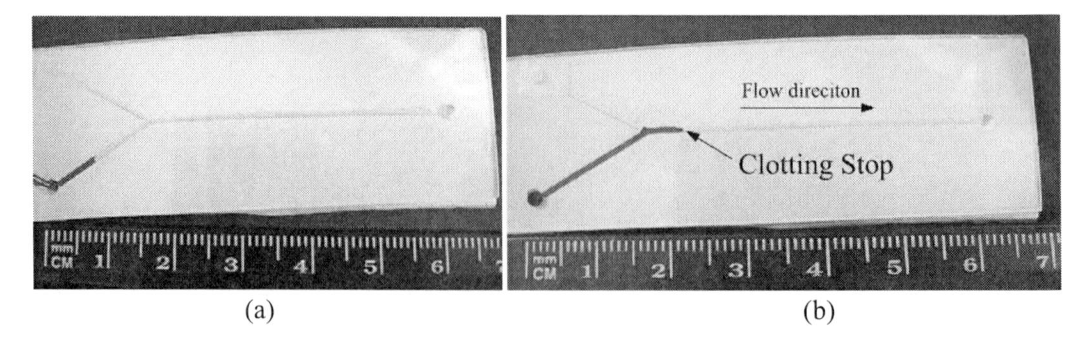

Figure 35. The whole blood clotting time test using the Y-channel glass-LCP chip (a) time = 1 sec (b) Time = 70 sec.

The crackless channel etching and hole drilling of glass have been demonstrated using liquid-assisted laser processing (LALP). It solves the problems of crack formation in traditional laser processing in air. The CO_2 laser ablated glass in air results in cracks due to the sharp temperature gradient and high temperature. Using LALP can cool down samples during ablation for less temperature gradient to eliminate the crack. The higher water depth can reduce more thermal stress together with the ablation rate. But deficiency in water depth for fast etching can lead to the crack at deep etching due to insufficient cooling. Therefore, a compromise between the water depth, glass thickness and etching rate of glass is desired. The ablated glass slide with native hydrophilic property is much suitable for fluidic chip application at low cost. Hydrophilic LCP was used an interlayer to bond two glass slides at low temperature for the rapidly capillary-driven low-viscosity DI water and high-viscosity whole blood without any additional surface treatment or application of external pumps. The clotting time testing under human whole blood sample in the $CaCl_2$ coated glass-LCP chip was achieved in the downstream of Y-shaped channel. The fluidic glass chip has the merit of hydrophilic behavior conquering the problem of traditional hydrophobic recovery for potential bio-fluidic application. It also simplifies the traditional complex fabrication procedure of glass chips.

3. Cover-Layer Protection Processing

3.1. PDMS Protection Processing for PMMA

An approach to fabricate microfluidic channel without bulges on PMMA material is proposed by adding a cover layer on the top of PMMA substrate. Different covering materials of un-exposed and exposed JSR photoresist and PDMS were used. A model for the bulge formation at the rim of polymer channel in laser micromachining based on the photothermal melting was proposed. Figure 36 shows the schematic process flow of CO_2 laser micromachining adding a PDMS or JSR cover layer on the top of polymethylmethacrylate (PMMA) substrate. First, a thick PDMS of 500 μm or a JSR photorsist of several ten μm was covered on the top of PMMA substrate as a protective layer. Then, CO_2 laser is used to pattern the desired surface profile or channel in the PMMA substrate through the JSR or PDMS layer. At last, the PDMS or JSR layer was removed. The sample size of PMMA substrate was 20 mm in length, 20 mm in width and 2 mm in thickness. At a constant power of the laser beam, the laser scanning speed was increased in 10% steps between 10% and 100%. This laser power of 10.5 W (30%) was performed at different cut numbers from 1 to 4 passes. Different cover layers of PDMS material and JSR photoresist which was unexposed or exposed by UV radiation of a wavelength of 365 nm were used to investigate the formation and elimination phenomena of bulges. A different number of passes were performed to examine the variation of bulge shape and its formation mechanism as well as the profile of channel. The laser scanning length for a channel was fixed at 20 mm. Traditional laser processing without any cover layer on the PMMA substrate was performed as a reference sample.

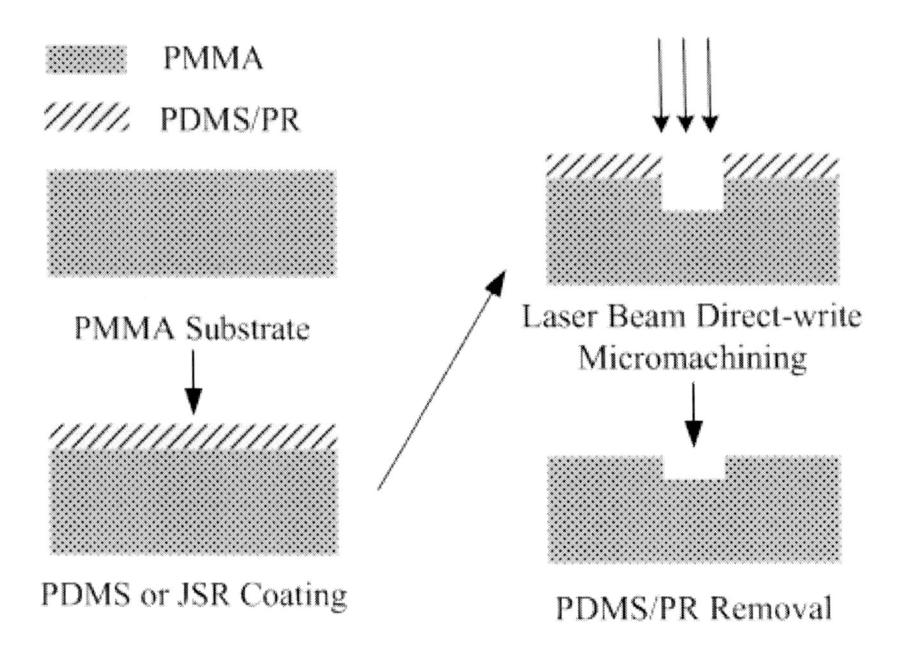

Figure 36. The schematic process flow of CO2 laser micromachining adding a PDMS or JSR cover layer on the top of polymethylmethacrylate (PMMA) substrate.

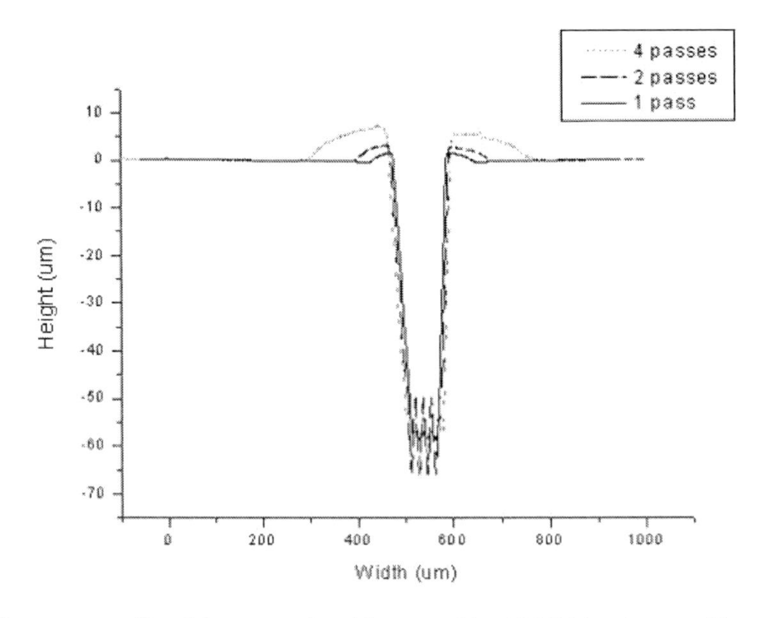

Figure 37. The α-step profile of the conventional laser machined PMMA polymer without any cover layer at by a laser power of 10.5 W at a scanning speed of 342.9 mm/sec and cut number of 1, 2 and 4 passes. The obvious bulges are formed on the rim of channel.

Figure 37 shows the α-step profile of the conventional laser machined PMMA polymer without any cover layer at by a laser power of 10.5 W at a scanning speed of 342.9 mm/sec. Three profiles at cut numbers are 1, 2 and 4 passes are arranged together for comparing the bulge formation. Obvious bulges are formed on the both rims of a channel at each cut number and the size of bulge increases with the increasing number of pass. The larger bulge in size at more cut passes also exhibits larger height and width of bulge on the rim of channel due to more laser energy to evaporate the polymer away. The height of bulge at cut number of 1, 2, and 4 passes are estimated to be 2.0, 3.1 and 6.8 μm, respectively. The width of bulge is about in the range of 50 to 200 μm at one rim of channel. The depth of machined PMMA channel is too deep to measure in the α-step profiler. Variation of width of PMMA channel increase little with the increasing number of pass in the α-step profile. Figures 38 (a)-(c) show the 3D confocal images of the laser machined PMMA without any cover layer at number of passes of 1, 2, and 4 passes, respectively. The laser power is 10.5 W and the scanning speed is 342.9 mm/sec. The increasing number of cut passes in laser machining results in the increase of both height and width of bulge on the rim of channel. It is in agreement with the result of α-step profile in the trend of bulge variation. Figure 39 shows the schematic model of the formation of the bulges, splashing and resolidification of polymer micromachined by a laser beam. The mechanism of CO_2 laser micromachining is the photothermal melting and evaporation due to the long wavelength of 10.6 μm of the laser. It results in the HAZ in solid, molten zone in liquid and evaporated gas with high pressure in the laser working area. The formation of bulge is attributed to molten polymer resolidified by atmospheric air cooling on the rim of channel after splashing from high-pressure gas or surface tension driven flow. The high-power CO_2 laser beam is used as the heat source and focuses on the surface of the PMMA material in a very short time to transfer energy to it. The PMMA instantly melts or vaporizes locally as the accumulated energy is high enough. One liquid layer in molten zone between the solid substrate, evaporated gas and the laser beam will be pushed out of the working zone by the high pressure of the hot evaporated gas and radiation beam. Furthermore, another driving force of pushing out liquid is probably from the surface tension due to the large surface property variation between the heat affect zone and the liquid-gas mixed zone with a sharp temperature gradient field and a volume expansion. So, the bulges will be formed as the ejected molten material meets with the atmospheric cold air, then to be resolidified and accumulated on the rim of channel. The more the cut number of pass, the more the molten polymer, and the larger the bulge in size.

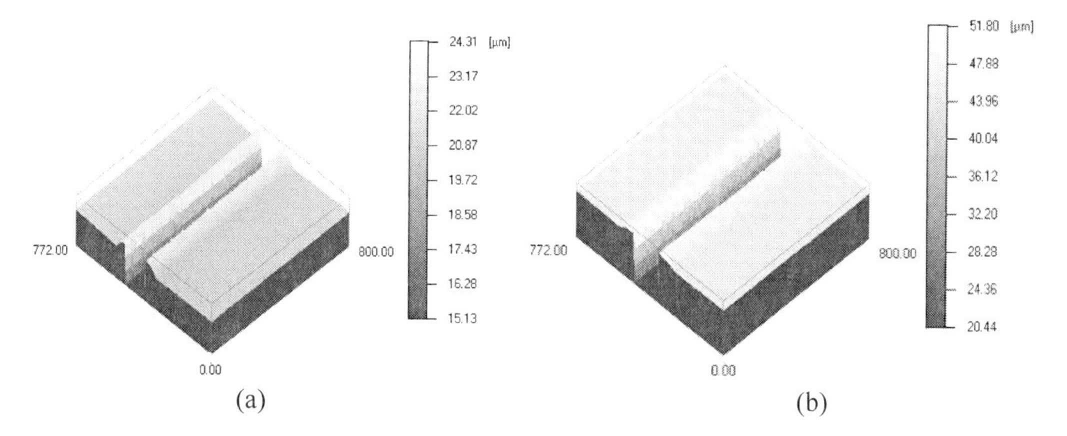

Figure 38. (Continued).

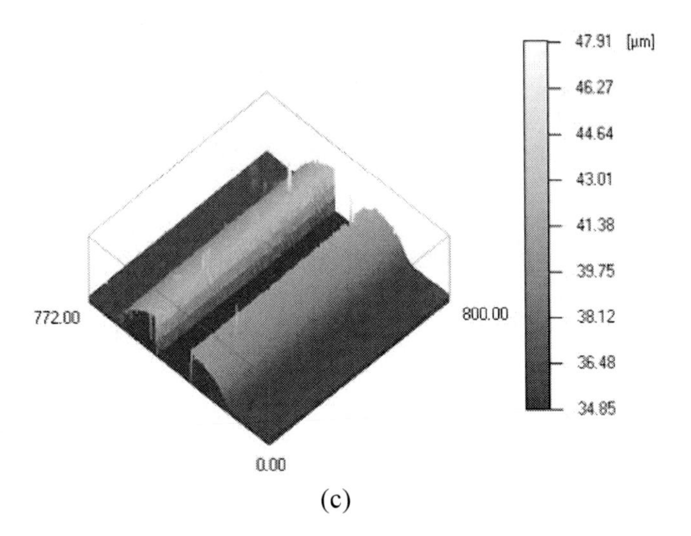

(c)

Figure 38. The 3D confocal images of the laser machined PMMA without any cover layer at number of passes: (a) 1, (b) 2, and (c) 4 passes. The laser power of 10.5 W and the scanning speed is 342.9 mm/sec.

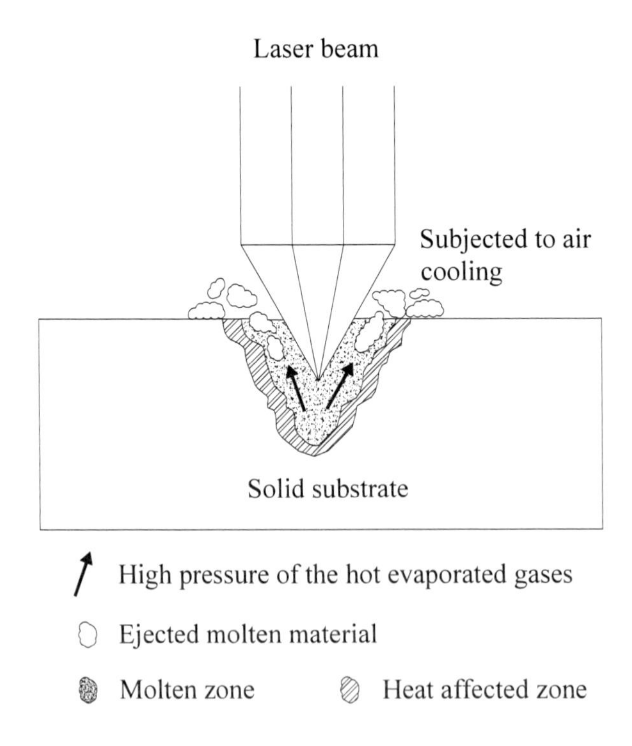

Figure 39. The schematic model of the formation of the bulges, splashing and resolidification of polymer micromachined by a laser beam.

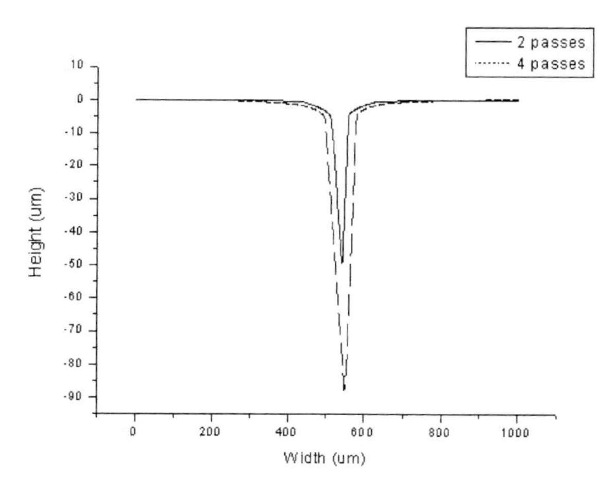

Figure 40. The α-step profile of the laser machined PMMA with a PDMS cover layer during processing and removed after machining. The laser power of 10.5 W is performed at a scanning speed of 342.9 mm/sec and cut number of 2 and 4 passes.

Based on the formation mechanism of bulge, we apply a cover layer on the polymer to protect it during laser processing and remove the cover layer to improve the machining quality. Figure 40 shows the α-step profile of the laser machined PMMA with a PDMS cover layer during processing and removed after machining. The laser power of 10.5 W is performed at a scanning speed of 342.9 mm/sec and cut number of 2 and 4 passes. The bulges are completely eliminated to get a smooth surface of the PMMA channel as expected. The depth of channel at cut number of 2 and 4 passes are about 50 μm and 88 μm, respectively. The cross section of channel displays a V-like shape subject to the Gaussian distributed laser energy decreasing in the PDMS cover layer of 500 μm thick during the processing and most absorbed by the PDMS layer. Figure 41 shows the 3D confocal image of the laser machined PMMA with a PDMS cover layer protection at cut number of 2 passes as an example. To get a clear picture on the rim of channel, the image about half of channel depth is taken. Smooth surface without bulge has been achieved. It is in consistent with the result of α-step profile (Figure 40(a)).

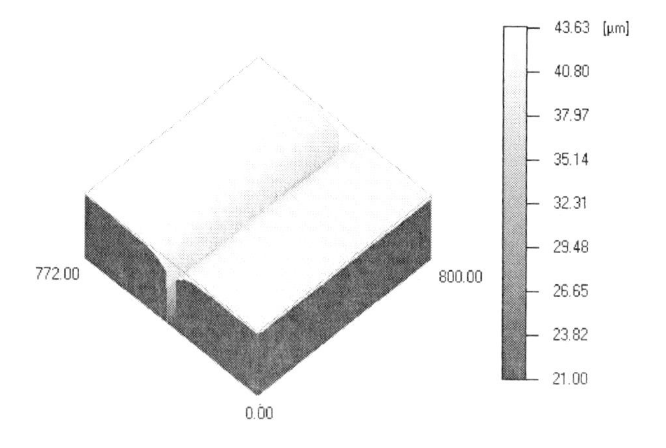

Figure 41. The 3D confocal image of the laser machined PMMA with a PDMS cover layer protection at cut number of 2 passes. Smooth surface without bulge has been achieved.

Another cover layer of JSR photoresist is also studied. Figure 42 shows the α-step profile of the laser machined PMMA with an unexposed JSR photoresist cover layer after processing and not removed at cut number of 1, 2 and 4 passes. Similar to the direct machined PMMA without any cover layer, obvious bulges are formed on the rim of channel but the shape variation of bulge is quite different. The height of bulge with an unexposed JSR cover layer decreases with the increasing number of cut pass while the width of bulge increases with the increasing number of cut pass. This is attributed to the difference of molecular weight and viscosity between the rigid PMMA substrate and unexposed JSR photoresist. The rigid PMMA substrate with high molecular weight and melting temperature is difficult to flow or diffuse in the molten state with high viscosity and the bulges resolidified and accumulated on the PMMA substrate are hard to be affected or re-molten by the lateral thermal energy from continuous laser processing. On the contrary, the unexposed negative tone JSR photoresist with low molecular weight and melting temperature is rather easy to flow with low viscosity and to be affected by followed thermal energy. So, the bulges resolidified on the top of unexposed JSR photoresist will change the position of accumulation to result in the increasing width and decreasing height of the budge with the cut pass. The amount of ejected molten PMMA for bulge formation decreases with the cut number of pass. As the unexposed JSR photoresist cover layer is removed, the bulges are eliminated to get a smooth surface of the PMMA channel as shown in figure 43. The depth of machined PMMA channel with an unexposure JSR cover layer is too deep to measure in the α-step profiler. The width of PMMA channel varies in little increase with the number of pass. As we use an exposed JSR photoresist for a cover layer, a different phenomena of bulge formation is observed. Figure 44 shows the α-step profile of the laser machined PMMA with an exposed JSR photoresist cover layer after processing and not removed at cut number of 1, 2 and 4 passes. In comparison with the direct machined PMMA without any cover layer, the shape of bulge formation is different each other. A "bulge-like or hump" surface is formed on the rim of channel with an exposed JSR cover layer. The height of "bulge" at cut number of 1, 2, and 4 passes are estimated to be 0.4, 1.5 and 4.5 μm, respectively. It is smaller than that formed without any cover layer. The width of "bulge" is about in the range of 50 to 175 μm at one rim of channel. It is similar to that without any cover layer. The total volume of "bulge" increases with the increasing number of pass. The exposed JSR cover layer for protection of PMMA can not be completely removed after laser processing. If the removal process is performed in ACE solution by ultrasonic agitation too long, cracks on the surface will be found. Figure 45 show the schematic model of bulge/hump formation mechanism in the laser machined PMMA with an exposed JSR photoresist cover layer. Bulges on the rim of channel are formed from two mechanisms. One called "conventional bulge" is formed from the thermal molten polymer resolidified by atmospheric air cooling (Figure 39). The other call "hump" is formed on the rim of channel of PMMA with the exposed JSR cover layer due to the thermal distortion from thermal stress or residual stress in a large temperature gradient induced by the difference of thermal expansion coefficient and cooling rate between the exposed JSR cover layer and PMMA substrate with strong bonding each other. The exposed negative-tone JSR photoresist has large molecular weight and becomes stable after UV activated cross-linking of molecules. It is easy to bond with PMMA due to over-exposure UV light transmitted to PMMA and increase the bonding strength with the increasing temperature during the laser processing. The thermal stress in a large temperature gradient is induced from the different thermal properties between the thermal unaffected zone, thermal affect zone (HAZ), molten liquid and exposed

JSR photoresist. It is very complicated during laser processing. In overall, the more the laser processing passes, the more the heat and temperature, and the more the thermal stress and hump formation. The cracks will occur on the surface if the large thermal stress is not able to release.

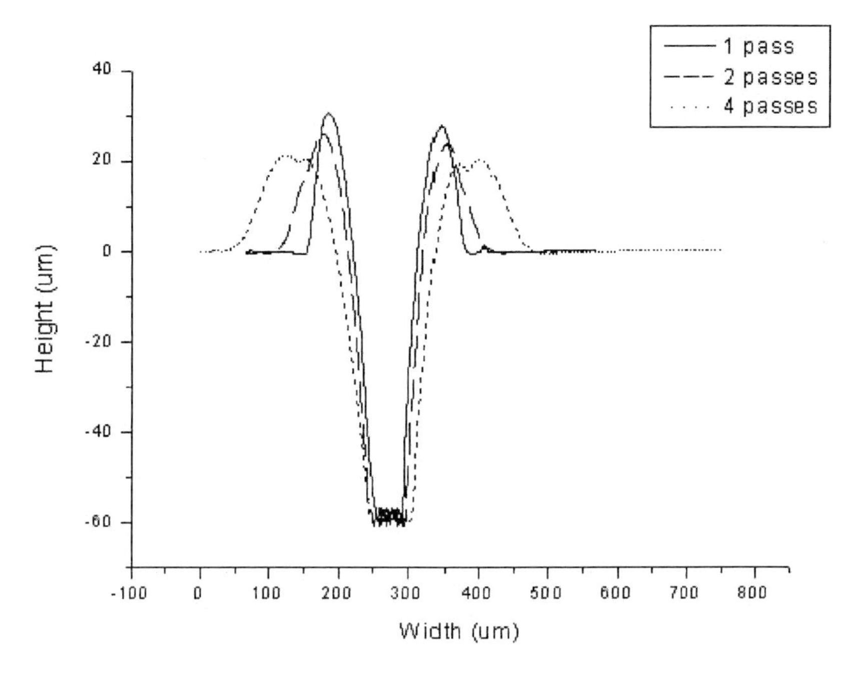

Figure 42. The α-step profile of the laser machined PMMA with an unexposed JSR photoresist cover layer after processing and not removed at cut number of 1, 2 and 4 passes.

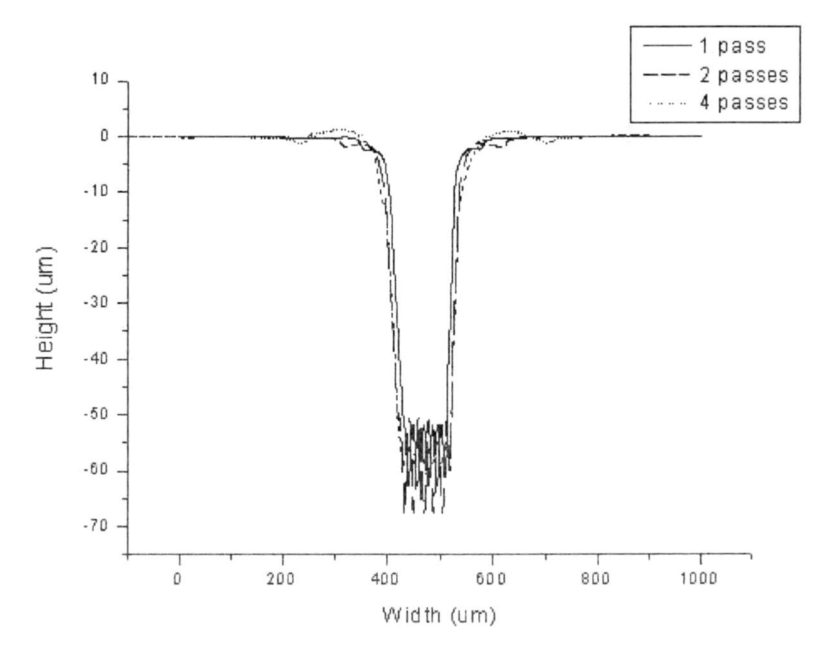

Figure 43. The α-step profile of the laser machined PMMA with an unexposed JSR photoresist cover layer after processing and removed at cut number of 1, 2 and 4 passes.

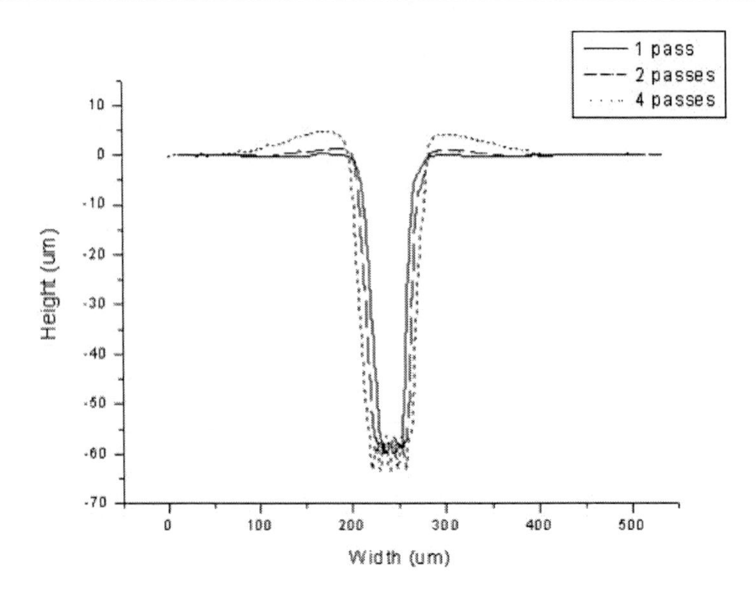

Figure 44. The α-step profile of the laser machined PMMA with an exposed JSR photoresist cover layer after processing and not removed at cut number of 1, 2 and 4 passes. The shape of bulge formation is different from that without any cover layer. A "bulge-like or hump" surface is formed on the rim of channel.

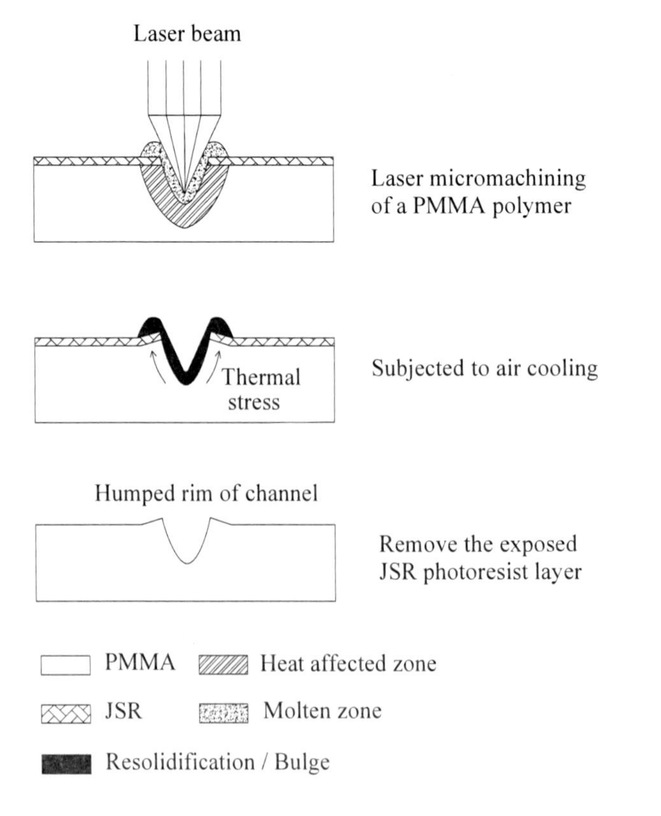

Figure 45. The schematic model of hump formation mechanism in the laser machined PMMA with an exposed JSR photoresist cover layer.

It has successfully demonstrated a novel approach to eliminate the bulges on the rim of micromachined channel in a PMMA substrate by adding PDMS or unexposed JSR layer on the top of polymer substrate during CO_2 laser processing. Bulges on the rim of channel are formed from two mechanisms. One is called "conventional bulge" resulting from the molten polymer resolidification in the cool-air atmosphere. This bulge can be eliminated by removing the cover layer. The other is call "hump" due to the thermal distortion from thermal stress or residual stress in a large temperature gradient which is formed on the rim of channel of PMMA with the exposed JSR cover layer. This bulge or hump can not be removed and subject to induce cracks in the following process. The shape variation of bulge at different cut passes is quite different between the laser machined PMMA without any cover layer and with an unexposed JSR cover layer before removal. Total volume and width of bulge in both cases increase with the increasing cut passes. But the height of bulge increase with the increasing cut passes in the former while that decreases with the increasing cut passes in the latter. This is attributed to the difference of molecular weight and viscosity between the rigid PMMA substrate and unexposed JSR photoresist. Good quality of smooth surface of the polymer microstructure can be achieved by adding suitable cover layer for the different application.

3.2. PDMS Protection Processing for Pyrex Glass

On the glass processing, the traditional glass forming using laser processing in air would produce many kinds of defects such as bulge, debris, crack and scorch. Glass materials have been widely used in the application of optical and optoelectronic devices, microfluidic and bio MEMS. Many publication discussions were made on improving CO_2 laser ablation method recently. Here, we have reported a poly-dimethylsiloxane (PDMS) assisted CO_2 laser process method to improve the drawbacks of transitional laser process for glass material in air. The ANSYS simulation was also used to analyze temperature distribution and thermal stress field in air and PDMS ambient during glass ablation. Figure 46 shows the schematic process flow of PDMS film assisted CO_2 laser ablation. Pyrex glass with an average thickness of about 500 μm was used and spin-coated a PDMS film at room temperature (RT). Then the direct ablation of CO_2 laser beam was performed to scan on the PDMS coated glass to get the desired shape at proper parameters. This laser power of 15 W was performed at different cut numbers from 1 to 10 passes. The PDMS thickness was controlled at about 150 and 250 μm on glass surface. Traditional laser processing in air was performed as a reference sample. The ANSYS software was employed to analyze the temperature distribution and thermal stress field. Figure 46 the schematic process flow of CO_2 laser micromachining adding a PDMS cover layer on the top of Pyrex glass substrate.

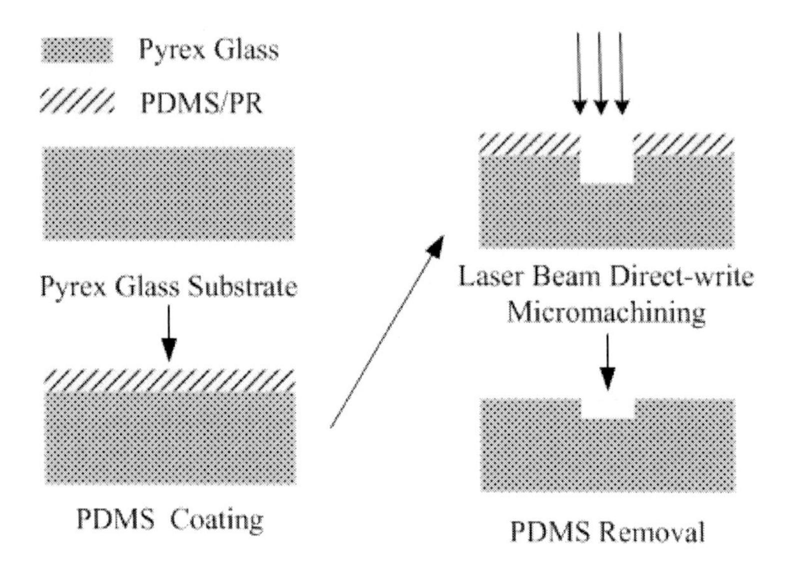

Figure 46. The schematic process flow of CO2 laser micromachining adding a PDMS cover layer on the top of Pyrex glass substrate.

The traditional laser processing in air would produce many kinds of defects such as bulge, debris, crack and scorch as show in Figures 47(a) and (b) under laser power of 15 W, scanning speed of 228 mm/s and scanning passes of 5 and 10, respectively. The cracks occur commonly in air condition due to the high thermal stress while the scorches formation are mainly from great accumulated heating energy for burning. The higher temperature gradient leads to the larger heat-affect zone region even cracks. Figures 48 (a) and (b) show the optical micrographs of 150 μm PDMS assisted CO_2 laser process at laser power of 15 W, scanning speed of 228 mm/s and 342 mm/s, respectively. It improves the traditional defects of laser machining with clear surface, reduced bulge, no crack and scorch. Figures 49 (a) and (b) show the optical micrographs of 250 μm thicker PDMS assisted CO_2 laser process, which have similar improved surface of ablation as Figures 48 but smaller cutting width. The variation of bulge height was measured by alpha-step profiler as shown in Figure 50. Figure 50(a) was the result of traditional laser process in air with bulge in maximum of 15.1 μm. Figures 50(b) and (c) were the results of PDMS assisted CO_2 laser process in 150 and 250 μm, respectively. The maximum bulge measured was 1.2 μm in case of 150 μm PDMS thickness for improving the defect clearly. As the thickness of PDMS material increases to 250 μm, the maximum bulge height is 2.94 μm due to the resolidification effect of some PDMS material. Figures 51 (a) and (b) show the simulation of temperature distribution and thermal stress field, respectively, using the material property of Pyrex 7740 glass and PDMS listed in tables 1 and 2. The higher PDMS thickness or scanning speed, the lower surface temperature and thermal stress are. That higher temperature gradient in air induced higher stress for crack formation while the smaller temperature gradient in PDMS had the less heat-affect zone range and diminished the formation crack during ablation.

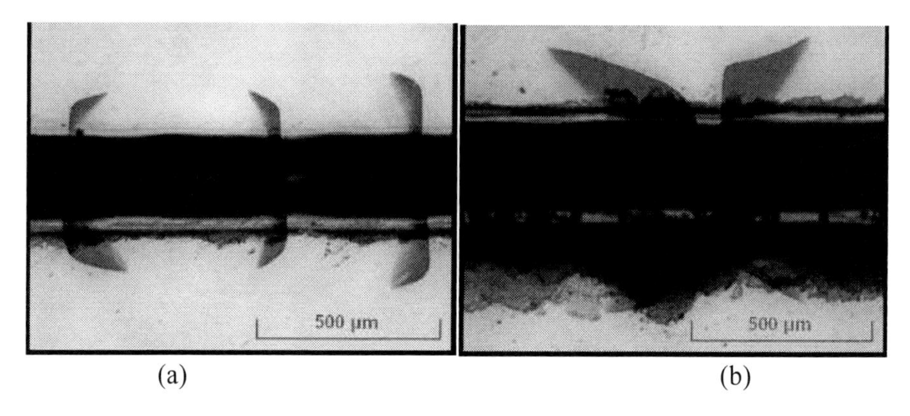

Figure 47. The optical microscope image of traditional laser machining in air condition; (a) 5 passes, (b) 10 passes.

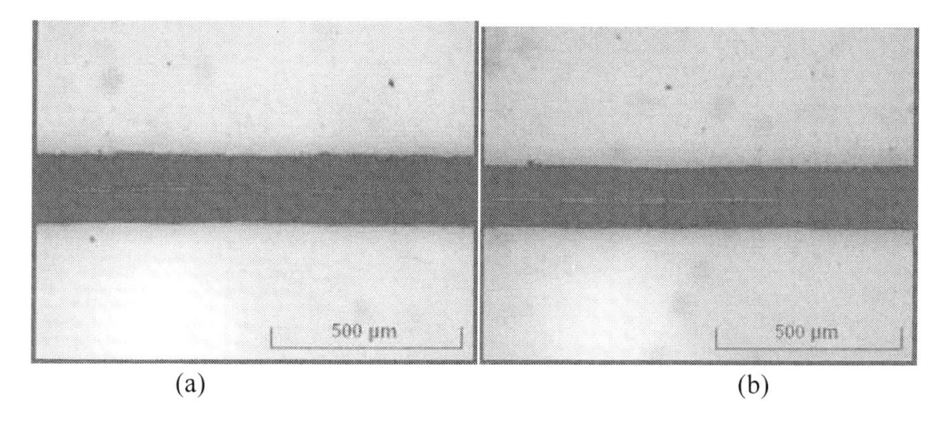

Figure 48. The OM image of PDMS assisted CO2 laser machining in thickness of 150 μm; (a) scanning speed 228 mm/s, (b) scanning speed 342 mm/s.

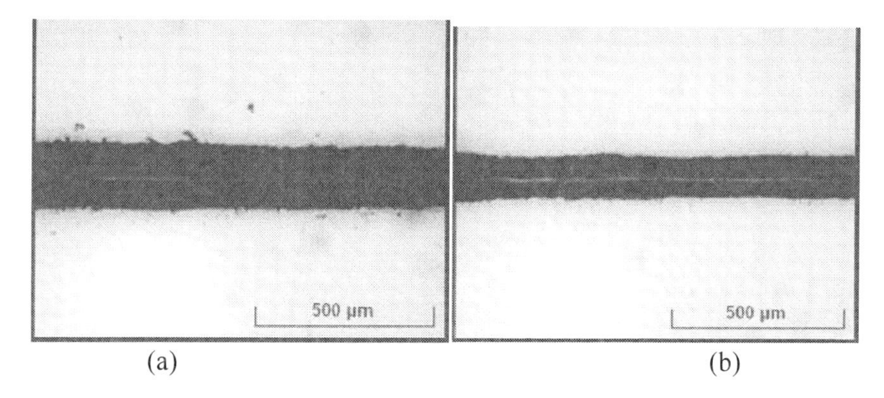

Figure 49. The OM image of PDMS assisted CO2 laser machining in thickness of 250 μm; (a) scanning speed 228 mm/s, (b) scanning speed 342 mm/s.

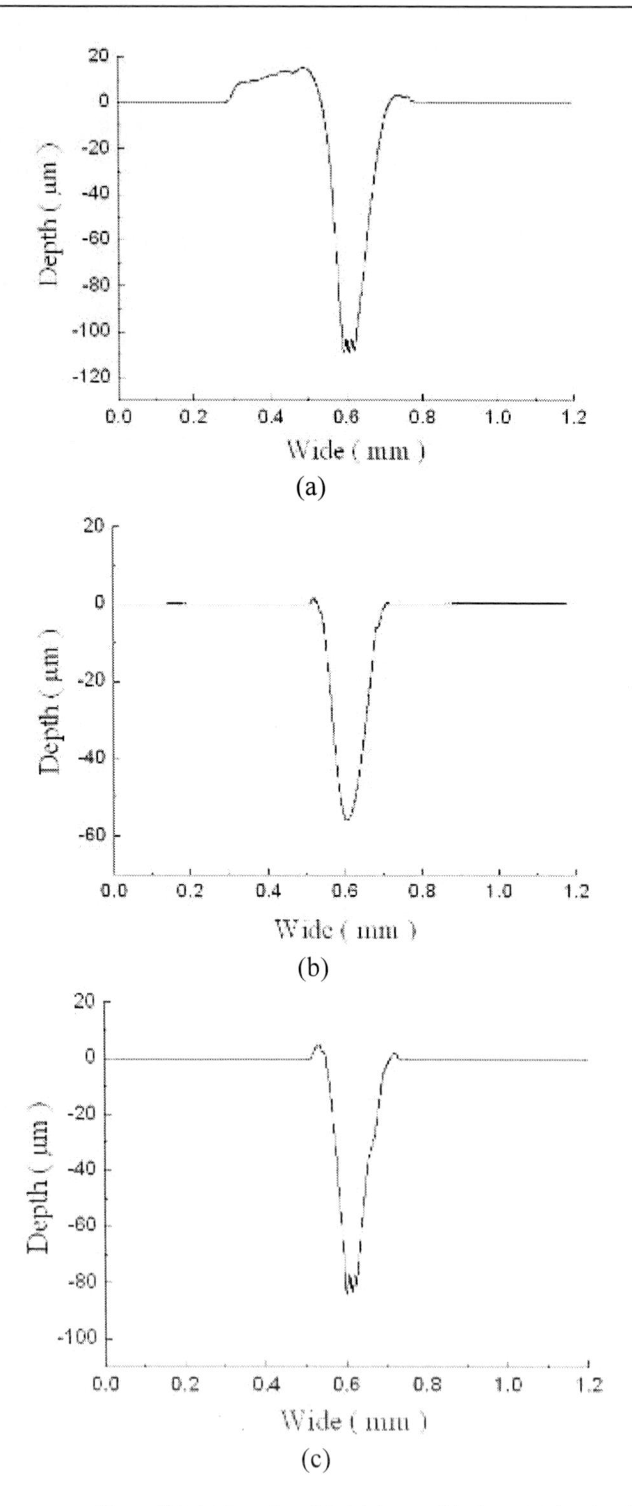

Figure 50. The alpha step profiles of PDMS assisted CO2 laser ablation: (a) in air, (b) in PDMS thickness of 150 μm, and (c) PDMS thickness of 250 μm..

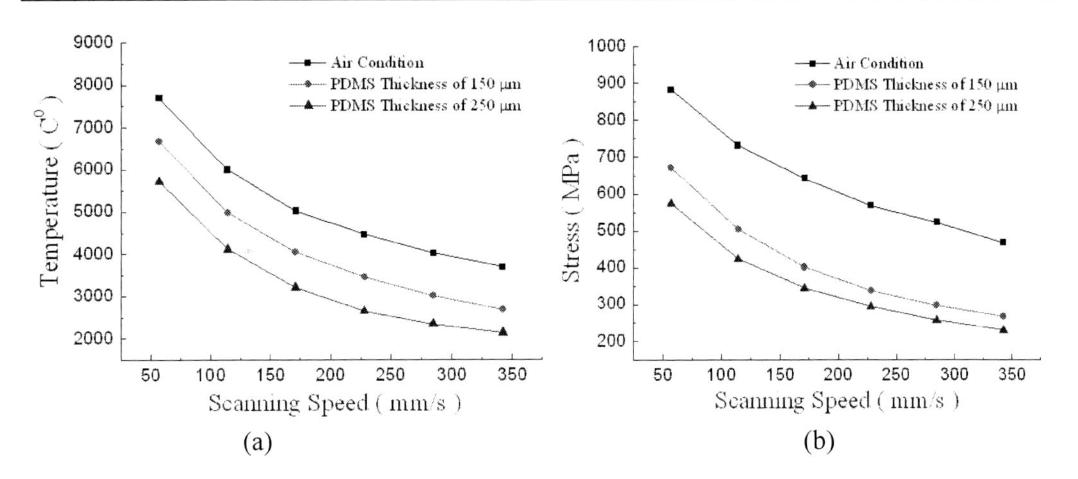

Figure 51. The results of ANSYS simulation; (a) temperature distribution, (b) thermal stress field.

The traditional glass forming using laser ablation in air would produce many kinds of defects such as bulge, debris, crack and scorch. In this article, we have applied the method of PDMS assisted laser processing to reduce the temperature gradient, heat-affect zone region for achieving crack-free glass ablation. The laser power of 15 W, PDMS thickness of 150 μm and scanning speed of 228 and 342 mm/s were used for ablation to obtain a good quality of cutting surface which was clear, less bugle, no crack and scorch in this technique. The alpha-step measured profile showed that the much reduced bugle height around the rims of groove was about 1.0 ± 0.5 μm in case of 150 μm poly-dimethylsiloxane (PDMS) thickness while that in air was 15.1 μm. The ANSYS software was also used to analyze the temperature distribution and thermal stress field in air and PDMS ambient during glass ablation. The smaller temperature gradient observed in PDMS-assisted ablation had the smaller heat-affect zone and diminished the crack formation during processing. In contrast, the higher temperature gradient in air induced higher stress for creating crack and scorch on the surface.

4. Glass-Assisted CO_2 Laser Processing

Conventional Si etching can be performed by dry plasma etching, wet anisotropic etching, laser ablation using Nd:YAG, Excimer or femtosecond lasers with wavelengths between 1.06 μm and 193 nm. No CO_2 laser with a 10.6 μm wavelength was directly used for Si etching. In terms of plasma or wet etching process, they all need photolithography to define the pattern geometry and select the suitable mask materials of the resist or hard mask to meet the different requirements of chemistry endurance and etching depth. Their etching rates are generally in the range of 0.5-3 μm/min, dependent on the chemistry, process conditions and sample geometry. Such etching rates are still low, especially for a through Si wafer etching of 525 μm, and it will take several hours. The equipment cost of dry plasma etching is high and the etched geometry for wet etching is limited to crystallography and directionality. In terms of laser ablation, Nd:YAG lasers with wavelengths of 1.06 μm, 532 nm and 355 nm, or KrF, ArF Excimer and femtosecond lasers with shorter wavelengths were used for direct Si ablation due to the adsorption in Si. But the cost of above lasers is much higher than CO_2 laser at the same laser power. In this article, we present a Glass Assisted CO_2

LAser Processing (GACLAP) for CO_2 laser direct processing silicon material at random cutting, drilling and marking. Figure 52 shows the schematic diagram of GACLAP setup. It includes a computer, a continuous wave (CW) CO_2 laser source, a reflective mirror, a focal lens and an experimental sample with a 525 μm thick Si wafer or chip on the top of a glass wafer or plate. The single-side polished p-Si(100) wafers were used with an average roughness around 1 nm and resistivities of 1- 20 x 10^{-2} ohm-m, corresponding to a boron doping concentration around $7x10^{20}$- 10^{22} m^{-3}. The glass can be a Pyrex glass or others which can absorb CO_2 laser energy. A pure Si sample was used as a reference for comparing the Si etching behavior with the glass assistance. The polished surface of the Si wafer was contacted to the bright smooth glass surface with the help of another heavy glass on the top of Si edges. The sample preparation for the contact of Si and glass is easy in this study. We used the clean-room class Kimberly-Clark wiper to clean the surfaces of Si and glass, and then slid two samples slowly to contact by hand. Thus both Si and glass could be closely contacted by electrostatic force and easily separated after experiments. The purpose of another heavy glass put on Si edges was used to fix the sample for the following laser processing. The laser spot was focused on the top surface of Si during GACLAP. A computer aided design program of CorelDraw software was available to set the experimental parameters of laser power, scanning speed and number of beam passes for auto-controlled processing. Two primary kinds of linear and circular patterns were drawn to study the Si etching behavior. The linear length of 5 mm and circular diameter of 300 μm were used in the experiments. At a constant power of the laser beam, the laser scanning speed could be adjusted between 0% and 100%. Because higher laser power and slower scanning speed were good for the observation of GACLAP phenomena, we controlled the laser power between 15 and 30 W and scanning speed between 2.3 and 11.4 mm/s. In order to see the pronounced GACLAP effect on Si etching, a nearly maximum laser power of 30 W and a low scanning speed of 2.3 mm/s were used to perform the etching of six linear patterns respectively with 10, 20, 30, 40, 60, and 80 passes. The sample was then cross-sectioned in order to evaluate the etching depth. The circular pattern was used for examining the hole drilling capability. Each condition was done at least two times for the repeatability.

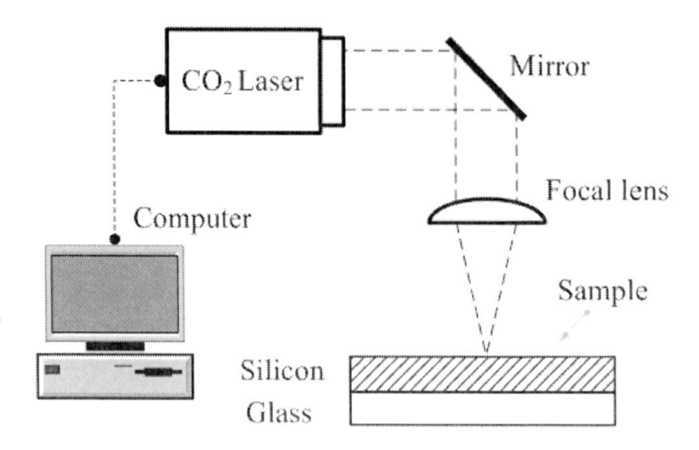

Figure 52. The schematic diagram of glass assisted CO2 laser processing setup.

Figure 53 (a) shows the optical micrograph of the Si sample on a glass with six evident etching lines from right to left using the CO_2 laser at a fixed 30 W power and a 2.3 mm/s speed for 10- 80 passes. The Si etching behavior by CO_2 laser is changed as it is put on a Pyrex glass plate compared to pure Si normally transparent to CO_2 laser where no etching occurs even at more passes input [70]. The sample was cleaved from the middle of trenches for the cross-sectional observation to examine the etching behavior. Figures 53(b)-(c) show the representative Si trenches etched at 10 and 80 passes, respectively. The cross section of the trench exhibits a V-like shape subjected to the Gaussian distributed laser energy. The occurrence of asymmetrical profile is due to the scattering effect of photon at high temperature in this new mechanism which leads to the non-uniform absorption of Si etching involved in a melting-and-evaporation process. The maximum roughness on the rim of each trench measured by the alpha-step profiler was in the range of 1 to 20 μm, roughly increasing with pass number. Figure 53(d) shows the overall relationship between the etching depth and laser passes. The etching depth measured was about 208, 236, 300, 354, 408 and 415 μm corresponding to 10, 20, 30, 40, 60, and 80 passes, respectively. The depth of trench non-linearly increases with increasing laser passes at the fixed power and scanning speed. The variation of depth in the repeated experiments exhibits the similar trend of etching, that is, the etching depth increases with the pass number. The depth may have tens μm difference in the repeated experiments attributed to the variation of contact between the Si and glass surfaces in our simple setup. The closer is the surface contact, the higher the energy transfer and the etching efficiency. The non-linear etching rate increases first at low-medium passes, then decreases at higher passes. It is because of the absorption of the laser beam at the sides of the trench near the surface that prevents absorption of the laser energy deeper inside the sample. This would also lead to the enlargement of the V-shaped trench near the surface at higher passes (Figures 53(b)-(c)).

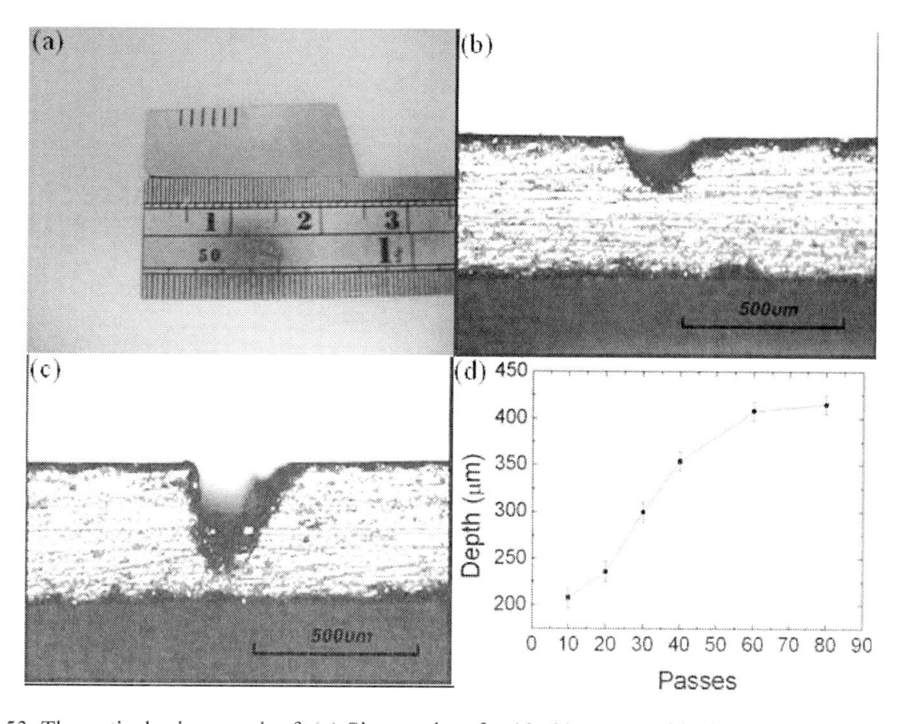

Figure 53. The optical micrograph of: (a) Si on a glass for 10- 80 passes with six evident etching lines. The representative cross-sectional images of Si trenches at: (b) 10 and (c) 80 passes. And (d) the etching depth as a function of laser passes.

Figure 54(a) shows the optical micrograph of the circular Si pattern in a 300 μm diameter etched by GACLAP technology at a 24 W power and a 5.7 mm/s speed for 20- 120 passes. Six visible holes including three blind ones had been achieved from right to left corresponding to the 20, 40, 60, 80, 100 and 120 passes drilling. The transparent holes with visible white light penetration indicate that Si had been etched through wafer of 525 μm thick after 80 or more passes drilling. All the magnified images of holes had been shown in Figures 53-54 of reference 12. The diameters of holes increase rapidly with pass number in the blind holes and then tend to a saturated value in the open holes. The range is 350-550 μm corresponding to the 20-120 passes. As blind hole is etched through and pass number is too large, the morphology around the hole becomes much rough [70]. The diameter is larger than the design value of 300 μm due to the lateral removal of Si from the accumulated energy in the continuous processing and the laser beam with a 76 μm spot size in a circular motion. The etching geometry affects the energy distribution and the etching rate. The accumulated energy enhances the vertical etching rate of a small-size circle (Figure 53) compared to the long-lines etching for trenches (Figure 52). That is, the 300 μm circular pattern is etched through 525 μm depth at a lower 24 W power and a higher 5.7 mm/s speed compared to the 5 mm long line which is only etched 415 μm at the 30 W power and 2.3 mm/s speed at the same 80 passes. Also, the accumulated energy influences the diameter variation of blind holes in the photothermal process. Figure 53(b) shows the diameter of the blind holes as a function of passes in the laser power of 15-27 W at a constant 5.7 mm/s speed. The diameter of hole roughly increases with increasing laser power and pass number. It is related to the total accumulated absorption energy of Si for the lateral removal, which is proportional to the laser power and passes and affected by the efficiency of lateral heat transportation. The variation is large at low pass number.

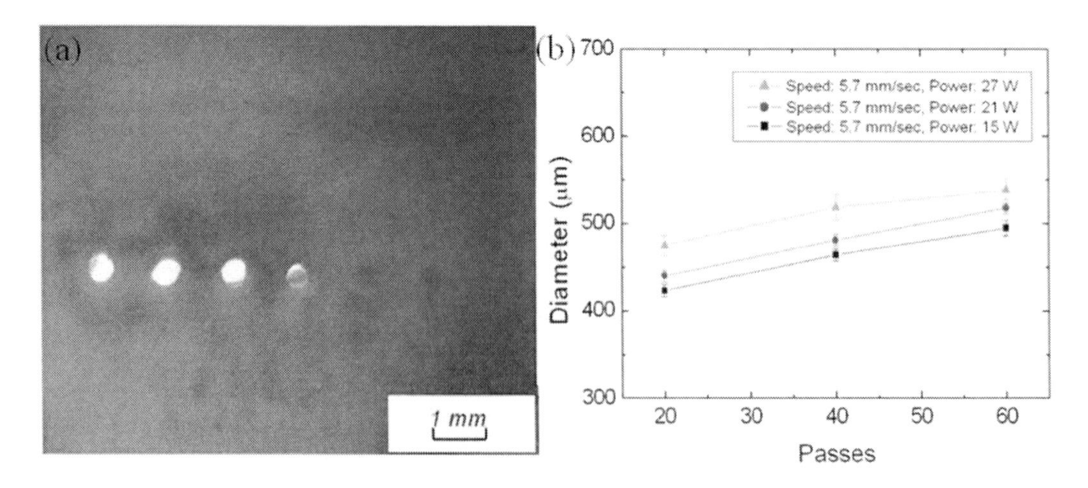

Figure 54. (a) The optical micrograph of the etched circular Si pattern by GACLAP for 20- 120 passes from right to left and (b) The diameter of the blind holes as a function of passes in the laser power of 15-27 W at a constant 5.7 mm/s speed.

From Figures 53 and 54, the Si etching from the top surface toward the interface implies that Si must absorb the laser energy with the assistance of glass to modify the Si absorption through some mechanisms to make etching possible. The phenomenal mechanism of Si etching goes through the following steps: (1) absorption modification before etching, (2) low-pass absorption and shallow etching, (3) middle-pass absorption and deeper etching, and (4) high-pass absorption to through-wafer etching. Figure 55 shows the proposed model for GACLAP mechanism. Glass initially absorbs CO_2 laser energy near the interface and then heats Si to change its absorption behavior extending to CO_2 laser wavelength of 10.64 μm for etching to start from the top surface. As we put our sample in water to etch Si like liquid-assisted laser processing, no etching occurs. It proves that the high temperature is a key factor for the modified high CO_2 laser absorption of Si. Since Si has a high thermal conductivity of 150 W/(m-K) for solid (300 K) and 450 W/(m-K) for liquid (1685 K) [71], the heat can rapid transport to the surface to result in high temperature at surface to change the Si absorption behavior. The microstructure will be changed at high temperature due to the thermal (/phonon) induced Si atom rearrangement from single crystal to polycrystalline or amorphous states with high entropy and defects in both the heat affected zone (solid) beneath surface and the molten zone at surface (liquid). The new mechanisms of the modified CO_2 laser absorption in Si with varied microstructure may be involved in two main reactions: one is the variation of Si energy band structure at high temperature [73],[74] for promoting the absorption of photon-electron interaction between the Si and CO_2 laser. The other is the increased Si oxidation at high temperature [[71],[72]. The thin oxide on the surface can directly absorbs the CO_2 laser for further etching, which may be the origin of modified absorption. The above two kinds of mechanism may occur simultaneously during GACLAP and interact each other for the high etching rate process, whose average rate larger than 240 μm/min at the condition in Figure 53. The thin oxide layer formation changes the light absorption from the interface to top surface of Si. The surface oxide is etched away after absorbing CO_2 energy and fresh oxide is formed immediately during continuous laser scanning to keep the Si at very high temperature with varied band structure to absorb CO_2 laser for deeper Si etching. The estimated temperature at the local surface of Si from the heating effect of glass and the isolation of the surface oxide will be between 1685 K (Si's melting point) and 3173 K (Si's boiling point) because the CO_2 photothermal etching mechanism is related to the melting and evaporation process for high etching rate.

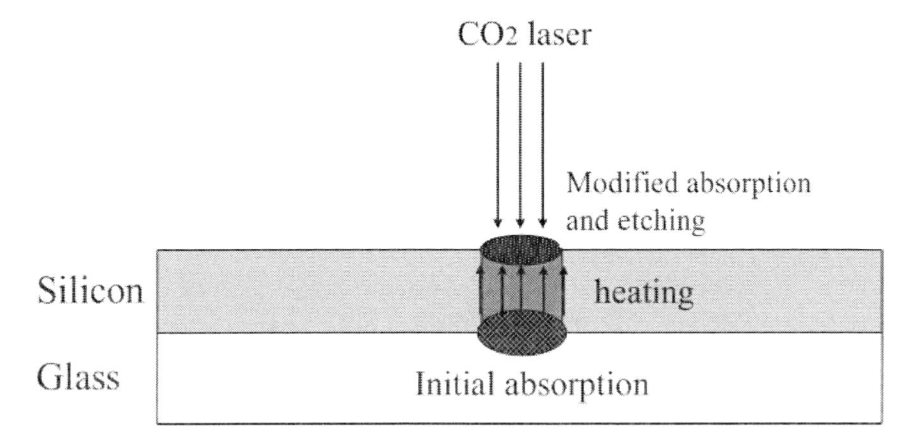

Figure 55. The schematic diagram of the proposed model for GACLAP mechanism..

In terms of the variation of Si energy band structure at high temperature, it may include the band gap narrowing [73], the formation of new defects states in the band gap [74] near/toward the valence band due to highly increased entropy of point defect configuration, electronic polarization [75] and the rapidly increased intrinsic free electrons concentration [73]. In principle, light radiation is absorbed in the nonmetal materials by valence band-conduction band electron transitions and electron transitions involved in defect levels lying within the bandgap and electronic polarization [75]. The bandgap energy of single crystal Si at 300 K is 1.12 eV corresponding to an absorbable wavelength of about 1.11 μm. However, the 10.6 μm CO_2 laser has a 0.12 eV photon energy. The occurrence of band gap narrowing and new defect states near the Si valence band at the local etching area at the high temperature will be very important for enhancing the photon absorption. The Si bandgap energy, $E_g(T)$, is a function of temperature in a universal formula (7) from 0 K to the melting point of material [73]:

$$E_g(T) = E_g(0) - \frac{\alpha T^2}{T + \beta} \quad (7)$$

Where $E_g(0)$, α and β constants for Si are 1.170, 4.73 x 10^{-4} and 636, respectively. The bandgap energy decreases with increasing temperature. For examples, $E_g(300)$ is 1.12 eV for 27 °C while $E_g(1685)$ equals to 0.59 eV for Si's melting point. If we extrapolate the data to higher temperature of liquid Si, E_g= 0.10 eV can be obtained at 2773 K. Although the value is just as a reference for the trend prediction, it implies that Si at very high temperature can absorb the CO_2 laser energy. In case the reduced E_g is not less than 0.12 eV, the bulk defect levels within the band gap may introduce valence band-defect level electron transition for enhancing the optical absorption [72],[73]. The absorption increases with increasing temperature. Also, the absorption by electronic polarization is important at the light frequencies in the vicinity of the atomic relaxation or vibration frequency [73],[74]. The CO_2 laser with a wavelength of 10.6 μm corresponds to a frequency of 2.8 x 10^{13} Hz. The average number of the occupied quantum state for photon or phonon particles following the Bose-Einstein distribution function ($\frac{1}{\exp(hv/kT)-1}$, h: Planck constant, k: Boltzmann constant, v: frequency of particle, T: temperature) at a frequency of 2.8 x 10^{13} Hz will be one at 1940 K [76]. The average number of occupation at 1685 K is 0.82. It is manifested that the electronic polarization contributes CO_2 absorption of Si at high temperature. The absorption increases with increasing temperature. In addition, the intrinsic concentration of Si will dominate the free carrier concentration at above 277-327 °C in our samples (10^{21}-10^{22} m^{-3} doping) and the free carrier concentration rapidly increases with temperature in log relationship [73] to 10^{25}-10^{26} m^{-3} at a temperature larger than 1685 K. These free carriers (electron and holes) make more collision with atoms to create excited electrons and holes, which increase the absorption probability. The above variations in energy band structure lead to the strong interaction between photons, electrons and phonons of materials for enhancing high absorption of CO_2 laser in Si.

Si etching has been achieved using GACLAP technology which is to put a Si sample on a glass plate for the CO_2 laser irradiation. The glass initially absorbs CO_2 laser energy near the interface of Si and glass and then heats Si to change its absorption behavior. The new mechanism may be involved in two main reactions: one is the variation of Si energy band structure at high temperature for promoting the photon-material interaction. It may include the band gap narrowing, new defect states formation, electronic polarization and rapidly increased intrinsic free carrier concentration. The other is the increased Si oxidation at high temperature. The thin oxide can directly absorbs CO_2 laser and keep high temperature for further Si etching. The above two reactions may occur simultaneously during GACLAP for high absorption of CO_2 laser to etch Si. The etching depth of Si increases with increasing laser passes and power. The etching geometry will affect the energy distribution and accumulation for the etching rate. The etching of the small-size circle is faster than the linear shape at the same laser energy.

5. Laser Treatment Technique

The crystalline titanium dioxide material has been intensively used in the photocatalyst or dye-sensitized solar cell application. Contrlling phase transformation of titanium oxide from initial amorphous to annealed crystalline phase is an important issue because the crystalline structure of titanium dioxide significant influences the film property. Three phases of rutile, anatase and brookite exist in titanium dioxide crystal structure, but both rutile and anatase are mostly discussed in optoelectronic application. Varied fabrication techniques, such as composite plating, sol-gel, anodic oxidation, evaporation, sputtering and CVD/PECVD have been performed for film deposition together with post annealing for crystalline titanium dioxide formation. The sol-gel solution was mixed by tetraisopropyl orthotitanate (TTIP), acetonylacetone, distilled water and alcohol at various molar ratios and coated on p-type silicon substructure for titanium oxide formation. Then the CO_2 laser annealing in air at powers of 0.5, 1.5 and 3.0 W in the defocus mode was performed on the coatings for studying crystallization of titanium oxide. The microstructure and phase transformation of titanium dioxide were examined by X-ray diffraction pattern. The ANSYS simulation was employed to calculate temperature distribution of films to correlate with phase transformation of titanium dioxide. Figure 56 shows the schematic diagram of CO_2 laser annealing processing combined with sol-gel coated film on the silicon substrate. It includes the CO_2 laser source, a reflective mirror, a focal lens and a p-type silicon wafer coated with sol-gel amorphous titanium dioxide. The sol-gel solution was made of mixing tetraisopropyl orthotitanate (TTIP), acetonylacetone (ACAC), distilled water and alcohol at various molar ratios. Table 3 lists the mixing ratios of sol-gel solution as groups $A_1(A_2)$, B and C. Table 4 lists the material property of titanium dioxide [77] for ANSYS software simulation to analyze temperature distribution on titanium dioxide thin film after CO_2 laser annealing in order to compare with XRD results.

Table 3. The molar ratio of TTIP, ACAC, DI water and alcohol mixing in sol-gel solution

Group	TTIP	ACAC	DI Water	Alcohol
A_1 (A_2)	1	1	3	20 (15)
B	1	1	6	15
C	2	1	3	15

Table 4. The material property of titanium dioxide [[77]]

Density (kg/m^3)	2100
Specific heat (J/kg.K)	790
Thermal conduction (W/m.K)	0.5

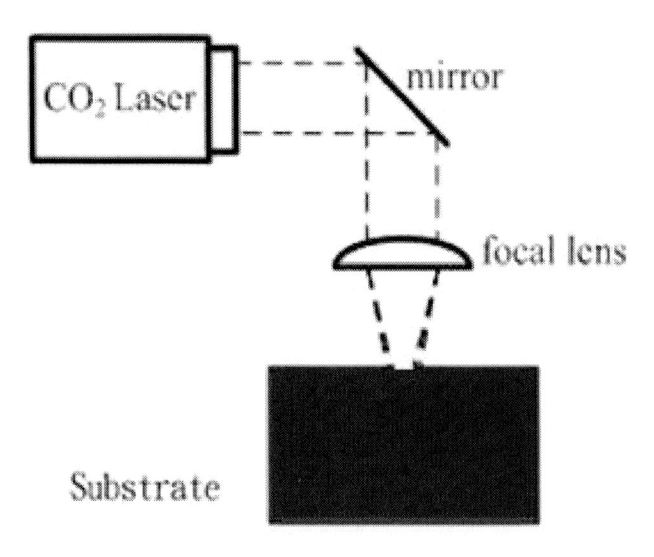

Figure 56. The schematic diagram of CO2 laser annealing setup.

The sol-gel solution was mixed by tetraisopropyl orthotitanate (TTIP), acetonylacetone, distilled water and alcohol at various molar ratios and coated on p-type silicon substructure for titanium oxide formation. The sol-gel solution was reacted on hydrolysis and condensation mutual reaction, that equation was shown in the following :

$$Ti(OR)4 + H2O \rightarrow Ti(OR)3(OH) + ROH \quad (7)$$

$$Ti(OR)3(OH) + Ti(OR)4 \rightarrow (RO)3Ti\text{-}O\text{-}Ti(OR)3 + H2O \quad (8)$$

Here, the equation (7) proposes a reaction step on hydrolysis process and equation (8) is the condensation process. After tetraisopropyl orthotitanate has reacted to titanium dioxide powder in the solution, that quantity of water is an important factor in sol-gel product or uniform property at powder size. In this study, we employ esterification reaction to revise produced ratio which can avoid the sedimentation during hydrolysis and condensation mutual reaction. Figure 57(a) and (b) show XRD patterns of groups A_1 and A_2 after CO_2 laser annealing at powers of 0.5, 1.5 and 3.0 W in air, respectively. In group A_1, the strong TiO_2 Rutile peak R(110) was observed at laser powers of 1.5 and 3.0 W. Compared with results of ANSYS simulation in Figure 58, the maximum temperatures of films annealed at 0.5, 1.5 and 3.0 W are 236, 842 and 1462 °C, respectively. It is consistent with the publication [77]. It indicates that our ANSYS simulation can be successful employed to calculate temperature distribution. In group A_2, the TiO_2 Rutile peak R(110) was also observed at laser powers of 1.5 and 3.0 W. Figure 59(a) shows XRD patterns of group B after CO_2 laser annealing at powers of 0.5, 1.5 and 3.0 W in air. As DI water quantity increases from 3 to 6 moles, better crystallization is also observed from R(110) peak together with R(111). The temperature distribution also exceeds the rutile formation temperature of 800 °C at laser powers of 1.5 and 3.0 W. Figure 59(b) shows XRD patterns of group C after CO_2 laser annealing at powers of 0.5, 1.5 and 3.0 W in air. The anatase peak A(101) was observed at laser power of 0.5 W as TTIP quantity increased. And the Rutile phase also has better crystallization from R(110), R(101) and R(111). Titanium dioxide phase transformation can be controlled in air at different parameters. Increasing TTIP quantity and decreasing laser power benefit the anatase phase formation. Seeing the results of thermal analysis, the temperature distribution catches the attention at laser output power of 0.5 W. When the temperature increases over 350 °C, that anatase crystallization was observed in XRD image. Compared with previously publication, the reaction temperature was reduced from common reaction temperature of 450-550 °C to lower reaction temperature of 350 °C. Figure 60 show the grain sizes of all titanium oxides at different sol-gel solution concentration ratio as a function of laser power. The crystallinity and grain size of titanium dioxide films increases with increasing laser power. The grain size calculated with Scherrer's formula was between 10 and 32 nm. This sol-gel processing combined with CO_2 laser annealing offers a good opportunity for controlling crystalline titanium dioxide formation and property for future optoelectronic application.

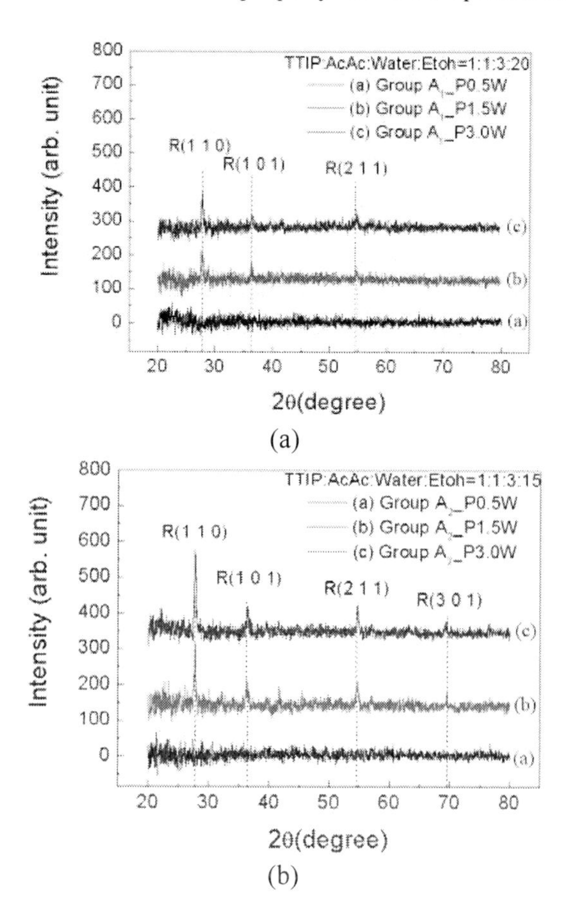

Figure 57. XRD patterns of: (a) group A1 and (b) group A2 after CO2 laser annealing at powers of 0.5, 1.5 and 3.0 W in air.

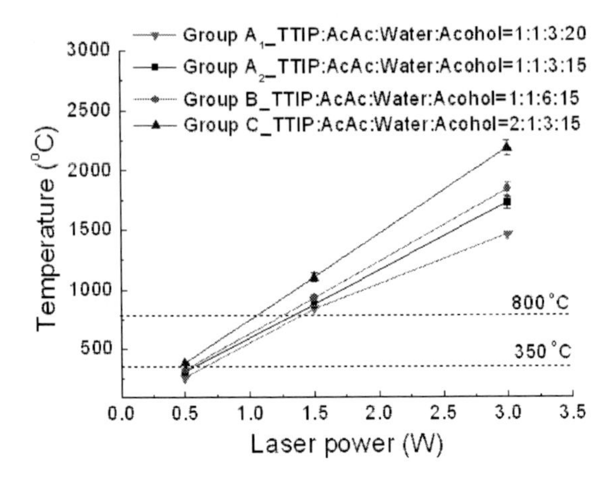

Figure 58. Simulation of annealing temperature of titanium oxide at different sol-gel solution concentration ratio as a function of laser power.

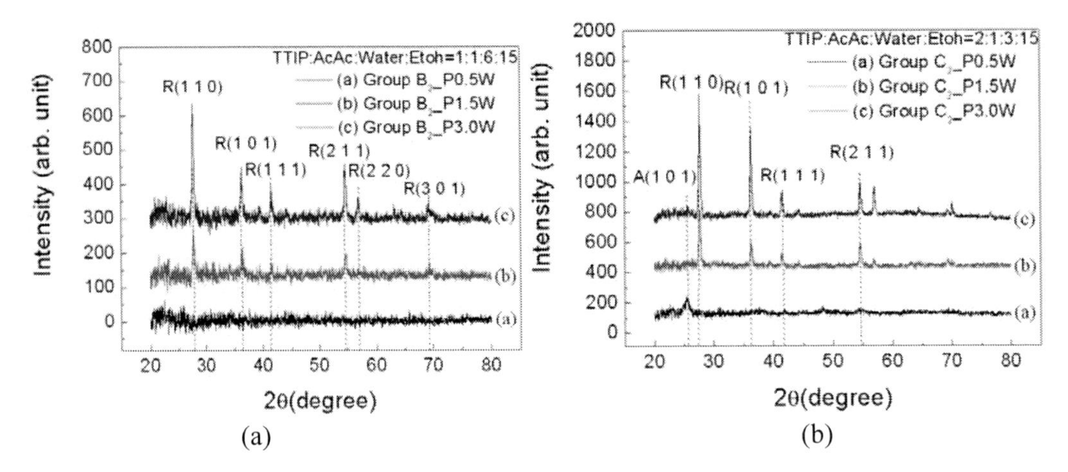

Figure 59. XRD patterns of: (a) group B and (b) group C after CO2 laser annealing at powers of 0.5, 1.5 and 3.0 W in air.

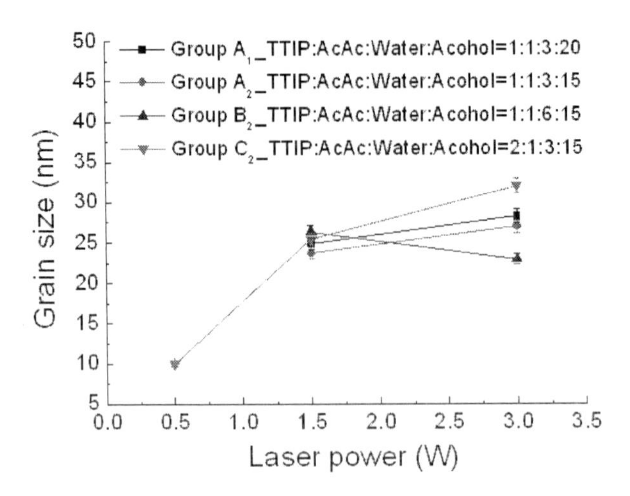

Figure 60. Grain size of titanium oxide at different sol-gel solution concentration ratio as a function of laser power.

6. Conclusion

Laser machining and processing have advantages of simple, maskless, fast and direct-write for different geometrical shapes and surface modification. In this article, the advanced CO_2 laser processing and its application have been presented to improve traditional defects. Liquid Assisted Laser Processing (LALP) can effectively reduce the temperature, heat-affected zone, thermal gradient and stress via water for hindering the crack and scorch formation together with the laser heating induced stronger natural convection in water for carrying debris away to reduce bulge height. Cover-layer protection processing employs a polymer thin film as protection layer against machining damage caused, it provides a simple approach for micro-groove or micro-hole manufacture in low energy processing. GACLAP can change light absorption behavior of Si and make Si be etched from the top surface toward the interface whose new mechanism is discussed in viewpoint of the variation of electronic band structure, surface oxidation and light absorption of Si at high temperature. A simple CO_2 laser annealing process for titanium dioxide treatment has also been developed instead of conventional expensive short wavelength laser annealing and non-selective high-temperature furnace annealing. Both crystalline rutile and anatase titanium dioxide transformation from amorphous titanium oxide can be controlled by the sol-gel composition and laser annealing parameters for material property adjustment which is potentially used for the photocatalyst and optoelectronic application. The proposed modified or advanced CO_2 processing approach has advantages of low-cost, simple composition, crackless formation, lager area machining and good quality of treatment, such as LALP method for microstructure manufacture and capillary-driven bio-fluidic application, GACLAP method for silicon material direct etching and defocus method for titanium dioxide treatment.

References

[1] T. Cormany, P. Enoksson and G. Stemme, *J. Micromech. Microeng.* 8, 84-87 (1998).
[2] C. Iliescu, J. Miao and F.E.H. Taya, *Sensors and Actuators A* 117, 286-292 (2005).
[3] C.K. Chung, *J. Micromech. Microeng.* 14, 656 (2004).
[4] M.A. Gosalvez and R.M. Nieminen, *New J. Phys.* 5, 100 (2003).
[5] X. Lia, T. Abe, and M. Esashi, *Sensors and Actuators A* 87, 139-145 (2001).
A. Goyal, V. Hood and S. Tadigadapa, *J. Non-Crystalline Solid* 352, 657-663 (2006).
[6] Baram and M. naftali, *J. Micromech. Microeng.* 16, 2287-2291 (2006).
[7] P. Gravesen, J. Branebjerg and O. Jensen, *J. Micromech. Microeng.* 3, 168-182 (1993).
[8] H. Makamba, J.H. Kim, K. Lim, N. Park and J.H. Hahn, *Electrophoresis* 24, 3607-3619 (2003).
[9] D. T. Eddington, J. P. Puccinelli and D. J. Beebe, *Sensors and Actuators B* 114, 170-172 (2006).
[10] K. Efimenko, W. E. Wallace and J. Genzer, *J. Colloid Interface Science* 254, 306-315 (2002).
[11] H. Hillborg and U. W. Gedde, *Polymer* 39, 1991-1998 (1998).
[12] J. A. Vickers, M. M. Caulum, and C. S. Henry, *Anal. Chem.* 78, 7446-7452 (2006).

[13] Daridon, V. Fascio, J. Lichtenberg, R. Wütrich, H. Langen, E. Verpoorte and N. F. de Rooij, Fresenius *J. Anal. Chem.* 371, 261-269 (2001).
[14] R. Prakash, KVIS Kaler, *Microfluidics Nanofluidics* 3, 177-187 (2007).
[15] Kaldos, H.J. Pieper, E. Wolf and M. Krause, *Journal of Materials Processing Technology* 155–156, 1815 (2004).
[16] William M. Steen, *Handbook of Laser Material Processing*, Springer, New York, p. 63 (1991).
[17] Y Xia, Q. Wang, L. Mei, C. Tan, S. Tue, B. Xu and X. Liu, *J. Phys. D* 24, 1933 (1991).
[18] G. Han and P.T. Murray, *J. Appl. Phys*. 88, 1184 (2000).
[19] J. Ren, S. S. Orlov and L. Hesselink, *J. Appl. Phys*. 97, 104304 (2005).
[20] M. C. Gower, *Opt. Express* 7, 56 (2000).
[21] H.C. Le, R. W. Dreyfus, W. Marine, M. Sentis and I. A. Movtchan, *Appl. Surf. Sci*. 96-98, 164 (1996).
[22] J. Ren, M. Kell and L. Hesselink, *Opt. Lett*. 30, 1740 (2005).
[23] J. S. Yahng, B. H. Chon, C. H. Kim, S. C. Jeoung and H. R. Kim, *Opt. Express* 14, 9544 (2006).
[24] W.J. Wang, Y. F. Lu, C. W. An, M. H. Hong and T. C. Chong, *Appl. Surf. Sci*. 186, 594 (2002).
[25] S. Nielsen and P. Balling, *Journal of Applied Physics* 99, 093101 (2006).
[26] Weck, T. H. R. Crawford, D. S. Wilkinson, H. K. Haugen and J. S. Preston, *Appl. Phys. A* 90, 537-543 (2008).
[27] J. Cheng, W. Perrie, M. Sharp, S.P. Edwardson, N. G. Semaltianos, G. Dearden and K. G. Watkins, *Appl. Phys*. A 95, 739-746 (2009).
[28] K. Venkatakrishnan, N. Sudani and B. Tan, *J. Micromech. Microeng*. 18, 075032 (7pp) (2008).
[29] Tan and K. Venkatakrishnan, *J. Micromech. Microeng*. 17, 1511-1517 (2007).
[30] B. Tan, S. Panchatsharam and K. Venkatakrishnan, *J. Phys. D: Appl. Phys*. 42, 065102 (9pp) (2009).
[31] Vogel, J. Noack, G. Huttman, G. Paltauf, *Appl. Phys. B* 81, 1015 (2005).
[32] M. Ohnishi, H. Shikata, M. Sakakura, Y. Shimotsuma, K. Miura and K. Hirao, *Appl. Phys. A* 98, 123 (2010).
[33] *http://iknow.stpi.org.tw/Post/Read.aspx?PostID=3020*
[34] M.J. Madou, *"Fundamentals of Microfabrication*", 2nd ed., CRC press, New York, p. 590 (2002).
[35] K. Zimmer, A. Braun and R. Böhme, *Applied Surface Science Volumes* 208-209, 199-204 (2003).
[36] Ben-Yakar, A. Harkin, J. Ashmore, R.L. Byer and H.A. Stone, *J. Phys. D* 40, 1447-1459 (2007).
[37] Y. Hanada, K. Sugioka, H. Kawano, I.S. Ishikawa, A. Miyawaki and K. Midorikawa, *Biomedical Microdevies* 10, 403-410 (2008).
[38] H. Y. Zheng, Y. C. Lam, C. Sundarraman and D. V. Tran, *Appl. Phys. A* 89, 559-563 (2007).
[39] R. Petkovšek, A. Babnik, J. Diaci and J. Možina, *Appl. Phys. A* 93, 141-145 (2008).
[40] Q. Sun, A. Saliminia, F. Théberge, R. Vallée and S. L. Chin, *J. Micromech. Microeng*. 18, 035039 (4pp) (2008).
[41] R. Bohme, K. Zimmer, B. Rauschenbach, *Appl. Phys. A* 82, 325-328 (2006).

[42] J. Y. Cheng, M. H. Yen, C. W. Wei, Y. C. Chuang and T. H. Young, *J. Micromech. Microeng.* 15, 1147-1156 (2005).
[43] J. Zhao, J. Sullivan and T. D. Bennett, *Appl. Surf. Sci.* 225, 250–255 (2004).
[44] C.K. Chung, Y. C. Lin, and G. R. Huang, *J. Micromech. Microeng.* 15, 1878- 1884 (2005).
[45] G. Allcock, P.E. Dyer, G. Elliner, and H.V. Snelling, *J. Appl. Phys.* 78, 7295-7303 (1995).
[46] H. Ogura, Y. Yoshida, *Jpn. J. Appl. Phys.* 42, 2881-2886 (2003).
[47] M. H. Yen, J. Y. Cheng, C. W. Wei, Y. C. Chung and T. H. Young, *J. Micromech. Microeng.* 16, 1143-1153 (2006).
[48] *Kruusing, Optics and Lasers in Eng.* 41, 307-327(2004).
[49] *Kruusing, Optics and Lasers in Eng.* 41, 329-352 (2004).
[50] K. Chung, Y. C. Sung, G. R. Huang, E. J. Hsiao, W. H. Lin and S. L. Lin, *Appl. Phys. A* 94, 927-932 (2009).
[51] Barnes, P. Shrotriya and P. Molian, *International Journal of Machine Tools & Manufacture* 47, 1864-1874 (2007).
[52] C.K. Chung, H.C. Chang, T.R. Shih, S.L. Lin, E.J. Hsiao, Y.S. Chen, E.C. Chang, C.C. Chen and C.C. Lin , *Biomedical Microdevies* 12, 107-114 (2010).
[53] R.E. Mueller, J. Bird and W.W. Duley, *J. Appl. Phys.* 71, pp. 551–556 (1992).
[54] G.A. Shafeev and A.V. Simakhin, *Appl Phys A* 54, 311–6 (1992).
[55] G.A. Shafeev and A.V. Simakhin, *Laser Phys.* 4, 631–4 (1994).
[56] F. Mafune′, J. Kohno, Y. Takeda and T. Kondow, *J. Phys. Chem. B* 105, 5114-5120 (2001).
[57] F. Mafune′, J. Kohno, Y. Takeda and T. Kondow, *J. Phys. Chem. B* 104, 9111-9117 (2000).
[58] Marc Madou, *"Fundamentals of Microfabrication"*, 2nd ed., CRC press, New York, p. 590 (2002).
[59] *http://www.engineeringtoolbox.com/overall-heat-transfer-coefficient-d_434.html.*
[60] G. M. Hale, and M.R.. Querry, *Appl. Opt.* 12(3), 555–563 (1973).
[61] Deborah D.L. Chung, *Handbook of Materials for electronic packaging*, Butterworth-Heinemann, UK, p. 218 (1995).
[62] K. Chung and M.Y. Wu, *Optics Express* 15, pp. 7269-7274 (2007).
[63] *http://chemicalland21.com/specialtychem/perchem/LAURAMIDOPROPYL%20BETAINE.htm*
[64] S. Kim, S. H. Lee, C. H. Ahn, J. Y. Lee and T. H. Kwon, *Lab Chip* 6, 794-802, 2006.
[65] P. K. Yuen, L. J. Kricka, P. Fortina, N. J. Panaro, T. Sakazume, and P. Wilding, *Genome Research* 11, 405-412, 2001.
[66] H. Sakai, S. Takeoka, S.I. Park, T. Kose, Y. Izumi, A. Yoshizu, H. Nishide, K. Kobayashi, and E. Tsuchida, *Bioconjugate Chem.* 8, 15 (1997).
[67] S. Bouaidat, O. Hansen, H. Bruus, C. Berendsen, N. K. B. Madsen, P. Thomsen, A. Wolff, and J. Jonsmann, *Lab Chip* 5, 827-883 (2005).
[68] T.P. Vikinge, K.M. Hansson, J. Benesch, K. Johansen, M. Ranby, T.L. Lindahl, B. Liedberg, I. Lundström and P. Tengvall, *J. Biomedical Optics* 5, 51–55 (2000).
[69] V. I. Furdui, J. K. Kariuki and D. J. Harrison, *Journal of Micromechanics and Microengineering* 13, pp. S164–S170, 2003.

[70] C.K. Chung, M.Y. Wu, J.C. Wu, Y.C. Sung and G.R. Huang, *IEEE Conference of Nano/Micro Engineered and Molecular Systems* (Nanotechnology Council, Zhuhai, China, 2006), 1445-1448 (2006).
[71] S.M. Sze, "*VLSI Technology, 2nd ed.*", (McGraw-Hill, New York, USA, 1988), p. 657 & Chap. 3 (1998).
[72] Koren, "Continuous wave laser assisted chemical material removal from Mo, W, and Si at faint red hot temperatures (700-800 °C)", *Appl. Phys. Lett.* 47, 1012 (1985).
[73] S.M. Sze, "*Physics of semiconductor devices, 2nd ed.*", (John Wiley & Sons, USA, 1981) Chap. 1.
[74] W. Schroter *et al.*, *"Handbook of semiconductor technology, vol. 1",* Jackson KA and Schroter W, ed. (WILEY-VCH, Weinheim, 2000), Chap. 10.
[75] W.D. Callister Jr., *"Fundamentals of materials science and engineering 2nd ed.*", (*John Wiley & Sons, NJ*, 2005) Chap 19.
[76] Chen, "*Nanoscale energy transport and conversion",* (Oxford, NY, USA, 2005) Chap. 4.
[77] S. G. Choi, T.J. Ha, B.G. Yu, S. Shin, H.H. Cho and H.H. Park, *Thin Solid Films* 516, 212-215 (2007).

INDEX

A

B

C

D

E

F

G

H

I

J

K

L

M

N

O

P

Q

R

S

T

U

V

W

X

Y

Z